AF477123

Mail Technology – Tomorrow's World

Conference Organizing Committee

R Annandale
TRW

P Barton (Chairman)
Royal Mail

S Bradford
IBM

D Brownsell
Royal Mail

J Dumper
IBM

F Gillet
Mannesmann Dematic Postal Automation

J Ivers
Pitney Bowes

G Lamb-Hughes
Lockheed Martin

G Merialdo
Elsag spa

R Paris
Siemens ElectroCom

J Schofield
Mannesmann Dematic Postal Automation

IMechE
Conference Transactions

I MECH E

International Conference on

Mail Technology – Tomorrow's World
Business Opportunities and Solutions in a Global Market

23–24 June 1999
Stakis Brighton Metropole, Brighton, UK

Organized by
The Manufacturing Division of the Institution
of Mechanical Engineers (IMechE)

In Association with Royal Mail

Sponsored by
Bell & Howell
Lockheed Martin
Mannesmann Dematic Postal Automation
Siemens

IMechE Conference Transactions 1999–5

**Professional
Engineering
Publishing**

Published by Professional Engineering Publishing Limited for the Institution of
Mechanical Engineers, Bury St Edmunds and London, UK.

First Published 1999

ISSN 1356–1448
ISBN 1 86058 220 6

A CIP catalogue record for this book is available from the British Library.

Printed by The Book Company, Suffolk, UK.

Front cover artwork – Ergonomic mail recirculation on the Mannesmann Dematic Postal Automation Mars Global Branch Office Solution.

Related Titles of Interest

Title	Editor/Author	ISBN
An Introductory Guide to the Control of Machinery	K Foster	1 86058 068 8
IMechE Engineers' Data Book	C Matthews	1 86058 175 7
A Practical Guide to the Machinery Directive	H P van Ekelenburg, P Hoogerkamp, D A Brown, and L J Hopmans	0 85298 973 3
Mail Systems 2000	IMechE Conference	0 85298 906 7

For the full range of titles published by Professional Engineering Publishing contact:

Sales Department
Professional Engineering Publishing Limited
Northgate Avenue
Bury St Edmunds
Suffolk
IP32 6BW
UK

Tel: 01284 724384
Fax: 01284 718692

Contents

Mail Generation

Mail Automation

Mail Generation

C568/013/99

End-to-end solution for mail or electronic messages

M FIELD and **K UEMOTO**
Pitney Bowes Inc, Connecticut, USA

SYNOPSIS/ABSTRACT

New electronic technologies have improved the quality and have increased the options available in the development of message creation and delivery systems. It is leading the way to better control over the end-to-end chain and to improve the quality of the delivery channel. It is significantly improving the value of paper-based communications. And it is also showing the way that paper-based mail as we know it can coexist with the newer electronic delivery forms.

The postal delivery channel for hard copy messages must evolve to use the capabilities provided by these new technologies. Through careful collaboration with all stakeholders, these technologies will enable more efficient delivery, while activating a large range of value added services.

1 INTRODUCTION

Message senders need end-to-end solutions that provide high quality customer communications with verifiable and timely delivery. They look for complete solutions that provide them a competitive edge, either in cost or in functionality.

The solutions are designed to take advantage of technology as it evolves, growing in capability and capacity with the needs of the sender. In the case of the end-to-end solution, the emerging challenge is to implement an integration of a company's information chain with multiple output processes, which range from traditional paper printing and finishing systems through multi-channel delivery systems which complement the Posts with other electronic delivery services. The key to success is integration that links all the parts of the end-to-end solution, from the company's information systems through the vendor control and reporting systems to the delivery tracking systems.

1.1 The End-to-End Definition

End-to-end needs a definition as it is actually a business cycle that begins with the customer and ends with the customer. In a typical business cycle, the customer initiates a business process that causes one or more communications to go from the company to the customer leading to one or more return communications. The classic example is a billing application where the customer buys something, the company bills the customer, the customer pays the bill and the company acknowledges the payment. Rarely are actual processes this simple, but, due to many reasons, most business process are built around the simple actions of posting transaction documents and processing remittances.

In practical terms this means the process does not end when the mail is deposited with the Post or other delivery service providers. It includes the Post and the other delivery service providers, **and** it also includes the recipient. These critical links in the loop must be considered in building successful end-to-end solutions. The company must always be seeking continuous improvement in control of the whole chain and will find that the careful uses of technology will provide them with this capability.

In an end-to-end solution, it is easy to see how cash flows are improved by timely delivery, but it is equally important to think in terms of life cycle value of a customer and the role of transaction documents in that life cycle. By increasing the quality of the communication, by providing needed information and by being more automated, end-to-end solutions increase the value of each customer generating higher loyalty and greater revenue potential.

1.2 The Value Proposition for the User

The vast majority of users look to their message systems suppliers to provide the highest level of document integrity and quality within a cost structure based on short-term financial goals. But long term value is based on repeated customer transactions arising from customer satisfaction. Therefore it is really the combination of the supplier and the Post/delivery service providers that create the value. So what is the value proposition?

- It is the printing and finishing of all the communication elements with no undetected errors that will lead to customer satisfaction.
- It is the envelope, correctly addressed per postal or other delivery channel rules that has been prepared to be fully automation compatible with the system of the delivery channel.
- It is the delivery to the correct location to the correct person in a timely manner.

2 THE MAILPIECE IS A MARVEL OF ENGINEERING

The mailpiece is a marvel of engineering, technology and communication; it is full of information. It is produced by an integrated system of information technology and paper-handling hardware.

2.1 The Envelope—A Very Critical Piece

The envelope contains the items that create the mailpiece. But most importantly it shows the instructions to the Post about how that mail should be processed and delivered. It contains several important parts:

- The recipient address is shown in human readable form for the use of the people handling and receiving the message. It is also represented as a barcode for the machine sorting of the mail by the Posts, or other identification of the message for the electronic delivery of the message.
- The return address can identify the sender, but it also specifically identifies the location where the mail piece will be returned for any reason when the message cannot be delivered.
- The postage indicia are classically thought of as a proof of payment – a limited type of currency. The current indicia are communication devices between the mailer's business systems and the postal systems. In the case of the Information Based Indicia Program or IBIP, this role is explicitly defined, but almost every Post worldwide uses the indicia as a means of carrying information compatible with their automated imaging systems.
- The envelope is often overlooked as a tool for carrying messages, but current technology allows personalized messages to be printed on the face of an envelope without conflicting with postal regulations governing the address block positioning and format. These messages can be targeted marketing teasers, regulatory notices, or simple callouts for important items inside the envelope.

2.2 The Envelope as an Interface Specification

In an End-to-End solution, interfaces are not important, they are critical. All solutions will consist of multiple components integrated from many sources: legacy components, internal development, vendor-supplied products, third party development, external service providers, etc. Each of these components will interface to one or more of the other components and each interface requires conventions as the parameters of that interface. Some interfaces are arbitrary and are developed on an ad hoc basis, but the majority will have pre-established conventions which are documented in an interface specification.

The Posts are critical service component providers in mailer's End-to-End customer messaging solutions. The envelope, besides physically protecting its contents, is the interface between the mailer and the Post and between the Post and the recipient. The conventions for this interface are well established, ex. Number 10 or C5 envelopes. The interface specification is also well documented – in the United States the specification document is the Domestic Mail Manual and the accompanying handbooks.

The Information Based Indicia Program, IBIP, is an interesting example of the evolution of the envelope as the postal interface as the Proof of Payment, tracking ID, originating location, and destination information are all contained within a machine readable symbol. In the development of the specification there has been a dynamic balance between the human readable and the machine readable information and the deployment of the IBIP is bringing forward new issues of the envelop material as it impacts mailpiece machinability.

2.3 The Mailpiece and All Its Parts

Inside the envelope are all the mailpiece elements and all the codes necessary for the accurate processing of the message.

- The variable messages designed specifically for the recipient.
- The address block information that assures accurate delivery, whether inside the envelope through a window or directly on the envelope.
- The inserter control codes that assures quality of the items inserted into the mail piece.
- The reply envelope with all the marks necessary for the Post to process this piece when the reply envelope is sent back.
- The reply envelope can carry marks bearing information about the recipient, who sent back the mail piece for the accurate processing of the message on return.
- Also included in the mailpiece may be a whole range of other inserts, such as regulatory messages, tear off return inserts and other company inserts—all of which can carry intelligent, information bearing marks.

3 THE MAIL/MESSAGE GENERATION SYSTEM

3.1 The Information System

A mailpiece is the output of a highly complex, continuous process, the parts of which are often managed in isolation. The result can be poor quality, production delays and customer dissatisfaction.

But, it is this process that produces the information about the recipient, the sender and instructions to the Posts. The process uses multiple databases and transaction files to generate the mailpiece. These databases or files are accessible on line to the staff developing the mail application. The information stream is then converted into a print stream that drives the printers and ultimately the finishing equipment.

The information systems will become increasingly important as the mailer to Post and to some extent the Post to recipient interface are based less on what is on the face of the envelope and more what is transmitted from one control and reporting system to the next in an integrated supply chain. In a tradition that goes back to the days of Ben Franklin's tenure as Postmaster General, it was expected that each mailpiece stood on its own and each person in the process could determine the information needed from their process step from the face of the envelope. In the near future, each message will be a data block with variable information within a tracking system irrespective of whether the media for the information is paper or electrons flowing over the internet.

3.2 The Document Generation System

New technologies have "freed" the document to be routed, stored, produced and sent in a dizzying array of alternatives. These technologies allow the process to manage the full range of documents and communications to be sent to a recipient – from simple messages to technical manuals and price books.

These technologies give a loud voice to the recipient of the mail/communication to choose how and when he wishes to receive his mail. All of a sudden, our customer's customer is important and his choices are noted and provided for, including the electronic delivery of the mail. As usual with new technologies, many companies today are not taking advantage of it to develop a customer-focused output strategy. As the customers insist upon digital document delivery-- there is the Digital Leak, which creates havoc for the controlled, high-quality document process. The outcomes are quality issues for all document delivery options that the senders use.

4 THE AUTOMATED DOCUMENT FACTORY

Pitney Bowes and many of our customers subscribe to a concept known as the Automated Document Factory (ADF). It is less a factory and more a powerful commitment that will bring quality back to the delivery of documents. This commitment leads to the deployment of an architecture and strategy that gives voice to internal and external customers to specify digital or hard copy delivery, with best practice levels of quality, integrity, security, control and productivity.

4.1 Excellent value of the Automated Document Factory

The Automated Document Factory adds an excellent value to the mailing systems of customers.

- It reduces costs per mail piece through an automated quality process
- It reduces cycle time for the development of a mail job since it takes advantage of the technology.
- It produces information that enables personalization.
- It controls the hard copy and digital streams
- It provides enterprise-wide visibility of the process.

4.2 The Automated Document Factory Building Blocks

The eight Automated Document Factory building blocks are focused on the whole process and defines the large chuck of work into cohesive work units, which are then logically sequenced. The blocks of work are print management and serving, print image manipulation and enhancement, printing, inserting, manufacturing floor controls and reporting, digital delivery, resource management, and archiving and regeneration. Each block can be handled in a number of ways and can even combine the operations of third party systems and equipment.

4.3 The Automated Document Factory Modules

These units of work are put into three modules: the input module, the transform module, and the output module. Each of these modules has a specific task. The input module takes information from any source and creates an input stream, which is combined with application programs to create the document in electronic form. In the transform module this stream is transformed into a form that can be delivered by the output module.

The whole process is controlled through the Operations Management System, which also provides reports on the functioning of the system and can develop the Statement of Mailing necessary for deposit with the Posts.

4.4 The Automated Document Factory Principles

Successful Automated Document Factories are built on seven fundamental principles:
- Take charge of the print-ready stream
- Condition that print-ready stream to perform efficiently

- Forgo data symbologies for file-based processing, until the data must be printed for others to use.
- Ensure control at the job and piece levels.
- Integrate control over all output streams.
- Use quality improvement production reporting
- Provide visibility of the ADF to the Enterprise.

5 AN EXAMPLE – A BILL PRESENTMENT SYSTEM

Automated Document Factories are used to generate a wide range of documents. A bill presentment application is an Automated Document Factory that produces bills for payment by the recipient, which incorporates plans for how the recipient will pay the bill. Thus, it integrates many different delivery channels, which facilitates the easy payment of the bill. This application usually requires the joint action by partners, who bring the capability of many different payment methods. Moreover it is an example of how diverse delivery options can become from the paper-based financial transaction through the Posts to the purely electronic transaction through the Internet. Bill Presentment is still largely a paper-based system, but it is quickly going to hybrid systems and to purely electronic systems. The ability to offer all forms and to migrate from one to the other is the value-added by such systems.

6 THE POSTAL DELIVERY SYSTEM AND THE USE OF CODE

Message delivery systems handle a very large volume. In order to provide quick and timely delivery, Posts and message delivery services have a very short window for the collecting, sorting and delivery of the messages. Delivering messages tend to be low-margin business, which requires very economical handling. This requirement poses a particular challenge for the delivery of hard-copy messages.

6.1 The Need for Universal Addressing and Directory Systems

Reaching the recipient is the goal of all messaging systems. The ability to effectively reach the recipient is dependent on addressing systems that are dependent on entities, which operate under different conditions, country by country. The Posts usually manage the addressing systems for physical locations. As the Posts become more liberalized access to the addressing system may become more and more restricted. However, among all the countries in the world, just 25 have addressing systems which encode addresses to the street level and another 33 have postal codes for geographical regions. Without address coding systems, the processing and delivery of hard-copy mail will remain uneconomically not automated.

In the Internet world addressing is conducted primarily through a network of computers that up until now have been managed under contract by private contractors to the US Government. The governance of the Internet will shift to a not-for-profit organization that will try to respond to the ever-growing need for IP addresses. The open accessibility to Internet addresses is critical to the growth of the Internet and the electronic delivery channels.

6.2 Machine-readable Codes are the Most Effective

The Automated Document Factory is an approach to converting information into machine-readable codes that can be read and used by the delivery channel to deliver the message accurately and quickly. All end-to-end solutions require the transfer of information from one link to the next. This transfer is most efficiently done if the information is in a machine-readable form or symbology. We may in the future see machines that can cost effectively read anything on a mail piece, but for now we are reliant on machines that can read symbols at a cost-effective speed.

6.3 The Post are Standardizing the Use of Machine-readable Codes
In the end-to-end solution, the envelopes play a critical role in assuring that the mailpieces are handled properly by the delivery channel. The Posts recognize that they need to standardize the symbols and their interpretation. So they have begun to establish standards for some of these codes for use within their countries. They require mailers to meet content, structure, placement and readability parameters. However, these codes have been developed from a postal centric view limiting the usability of these codes by anyone other than the Posts.

6.4 Codes Need To Be Standard Across Borders
Unfortunately, once these codes cross a country's borders they are rendered useless. Cross-border, global standardization would greatly assist in the development of the mailer-Post-recipient information chain. Today, the Posts are moving in a multitude of directions with no agreement on even a common symbology. These unique country-by-country codes create havoc for manufacturers and delay the development of a single Postal delivery network worldwide.

6.5 The Codes Contain Large Amount of Information in a Small Space
The search is on for the symbol that is the most efficient, i.e., getting the most information printed in a small space and that can be printed and read with currently available technology. The international postal community has begun to focus on the standardization of the symbols—what they contain and the placement of the symbols on the envelope.

The discussion and testing of these symbols are leading to a set of emerging requirements. These requirements are:

- The symbols need to have standard structure, especially for the data content and coding.
- The symbols need to be printed in a manner that can be read. No standards have been established in this field. The problem is very complex when considering the variety of printers and readers that are available and in use. The complexity is increased when including the variables of paper and ink. Furthermore, mail envelopes create a very wide range of items when considering the bulkiness of some envelopes and the uneven printing surfaces that are created. However, the sheer volume of mail would justify the effort to set standards in this area.

6.6 A Required Infrastructure to Support the Use of the Symbols
The use of codes is greatly enhanced if their use results in valued-added services to the customer. For example, the use of this mechanism can facility the tracking and tracing of letters in the mailstream. The codes enable an automation of this function across mailers, Posts and across borders to bring the cost of such service within acceptable bounds. In order to take advantage of this information, Posts and their partners will need to put into place an infrastructure. This infrastructure must have the capability to efficiently read the symbols to trigger the appropriate action by the system, whether it is to record information and store it in a database or to instruct a delivery employee to get a recipient's signature.

7 CONCLUSION

New electronic technologies have improved the flexibility and have increased the options available in the development of the message delivery system. It is leading the way to better control and improve the quality the end-to-end chain. It is significantly improving the value of paper-based communications. And it is also showing the way that paper-based mail as we know it can coexist with the newer electronic delivery forms.

Address management worldwide

S B O WORTH and **R W MASON**
QAS Systems Limited, London, UK

Synopsis

Whatever your business, maintaining accurate and up to date customer address data isn't just important, it's essential.

Every time you mail to an address, you mail to a person, and most people reason that if you can't get their name and address right, you're likely to fall down elsewhere.

That's why good Address Management is essential for any company that communicates with its customers.

But there is also significant benefits for the Postal Administrations throughout the world. The costs for the entire mail processing chain can be readily reduced when there is an accurate and clean address on the mail piece.

1 HOW TO KEEP YOUR CUSTOMERS

Knowing your customers is the key to retaining them and keeping them is a key factor in prospering in today's increasingly competitive market place. Databases containing important customer information are growing all the time and they provide the foundations from which customer relationship marketing grows.

According to the US Office of Consumer Affairs, it costs six times more to acquire a new customer than it does to retain an existing one. Research also shows that two thirds of customers who take their business elsewhere do so because no one bothered to communicate with them. (1)

The Henley Centre for Forecasting recently stated that from their surveys, the key to holding on to customers is keeping in touch. A business needs to develop relationships with its customers to make them feel valued and respected. Loyal customers will place frequent and consistent orders and are therefore less costly to service. Loyal customers will often recommend new customers to your company, eliminating extra costs in terms of advertising or promotions. Furthermore they are more likely to pay premium prices with a supplier that they know and trust. After all, we all know that the most profitable business is referral business.

The Henley Centre's figures reveal that 68 per cent of customers actively want more information from the companies that they deal with, and 60 per cent say they are more likely to buy from firms that stay in touch. Even more importantly, 70 per cent actively look forward to receiving their post. (2)

One obvious way of staying in touch with your customers is to send them regular communications – newsletters, information on new products and any new promotions. But when mailing your customers, it is essential that you get their address correct.

2 CORRECT ADDRESSES

From your own personal viewpoint, what do you think of a company that corresponds with you, and they get your name and/or address wrong? You are not likely to think very much of them. It is the same for your customers, every time you mail to an address; you are mailing to a person. If you get part of it wrong, it gives the impression that your company is not concerned with detail and has not taken the time or trouble to find out specific details about the individual customer.

Correctly addressed mail clearly presents a more positive image.

In a recent sample of direct mail recipients, 85 % agreed that "inaccuracies in the name and address reflect badly on the image of the company" (ref: "The Trends Survey" commissioned by the Direct Mail Information Service and carried out by the British Market Research Bureau every two years.) (3)

Today I therefore want to focus on the address aspect because that gets each item of mail delivered to the right address. Getting an individual's name spelt correctly is also very important, but that's the subject for a future presentation.

What goes through your mind if a letter is addressed to you as follows:

6 High Street
TUNBRIDGE
Kent
TN9 1EJ

Or

6 High Street
TONBRIDGE WELLS
Kent
TN1 1UX

These are perhaps fairly easy to see if you take the trouble to check them, that I've transposed the "o" and the "u" to make the point.

But what about this address:

Trottiscliffe House
Trottiscliffe
West Malling
Kent
ME19 5EA

Which is pronounced:

"Trosley"
and West Mawling"

It is unlikely that you would get this address right by just typing it into a database if you did not have either any local knowledge or someone spelling out each word letter by letter.

Hence ADDRESS MANAGEMENT - the business of maintaining accurate and up to date address files.

3 GOOD ADDRESS MANAGEMENT

So exactly what is Address Management and how do you go about implementing it.

3.1, Agree what your overall requirements are for the use of the data. For example - are you going to have just 1 file as a master reference, or will you want to maintain a number of separate files.

3.2, Ensure that every one in your organisation is aware of your plan for Address Management.

3.3, Clean up the mistakes and errors of the past. Previous addresses will have been entered by hand, with no level of validation at point of data entry. You can either send your files externally to a bureau or you can do the work yourselves using a batch software package. The Royal Mail can provide a list of their Value Added Resellers who take the Postcode Address File and incorporate it into their products. Royal Mail works closely with these licensed resellers to promote a range of products that can maintain your customer database accurately, easily and effectively.

Your customer details are perhaps one of your greatest assets, and it may be that automated processing using a bureau could change some addresses for the worse. Addresses where a precise or obvious close match cannot be achieved should come back to you for resolution anyway. You and only you may have some information on a customer that enables you to decide exactly what the correct address is.

3.4, Verify and validate EVERY new address that goes onto your database. For the UK as you enter a postcode, the data capture tool should be verifying that this is the correct one for the premise number. Hence TW11 9LY covers numbers 1 - 27 and 29 - 35. If someone gives you this postcode and 36 as their house number, then either the house number or postcode is wrong, because the postcode for number 36 in this street is TW11 9LU. Far better to resolve this problem right there and then when you have got the person on the telephone rather than finding the problem later on. You can also benefit from keystroke savings, a typical address being captured in about 10 keystrokes rather than 40+, and the correct address can then be returned to your underlying database in the correct format.

Make sure that you capture a full address. There are 20 High Streets in London, and this by itself would not get your mail delivered promptly, if at all.

When you've captured it, make sure that it is formatted to the requirements of the local postal administration. These requirements should be embodied in the underlying application that you are returning the addresses to.

GIGO – garbage in, garbage out also applies to address files!

3.5, Ensure that your standard address reference files are kept up to date. New houses are put up, old ones are demolished, and from time to time, postcodes actually suffer from exhaustion!. Where an area is expanding fast, then the Royal Mail actually run out of postcodes available and it becomes necessary to re-code the entire area. For example, this recently happened in Aberdeen and also Southampton. Hence SO1 1AA has now become SO14 3AA
i.e.
Maritime Chambers
Canute Road
Southampton
SO14 3AA

In the UK, PAF is dynamic and the Royal Mail are updating this master database on a daily basis. However for companies to be updating their database at such a frequency would be clearly impractical, and hence typically they will be supplied with quarterly updates, although monthly updates are available if required.
Its one thing being supplied with them, its another to implement them – so do see you have the latest one on the system!

3.6, Keep up to date with changes!
There's still more to be done. You've corrected the sins of the past!, you're validating each new address as its entered, but as well as postcode changes, there are locality changes and sometimes posttown name changes. Either once a year you should consider running your records through a batch routine again, or better still looking at using an update program which just updates any changes.

Both in terms of prompt delivery, and the perception of the recipient, a company can waste precious resources through poor Address Management.

4 MOVING PEOPLE

A whole separate task is tracking people moving, and in the UK up to 10 per cent of the population can change address. That means that one in ten people on your database could simply disappear. Of course some will let you know where they have moved to, but experience suggests that most will not. So not only are you losing contact with customers but you are also sending out mailshots to the wrong people and wasting money on print and postage. This is where the Royal Mail's NCOA – National Change of Address service can be of great assistance. One of the most comprehensive lists of its kind in the UK, NCOA data only includes the names and addresses of customers who allow their details to be passed on. To access the new address you compare your list of names and addresses and where you get a match, this database will provide you with their new address. It is as controlled as that, and for example you cannot list out all of the people who have just moved into an area over the past 6 months.

5 STANDARD ADDRESS DATABASE

One of the tools you will need is a standard address database, and that is where all of the Postal Administrations clearly have a key role to play. They are visiting every premise in their country on a daily basis. You could say that they have an in-built team of site surveyors - although they are usually referred to as The Postman or Postwoman!!.

In the UK, the standard address file is called PAF – Postcode Address File. This contains some 26 million addresses to sub-premise level and 1.7 million postcodes (an abbreviated form of the address to street level). (4)

Lets look at how powerful this data is. Take any postcode in the UK, e.g. TW11 9LY.

King Edwards Grove,
TEDDINGTON,
Middlesex
TW11 9LY

And where you do not know the postcode, just access the PAF data with part of the address, misspelt if necessary e.g. 24 kaw road,richmand will return:

24 Kew Road
RICHMOND
Surrey
TW9 2NA

The United States Post Office (USPS) have a range of standard address files, which link their addresses to either a 5 digit zipcode or for more precision to a zip + 4 code. e.g.

222 DABNEYS MILL RD
MANQUIN VA 23106 2506

So why should a postal administration produce such a master reference file:

5.1, With a correct address together with a postal code, they will be able to process the mail quicker and much more cost effectively. Royal Mail have stated that accurately postcoded mail goes through their automated system up to 20 times faster than when it has to be sorted by hand. (5) This means that they can keep their prices competitive. A clear example of this working in practice is the fact that from 1st April this year, the cost of 2nd class postage is to be reduced by a penny and the cost for 1st class letters remains unchanged.

5.2, You can improve the Quality of Service levels that you provide to your customers. Mail will be routed from its point of collection via the most direct route to its point of delivery, and not via totally unnecessary location(s). i.e. it will go A to B and NOT A to C and then to B.

5.3, You could make some additional revenue from sales of the address data files.

5.4, Effective use can be made of a Mailing Preference Service or Robinson list which will reduce the number of aggravated customers.

Overall it will lead to an increased level of customer satisfaction with the mail service, and an increase in the volume of mail traffic. In today's increasingly competitive market that must be good news.

6 ADDRESSES AND THEIR LOCATIONS

Every address has a geographical location, called a grid reference or x,y co-ordinate. Combination of the two datasets (addresses and grid references) can provide some additional benefits. You can then:

6.1, Plan, analyse and develop routing applications such as home deliveries of groceries and other retail goods.

6.2, Link in with vehicle navigation systems.

6.3, Analyse and plan utility services such as gas, electricity, water and telephone.

In the UK, Ordnance Survey have seen this potential with their dataset – ADDRESS-POINT, and An Post and Ordnance Survey Ireland have just announced GeoDirectory. This will contain over 1,200,000 accurate building addresses together with business address information. The database will be updated continuously with information coming from An Post and OSI surveyors. It's:

"The new standard that's right up your street" (6)

7 DELIVERY POINT IDENTIFICATION AND BARCODING

Something we are clearly going to see more and more of is the use of barcoding. Not surprisingly, the USA is trail blazing, where some 80% of the mail is already bar coded, but a number of other countries will soon be chasing it hard. In the UK, if you can identify the exact premise or sub-premise number then you can add a "delivery point suffix" or DPS to the address. For example, the DPS for the first address example I used, 35 King Edwards Grove, is 1S. If you send the postcode with the DPS to a printer, then this will generate a barcode.

But what's special about a barcode – in addition to sorting the mail to the correct delivery office, it can all sort the mail into delivery sequence. i.e. all the letters for 1 London Road can be put together, etc. The postman then receives his mail for delivery in the correct order. Of course there will always be some mail, mainly personal letters, that doesn't have a barcode on them which will need to be inserted manually into the delivery pack, but over time these items will be a decreasing portion of the total.

One country that has recently latched onto this is Australia, where they have just made an investment of some $500 million in new sorting equipment. This will read the barcode on the mail, sort it into delivery areas, and then sort it into delivery sequence. From the mailers point of view, all they have to do is include a barcode on the mail piece. They do not have to do any level of pre-sorting. But here's the key point – if you do NOT know the precise delivery address that matches the master reference database you cannot produce a DPID – a delivery point identifier and hence a barcode. (7)

For example:

4 Dalley St
SYDNEY NSW 2000

produces a DPID of 8321436

2 Dalley St
SYDNEY NSW 2000

does not return a DPID, so you cannot get a discount on this mail piece, nor will you get the advantages of quicker sorting and delivery.

8 ADDRESS STRUCTURES THROUGHOUT THE WORLD

The above addresses for the UK, America and Australia have some level of similarity in that the broad structure is:

Premise number, street name, street descriptor
Town name, County/state code and finally postcode/zipcode

Even the abbreviations for Street and Road look similar.

But what about:

Viale Europa 77
00144 ROMA RM

or

22 rue Dabray
06000 NICE

In the case of the Italian address, the street descriptor, Viale, is at the beginning of the first line and the premise number is at the end. The French example has the premise number back in its familiar position at the beginning of the address, but its then followed by the street descriptor – "rue" and this time the first letter is in lower case.

On the next line the postcode comes first before the town, and in fact in most cities there is only one postcode per town name.

In fact, according to the Global Source Book for Address Data Management by Graham Rhind, at the last count there were some 91 different formats from around the world. (8) The UK can be one of the most complex, where there are up to 12 different address elements, and 6 different formats for the postcode.

1 Gorse View,
Peak Lane
Kingston Gorse,
East Preston
LITTLEHAMPTON
West Sussex
BN16 1RW

Is an example with 7. Not surprisingly the Royal Mail is looking to see whether this is really the way forward and to whether they can reduce the number.

However for Europe there are in fact only 3 main address styles:

8.1, "continental" where the street name appears first (with the street type attached to the street name) followed by the building number e.g. Blomstervej 28. Germany and the Scandinavian countries generally use this format.

8.2, "modified continental" where the street type appears first, followed by the street name and then the building number, e.g. via Sacchi. Italy and the other Mediterranean countries generally use this format.

8.3, "modified UK" where the house number appears first, followed by the street type and then the street name separately. France uses this format. (9)

Details on every country's precise requirements can be obtained form the UPU – Universal Postal Union in Switzerland, (10) or from Graham Rhind's book. (8)

9 THE INTERNET

But maybe we do not need to concern ourselves with any of this. Surely the Internet is going to replace "snail mail" and the Postman. Independent studies have shown that the use of mail, particularly direct mail is still growing and that this will carry on for the next 10 years at least. And useful as the Internet is, theres many things it cannot do. If you order a book using a web site, you still need a physical delivery address to send the book to.

10 BENEFITS OF ADDRESS MANAGEMENT

and what benefits does Address Management bring you. There are eight main ones:

10.1, You will save increasing amounts of money on postage. The Royal Mail have a mailing cost analysis model "Mailmax" which demonstrates that a company which sends out a monthly mailing of 200,000 items with a 5 per cent error rate could be wasting over £100,000 per year. Or if you are just sending out 2,000 items per month with the same error rate, you could be wasting over £1,000 per year. (11)

10.2, You'll save increasing amounts of money on capturing addresses. Bradford and Bingley Building Society recently confirmed that on a typical mortgage application they may require up to 12 addresses – the applicants; their bank's; their new property; the estate agent's; two references; 2 insurance companies; their doctor etc. By using an on-line address capture and validation tool they saved 10 minutes per application!. Problems with poor quality addresses and regional accents were eliminated.

Get it right, first time, every time

10.3, You will enjoy increased customer satisfaction leading to additional business from the same customer, and overall higher retention levels (No more misspellings of customer addresses!)

10.4, You can eliminate non-deliverables. If all of your addresses match the standard address database, you can be sure that they will all be delivered safely.

10.5, Increased levels of de-duplication can be achieved. (or "double-clean" as its very aptly put in Germany) which leads to not only a reduction in your costs, but would also a reduction in aggravation from your customers and prospects.

Using the previous example, a mailing piece with the wrong spelling of Tonbridge Wells written on it would probably get delivered to the correct address in Tunbridge Wells. In fact, the Royal Mail may see it as a challenge to deliver it there, but unless you were using very clever de-duplication software, you would end up sending the same item of mail twice to the same person.

10.6, More efficient and faster delivery will be achieved, and you will secure a higher probability of the mail-piece being delivered the next day where you are using first class post.

10.7, You can undertake geo-demographic and statistical analysis of your customers and prospects. This can be done at individual household level.

10.8, And finally you will get access to mailing discounts with such schemes as Mailsort, Walksort and Customer barcoding. These discounts can go as high as 40% in the UK. Other countries offerings are different, but in every case there are rules on what needs to be included within the address and how it needs to be formatted.

11 SUMMARY

In summary:

Your customers are your biggest asset.

Contact them regularly and address them correctly

Use a standard reference address database, preferably provided and maintained by the local Postal Administration.

Enjoy the benefits and cost savings that good Address Management will bring you.

REFERENCES

(1) and (2) mail*plan* Spring 1999 published by the Post Office. This information correct at the time of printing. This information appears courtesy of the Post Office.

(3) and (5) Postcode Update 27 published by Royal Mail Address Management Products, Edinburgh.

(4) PAF is a registered trademark of the Royal Mail.

(6) GeoDirectory by An Post and Ordnance Survey Ireland

(7) Australia Post

(8) Global Source Book for Address Data Management by Graham Rhind

(9) European Postcodes by ALLM Systems and Marketing

(10) UPU – Universal Postal Union, Case Postale, 3000 BERNE 15, Switzerland.

(11) mail update – Managing Address Data Supplement published by Royal Mail. This information appears courtesy of the Post Office.

The security of next-generation postage meter devices

D COWARD
Royal Mail, UK
D PERKINS
Logica, UK

SYNOPSIS

The use of new technology in the postal businesses is developing rapidly, and improvements in the security of devices which contain or control monetary value, has provided a technology enabler so that step changes in the way in which postage meter devices can be used will be implemented in the near future.

This paper describes the background to the use of postage meter devices in the UK postal environment, discusses the existing and emerging technologies, and describes the security system that Royal Mail is proposing to use in the future.

1   INTRODUCTION

Postage meter devices (also known as Meters or Franking Machines) have been used in the UK by commercial enterprises for a significant part of the 20th Century for applying franking marks to envelopes as evidence of postage paid.

This paper discusses postage meter devices, the technological changes that have occurred, and the likely shape of the Royal Mail system which will utilise modern technology.

2   HISTORY

The benefits of postage meter devices to customers are to enable items of any size or shape to be prepared for posting at any time, by allowing customers to generate their own postage marks and affix them onto envelopes, because the postage meter is able to generate a variety of transaction values. For large posters, the postage meter device can form an integral part of a larger mailing system which may include folders and "stuffers" etc. Small business users have a perception that franking marks lend them an air of "respectability" which may or may not be accurate. The benefit to the commercial customer is offset against the cost of owning or leasing such devices, and the cost of consumables.

The technology used for postage meter devices when they were first manufactured was quite simple, and for the majority of their use this century the technology of using a pre-inking process and an image generated using some form of impact print, has not changed.

As electronics became more prevalent in manufacturing products the control systems for printing moved away from the electro-mechanical approach to micro-processor or micro-controller systems. However, the image generation was still created in essentially the same way; a pre-ink and an impact print. It was not until relatively recently that step changes in technology have allowed the use of alternative printing methods such as thermal transfer, ink jet and bubble jet printing.

3   SECURITY

From the perspective of the Postal Authorities, the primary security concern is the protection of postal revenue, viz., to ensure that all franking marks generated by a postage meter device are legitimate, which implies that

(i)   the user (or customer) has a contract with Royal Mail to allow the use of a postage meter device,

(ii)   the machine has sufficient credit stored upon it,

(iii)   credit has been paid for with some form of acceptable funds transfer from the customer to the Royal Mail,

(iv)   the machine model or version has been approved for use by Royal Mail

(v)   the machine produces, or prints, a legible, acceptable, mark.

Other features that would normally be part of a traditional machine design would be a cumulative total, often referred to as the "ascending register" (the credit total is often referred to as the "descending register"), physical protection of the sensitive or vulnerable parts of the machine, and provision for a Royal Mail security seal for identification or as evidence of tampering.

Resetting centres currently have a significant influence on the risks to Royal Mail's revenue. The protocol discussed in this paper limits the risk at resetting centres to an acceptable level; determined by the Key Distribution Centre which would be directly under Royal Mail's control.

4   CHANGES IN TECHNOLOGY

4.1   Electronics

The electronics industry has developed rapidly in recent years, and, consequently, more and more functionality has been installed on traditional style postage meter devices. Often, improvements have been made that benefit the end user , such as incorporating accounting facilities, but the more important improvement from the security perspective was the development of the ability of the machine to store, and protect from manipulation, the sensitive register information. The sensitive information has been held on a variety of electronic storage media both standard memory devices such as, for example, battery backed SRAM (or NVRAM) and in other cases company specific, bespoke, silicon devices. The robustness of such storage is obviously important to any postal authority, as is the need to prevent both unauthorised access to the secure memory and any manipulation of the information stored.

4.2   Encryption

In common with other industries that store and transfer sensitive monetary information, the postage meter market has been looking towards encryption to solve some of the problems associated with the transfer and storage of sensitive information. Encryption has been used on discrete storage devices within a postage meter, communications between sensitive areas of a postage meter and in communications with external devices, such as a re-crediting facility.

4.3   Smart cards

Once encryption was an accepted part of postage meter design, the obvious next step in the development of acceptable products, was to consider the use of industry standard technology that was accepted and used by the banking and finance industry. The smart card has been through a number of iterations, and the most recent versions appear to offer acceptable security, and adequate memory space for useful applications. Smart cards, either in their normal card size guise or as extracted and embedded chips, are likely to play a more significant role in this market as the technology develops, and the security improves.

5   The indicia

Most postage meter machines approved up until recently have produced an indicia that has changed little since the initial design. Each indicia consists of the town circle, which has information concerning the date and the registered sorting office, and the Crown die, which has the value of postage printed, and the serial number of the machine.

The information contained within the existing design of crown die is limited and relatively static. The dynamic elements are the date and the value. The changes in printing technology, and the increasing use of cryptographic methods to protect sensitive information has led to the investigation of the use of far more dynamic information printed in the indicia. Once the principle of using dynamic information as an integral part of the revenue protection system has been established then the opportunities to exploit the information flows for security, and other reasons, present themselves.

The current sorting equipment in use by Royal Mail has the ability to scan the whole of the envelope area, and then to focus on areas of special interest, such as the indicia area. Developments in "two dimensional" bar codes in recent years, when combined with the scanning ability of machines, have enabled the principle to be established of a significant amount of dynamic information to be printed in a relatively small area on the envelope and to be swiftly decoded by the automation equipment.

A number of international standards have been developed recently which address this area see Appendix I.

6   "PC" POSTAGE

Once the principles of dynamic information printing and the use of two dimensional bar codes have been established, an obvious next step in the development of postage meter devices is to utilise the ubiquitous personal computer (PC) to control most of the aspects of the generation and printing of an indicia. The use of existing equipment, the PC, to utilise the indicia generating software, and the use of the attached printer to produce the final printed envelope, has demonstrable benefits in terms of cost- effectiveness and functionality.

The PC approach can also be used by businesses whose envelope print runs may be large, by using the PC as the data control hub connected to a specialised high speed printer to print the dynamic information onto the envelopes or items to be posted. However, the use of such distributed technologies brings with it risk; the prevalence of the PC enables a large number of customers easy access to the ability to print postage, but also enables the small proportion of criminally inclined in our society the opportunity to attempt fraud using the same technology.

The expanding use of the Internet highlights the opportunity to produce variations on the PC postage theme, by, for example, producing a server version of the PC and connecting to an Internet Service Provider, and hence allow on-line production of legitimate indicia. In a similar way to the PC postage approach, Internet postage also has a number of risks associated with it which also need addressing.

7   SECURITY ASPECTS OF EMERGING TECHNOLOGIES

The section above outlines the likely expansion of PC and Internet postage. The list below highlights (but not exhaustively) a number of security concerns such as;

(i)   attacks on the software for generating indicia,

(ii)   generating replica indicia using legitimate software, but modifying small elements of the information printed to, for example, increase the value

(iii)   attacks on the hardware elements containing the revenue sensitive information

(iv)   copying the indicia, either digitally, or using a photocopier,

(v)   obliteration of elements of the indicia to prevent reading and decoding

The approach to dealing with such security problems is multi-faceted, and among the elements currently being considered are:

7.1   Hardware - the Postal Security Device (PSD)

The security requirements for the PSD are likely to be significantly different to the current testing and approval requirements in place in most postal authorities for franking machines. The step change in technology requires a different approach and a different set of tests to be conducted before a product gains approval from a Postal Authority. The international postal community has recognised the need for a change in approach and under the auspices of the Universal Postal Union, Technical Standards Board, a document has been produced which identifies the requirements for postage meters; " S30-1 International postage meter approval requirements". S30-1 borrows heavily from the American FIPS 140-1 standard.

When cryptographic keys are unique to a PSD, a successful attack on a single PSD does not affect the cryptographic keys of the other PSDs. A well designed PSD should have the ability to delete the cryptographic keys in response to an attack, which would render it unusable, however, there are other security considerations to take into account if they keys were to be deleted.

A PSD would be expected to have a secure real time clock so that timing constraints could be utilised when required.

7.2   Encryption

Encryption of the dynamic data in the indicia is likely to be employed, so that security checks can be utilised. There are a large number of cryptographic algorithms that are available at the present time, some of which are freely available, and some of which have licensing and export issues. Each cryptographic algorithm has advantages and disadvantages and the task for system designers is to understand the features of each approach and to reach an appropriate compromise.

7.3   Key Management system

Any security approach that uses encryption as part of the system has a further requirement to manage and maintain the key load and distribution network. An encryption system ceases to be secure if the keys have been compromised, thus the design and use of the key management infrastructure is crucial.

7.4   Postal Infrastructure requirements

Secure generation of encrypted dynamic information, and the subsequent printing of that information onto an envelope, is not, in itself, a complete secure system. At some stage the information needs to be decrypted and reconciled with expected or known customer and item information. The reconciliation process requires that the infrastructure for scanning is in place, and that the information flows required for decryption and customer account checking are enabled. The encryption keys would be generated and distributed by Royal Mail, so the communication infrastructure to support the key management system described above would also need to be enabled.

8   PROPOSED ROYAL MAIL SYSTEM

The system that Royal Mail is investigating at present is described below, and is shown as a schematic block diagram in figure 1.

Each mail customer possesses one or more PSDs which can be used to produce indicia in conjunction with either a PC and a printer or with a dedicated postal franking machine (PFM). As discussed above, the PSD is the only piece of customer equipment that is relied upon for providing security.

A customer uses the equipment to print 'proof of payment' indicia onto envelopes or labels. When the mail reaches a Mail Centre the indicia are scanned and payment is validated.

Each customer obtains postal value, which is used for generating indicia, from a Re-crediting Centre. This is transferred on-line onto the PSD. Payment for postal value is through means outside the Royal Mail system. Re-crediting Centres may be run by organisations other than the originator of postal value. The system permits them to be given a limited amount of trust.

Re-crediting Centres are given authorisation for distributing credit to customers from a Key Distribution Centre (KDC). KDCs are operated in highly secure conditions by a trusted organisation.

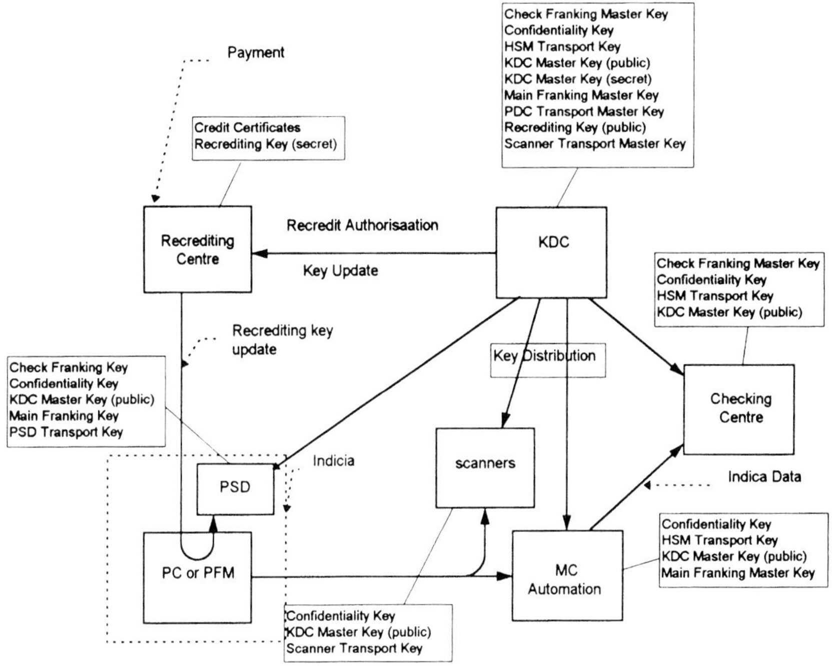

Figure 1
A schematic block view of the postal franking system.

The system components are described in the following sections

8.1   PSD

Each PSD contains:

(i)   a PSD Identifier, a unique identifier for the PSD,

(ii)   a number of operational keys, which are used for purposes such as producing indicia

(iii)   a key for decrypting encrypted keys

(iv)   Key Versions, which are identifiers for the versions of these keys and form an increasing sequence (modulo k, where k is a constant in the system)

(v)   the public part of the KDC Master Key, which is used to verify signatures of Credit Certificates (see section 3.2).

The PSD also contains two registers:

(i)   the Total Value Credited, the total value that has ever been credited to the PSD (known in the past as the ascending register).

(ii)   the Total Value Spent, the total value of indicia that have ever been produced by the PSD.

The value available to the PSD is the difference between the Total Value Credited and the Total Value Spent. Creation of an indicium adds to the total value spent. A re-credit operation adds to the total value credited. Neither of these values ever decreases.

A PSD's keys are updated on-line on a regular basis and on suspected key compromise. The Main Franking Key and Check Franking key are unique to the PSD that holds them, and may be derived for each PSD from master keys, the Main Franking Master Key and Check Franking Master Key. Both Main and Check Franking Keys are 128-bit Triple DES symmetric keys.

Additionally, each PSD has a unique 128-bit Triple DES Transport Key. The PSD Transport keys are derived from the PSD Transport Master keys held by the KDC.

8.2   Re-crediting Centre Keys and Data

Re-crediting centres contain:

(i)   a set of Credit Certificates (see below), at least one for each PSD for which they are authorised to re-credit

(ii)   one or more Re-crediting Keys which are unique to the Re-crediting Centre and are used to sign re-crediting data sent to PSDs

(iii)   the Maximum Total Value Credited for each PSD that may use the Re-crediting Centre, which includes the amount that the re-crediting centre is willing to distribute to the PSD.

(iv)   appropriate accounting data.

8.3   A Credit Certificate includes:

(i)   a Credit Certificate Indicator

(ii)   a PSD Id

(iii)   Key Versions of the current set of keys that must be held by the PSD

(iv)   a Limit of Total Value Credited that the Re-crediting Centre can assign to the PSD using this Credit Certificate [1]

(v)   a Re-crediting Public Key, the public key corresponding to the private key used by the re-crediting centre to sign the re-crediting data

(vi)   a signature covering the data in the Credit Certificate using the KDC master private key.

A credit certificate is a limited authority from the KDC to a re-crediting centre to re-credit a PSD. Credit certificates allow the KDC to limit the risk of trusting re-crediting centres, to some extent also permitting monitoring of credit usage on specific PSDs and can enforce the requirement that a PSD's keys are kept up to date.

8.4   KDC Keys and Data

The KDC contains:

(i)   KDC Master Keys which it uses to sign Credit Certificates

(ii)   Transport keys used for securing confidential data for each PSD

(iii)   Key Versions.

The KDC Master Key is a 1024-bit RSA key. The key needs to be considered generally secure, and yet be readily usable in the sort of limited functionality devices that are expected to be used for PSDs.

9   RE-CREDITING PROTOCOL

The re-crediting protocol adds new value to a PSD, and possibly also transmits control messages and changes to PSD keys. The protocol is illustrated in figure 2. It is initiated by the user of the PC or PFM and communicates with the Re-crediting Centre.

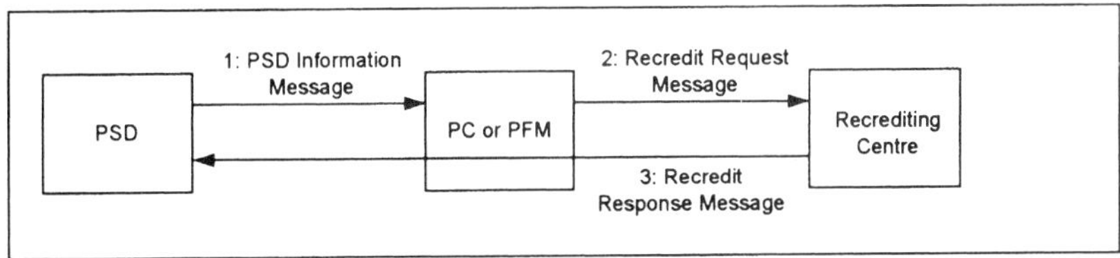

Figure 2: The Re-crediting Protocol

---

[1] There is a "loose" relationship between the reset requested by the user and (iv) above, i.e. the Re-crediting Centre does not have to communicate with the KDC for every transaction- only those which exceed (iv), except where the transaction is performed in advance.

9.1   Message sequence

The aspects of the re-crediting protocol that are relevant to cryptography are contained in a Re-credit Request Message from the PC or PFM to the Re-crediting Centre and a Re-credit Response Message from the Re-crediting Centre to the PC or PFM. The PC or PFM requests some basic information about the PSD's state which it combines with the value requested and sends to the Re-crediting Centre. The response from the Re-crediting Centre is passed straight to the PSD.

In addition to a re-crediting instruction for the PSD the Re-credit Response Message can contain PSD control instructions and key update messages. The control commands can be used to disable a PSD. The key updates must be entered into the PSD before the re-crediting instruction.
Key updates should take place from time to time and new keys are distributed to PSDs during re-crediting operations. This ensures that any PSD that needs re-crediting will have the latest keys (helping to control key lifetimes) and can also allow timely key update in the case of suspected key compromise.

The re-crediting protocol has the following properties:

(i)   Multiple re-credits are prevented since the PSD checks the Total Value Credited in an incoming message against the Limit of Total Value Credited in the Credit Certificate.

(ii)   Re-crediting value cannot be lost because a request can be reissued.

(iii)   Re-crediting cannot take place if a PSD holds keys that need to be replaced.

Note that the protocol relies upon knowledge of the Maximum Total Value Credited at a Re-crediting Centre. The Maximum Total Value Credited is incremented as a result of a payment associated with the PSD or an extension of credit. If PSDs possess an initial credited value, this must be known by the Re-crediting Centre.

9.2   PSD information message

On request from a PC or PFM a PSD returns the following data:

(i)   the PSD Id of the PSD

(ii)   Key Versions held by the PSD

(iii)   KDC Master Key Version held by the PSD

(iv)   Total Value Credited held by the PSD.

This message is used by the PC or PFM to construct a Re-credit Request Message. No encryption is provided because without an element of variable data, the PSD could be vulnerable to attack from its users.

9.3   Re-credit Request Message

The PC or PFM appends information to the PSD Information Message about the value to be credited, to give:

(i)   PSD Id, from the PSD Information Message

(ii)   Key Versions, from the PSD Information Message

(iii)   KDC Master Key Version, from the PSD Information Message

(iv)   Total Value Credited, from the PSD Information Message

(v)   Total Value Spent, from the PSD Information Message

(vi)   Requested Value, a positive financial value to be added to the PSD's Total Value Credited.

The Re-crediting Centre accepts a Re-credit Request Message if there is sufficient credit associated with the PSD Id, i.e. if Requested Value + Total Value Spent $\leq$ min(Maximum Total Value Credited, Limit of Total Value Credited). Note that a Re-credit Response Message can be sent out any number of times, even for the same Total Value Credited and different Requested Values. The processing is illustrated in figure 3. The Re-crediting Centre has the total value paid since licensing (the maximum TVC), thus it can ensure that it never accepts requests with a forged TVC, so the request message doesn't need to be encrypted.

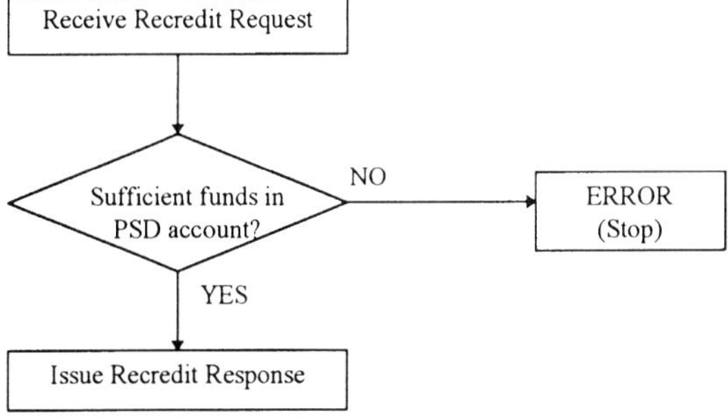

Figure 3: Processing a Re-credit Request

9.4   Re-credit Response Message

In response to a valid Re-credit Request Message a Re-crediting Centre returns a Re-credit Response Message to the PC or PFM. A re-credit Request Message consists of one or more blocks:

(i)   Key Update Blocks (a possibly empty sequence)

(ii)   PSD Control Block (optional)

(iii)   Re-credit Response Block.

The PC or PFM receiving this message must process each part of the message where present in the order in which they appear in the message. The order of processing is illustrated in figure 4.

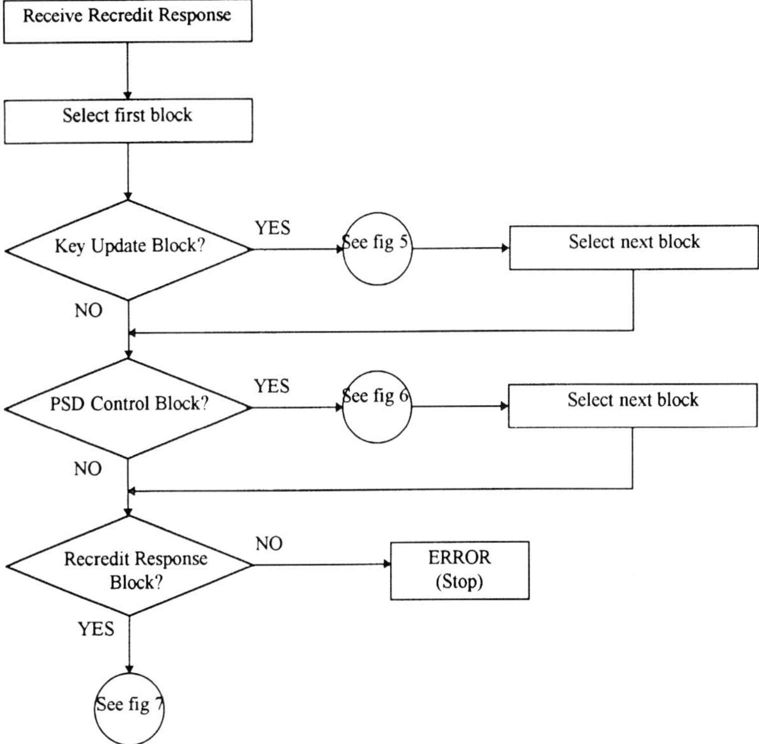

Figure 4: Processing a Re-credit Response Message in a PSD

9.5   Key Update Block

The Key Update Block contains a new value for keys held by the PSD:

(i)   Key Update Indicator, a unique identifier

(ii)   the PSD Id of the intended destination PSD

(iii)   new Key Versions, for the updated keys

(iv)   the Encrypted Keys, to be the replacements

(v)   a signature using the KDC master key[2] covering the above data.

On receiving a Key Update Block the PSD:

1   checks the Key Update Indicator, then

2   checks that the PSD Id is correct, then

> 3   checks that the new Key Versions do not cause old keys to be reinstalled, i.e. that $(new\ Key\ Versions - Key\ Versions)$ mod $k < x$,
>
> where x is a value configured into the PSD and $x << k$

4   verifies that the signature on the message is correct, then

5   updates all of its keys with those in the Key Block and replaces the Key Versions indicator with the new Key Versions indicator.

This is illustrated in figure 5.

---

[2] A key update block may contain a new KDC master key. The version of the KDC master key used to sign a key update block should be the one expected before the update.

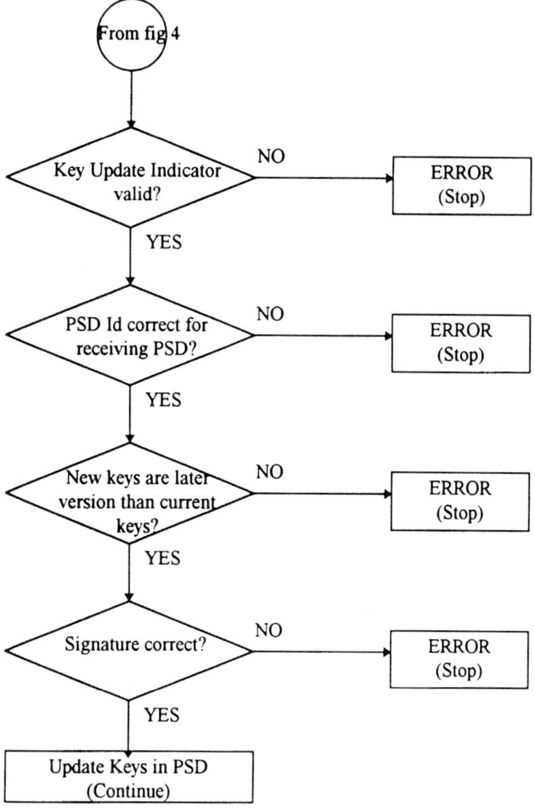

Figure 5: Processing a Re-credit Response - the Key Update Block in a PSD

9.6 PSD Control Block

A PSD Control Block is a facility initially intended for controlling PSDs remotely, for example disabling the PSD. The PSD Control Block contains

(i)   PSD Control Block Indicator, a unique identifier.

(ii)   the PSD Id of the intended destination PSD

(iii)   an Instruction, a value indicating the action that the PSD should perform, e.g. to cease issuing indicia

(iv)   a signature using the KDC master key covering the above data.

On receiving a PSD Control Block the PSD:

1   checks the PSD Control Block Indicator, then

2   checks that the PSD Id is correct, then

3   checks that the Instruction is recognised, then

4   verifies that the signature on the message is correct, then

5   performs the instruction.

This is illustrated in figure 6.

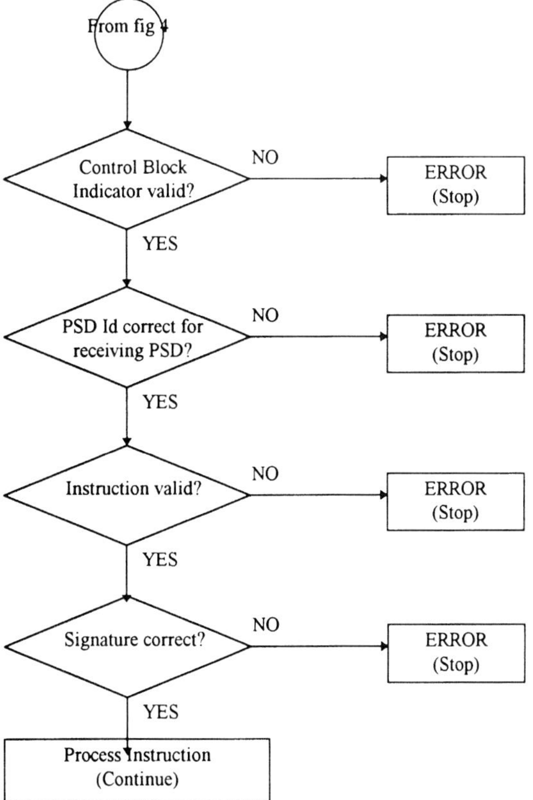

Figure 6: Processing a Re-credit Response - the PSD Control Block in a PSD

9.7   Re-credit Response Block

A Re-credit Response Block contains

(i)   Re-credit Response Indicator, a unique identifier

(ii)   the PSD Id of the intended destination PSD

(iii)   the Key Versions after all key update operations have been performed

(iv)   New Total Value Credited, equal to the sum of the supplied Total Value Credited and the Requested Value

(v)   a signature using the Re-crediting Key covering the above data

(vi)   a Credit Certificate for the Re-crediting Key used in the above signature.

The PSD only accepts the Re-credit Response Block if:

(i)   the block has a Re-credit Response Indicator

(ii)   the PSD Id field in the Re-credit Response Block matches that held in the PSD

(iii)   the Key Versions field from Re-credit Response Block matches that held in the PSD

(iv)   the Total Value Credited is greater than the equivalent field held in the PSD

(v)   the Limit of the Total Value Credited in the Credit Certificate is not less than the New Total Value Credited in the Re-credit Response Block

(vi)   the Re-crediting Public Key can verify the signature in the Re-credit Response Block

(vii)   the KDC Master Public Key held by the PSD can verify the signature in the Credit Certificate.[3]

If so, the PSD adds the Credit Requested to its Total Value Credited.

This is illustrated in figures 7 and 8.

---

[3] If a KDC Key Update is contained in the message, the new key should be used to check the signature.

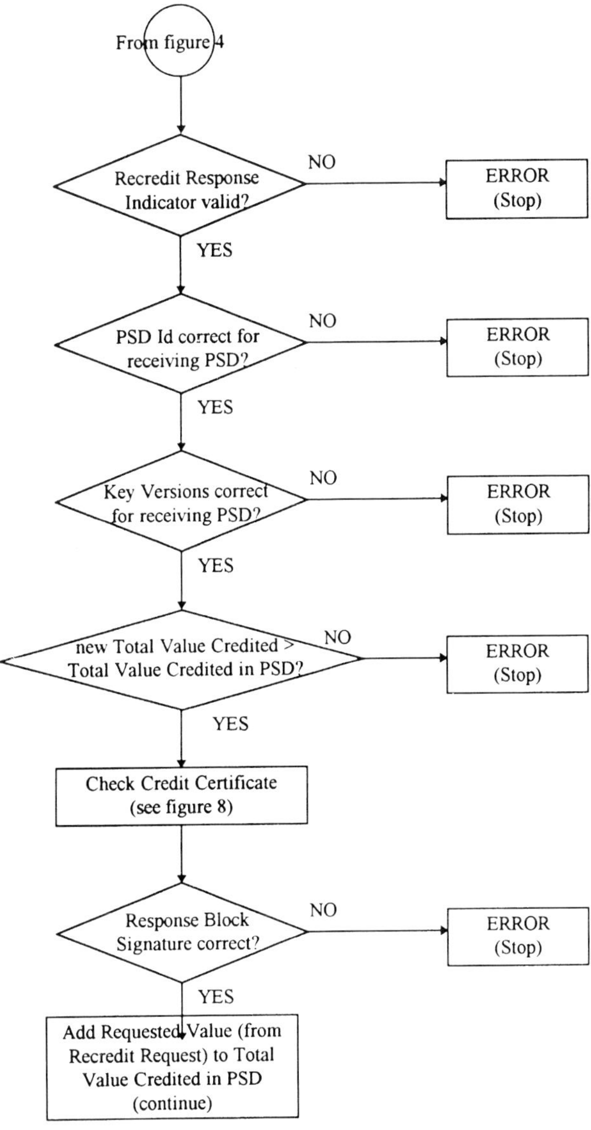

Figure7: Processing a Re-credit Response - the Re-credit Response Block in a PSD

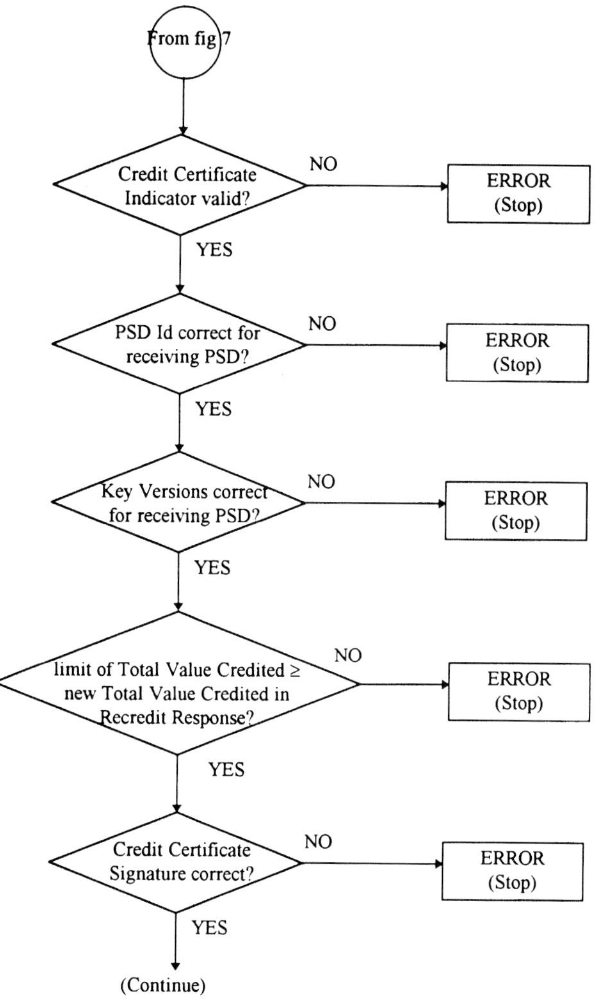

Figure 8: Processing a Re-credit Response - the Credit Certificate in a PSD

10 RE-CREDIT AUTHORISATION PROTOCOL

The re-credit authorisation protocol enables Re-crediting Centres to obtain Credit Certificates for new PSDs, periodically to obtain increased limits on their Credit Certificates or to update Credit Certificates for use with a new KDC Master Key. The protocol is illustrated in figure 9.

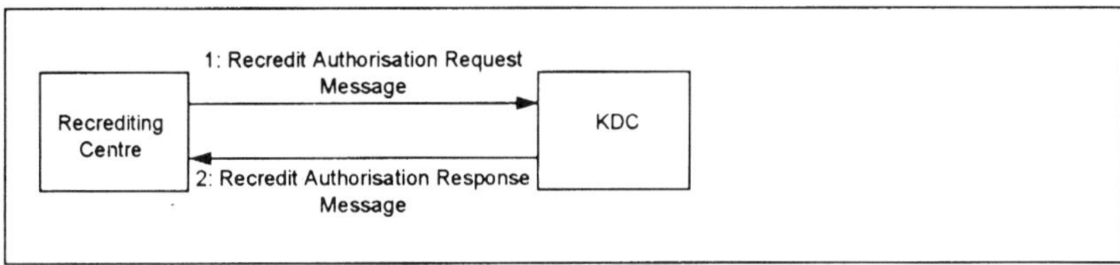

Figure 9: The Re-credit Authorisation Protocol

10.1 Message Sequence

The re-credit authorisation protocol consists of two messages.
Before this protocol can be used the KDC needs to possess an authentic copy of the Re-crediting Public Key. This is by means outside the scope of this protocol.

10.2 Re-credit Authorisation Request Message

A Re-crediting Centre sends a Re-credit Authorisation Request Message to the KDC to request a new Credit Certificate. A Re-credit Authorisation Request Message contains:

- Re-credit Authorisation Request Indicator, a unique identifier

- Re-crediting Centre Id, the source of the request

- PSD Id, the Id of the PSD to which the Credit Certificate is to apply

- a signature using the private Re-crediting Key covering the above data.

The KDC must check:

- the Re-credit Authorisation Identifier

- that the PSD is associated with that requesting Re-crediting Centre

- that it wishes to provide a Credit Certificate for the identified PSD and Re-crediting Centre

- that the signature for the message is valid.

10.3   Re-credit Authorisation Response Message

The KDC sends a Re-credit Authorisation Response to the Re-crediting Centre. The Re-credit Authorisation Response Message contains a Credit Certificate and possibly also PSD control and key update instructions to be passed on to the PSD.

- Key Update Blocks (a possibly empty sequence)

- PSD Control Block (optional)

- Credit Certificate.

The Re-crediting Centre stores the information from this message. When any of the optional fields is present they must be included in the next Re-credit Response Message sent to the PSD.

11   SUMMARY

In summary, it is clear that recent changes in technology have opened up a wide range of options for customers, manufacturers and Postal Authorities, for the generation and capture of indicia which contain dynamic information. However, the opportunities and benefits do not come without risk, and that risk must be carefully evaluated, understood, and managed.

With the system described in this paper, Royal Mail largely determines the overall level of risk to their cryptographic keys, and hence to their revenue.

APPENDIX I

References

S30-1 International postage meter approval requirements. Release date 28/10/98

    UPU Technical Standards Manual

S25-1 Data constructs for the communication of information on postal items, batches and receptacles. Release date 15/7/98.

    UPU Technical Standards Manual

S27-1 Framework for Communication of information about postal items, receptacles and batches. Release date 15/7/98.

    UPU Technical Standards Manual

S28-1 Communication of postal information using two-dimensional symbols. Release date 15/7/98.

    UPU Technical Standards Manual

FIPS 140-1 Security requirements for cryptographic modules. Issued 11/1/94

    National Institute of Standards and Technology, USA.

UK IT Security Evaluation and Security Scheme

    Department of Trade & Industry, UK.

C568/009/99

Print quality for machine-readable postal indicia

C ZELLER, D MACKAY, W BROSSEAU, and **J AUSLANDER**
Advanced Concepts and Technology, Pitney Bowes Inc, Connecticut, USA

ABSTRACT

Data Matrix two-dimensional bar code symbols were printed with various mailing machines and printing systems on a set of envelopes representative of the U.S. mail spectrum. These specimens were used to compare readability measurements made using a commercial bar code reader, a bar code verifier, and a machine vision instrument. The statistics of the principal image quality attributes and symbol quality parameters were compiled for the various readers and envelope/printer combinations. Analysis of the results permits quantification of the effects of paper/ink interactions and substrate dependence on the readability success rate. The results of this study can be used to develop print quality specifications to assure machine readability of postal indicia.

1. INTRODUCTION

To automate mail processing, Postal Services worldwide are introducing technical specifications for digital franking machines and new electronic postage standards. The UPU (1) and USPS (2) require printing a two-dimensional bar code in the Postal Revenue Block with a successful readability rate of 99.5% to 99.8%. To achieve these high readability rates, certain print quality attributes must be within acceptable limits.

The bar code industry has little experience using 2D bar codes in a non-controlled (open system) environment such as mailing. Printing on filled mailpieces with their uneven surfaces presents a specific print quality concern, which could be overcome with the use of an application standard. No international standards exist for print quality attributes of machine-readable postal indicia (or franking marks). A quantitative specification of these parameters is essential for the development of products that meet the needs of the Postal Services.

The objective of this project was to develop specific limits for the values of print quality attributes that will assure that a bar code image can be scanned and successfully read with

available scanning systems. The process we used to evaluate print quality and machine readability was:
1. Generate print samples.
2. Measure machine readability.
3. Measure print quality.

2. GENERATING PRINT SAMPLES

2.1 Sample image
Samples were printed under a variety of conditions including different print speeds and different gaps between the printhead and the paper to produce images with a range of print quality. The test images contained a number of graphic elements to facilitate measurements. The elements of interest are described below.

2.1.1 2D bar codes
A Data Matrix two-dimensional (2D) bar code symbol was created with an element size of approximately 0.020" (0.5 mm). The dimensions of the symbol were determined by the printhead height. All symbols were created according to the ANSI/AIM specification (3).

2.1.2 Solid black square patterns
Two 0.25" solid black squares were used for measuring large-area print quality attributes such as mottle, graininess, reflectance, and fill (or void area), and to automate sequences of the print quality measurements.

2.1.3 Lines
The print quality pattern included both horizontal and vertical lines of different thickness. The vertical lines were perpendicular to the direction of motion of the mail. The thick lines were approximately the width of an individual module of a 2D symbol. Of primary interest were the trailing edges of the vertical lines; these showed the worst print quality due to satellite phenomenon (small, slow-moving ink fragments that land on the paper some distance away from the intended target). The print quality data from the thick vertical lines were plotted versus the machine readability data in figures 5 to 8.

2.2 Printing technology, printheads, and ink
A number of different printing technologies and printing conditions were used to print the sample images. Ink jet printing technologies studied included thermal ink jet, piezoelectric ink jet, and solid-ink piezoelectric printheads. Commercial laser and ink jet printers were used to establish a baseline for performance comparisons.

2.2.1 Thermal ink jet technology
Two thermal ink jet technology printheads were used, one with vertical resolution of 600 dots per inch (dpi, nozzle spacing) and the other with 360 dpi. The 600 dpi head printed at a resolution of 600V x 300H dpi (300H dpi being the resolution in the direction of the mail transport). The 600 dpi printhead used a black pigmented ink (see figure 1 for the ink spectral characteristics). The 360 dpi printhead generated images at a resolution of 360V x 240H dpi with a black dye-based ink (figure 1). The primary consideration in selecting the operating points for the printheads was that the print quality was degraded enough to produce defects that require the use of Error Correction Codewords (ECC) when decoding the 2D symbols.

2.2.2 Piezoelectric printing technology
Two piezoelectric ink jet printheads were included in the study. One used a low optical density water-dye-based black ink and printed at 160V x 430H dpi; the second printhead used a black solid ink and printed at approximately 300V x 300H dpi. As with the thermal ink jet printheads, operating points were selected that required ECC to decode the bar codes.

2.3 Printing platforms

The different printing platforms used to evaluate the printing technology are described below.

2.3.1 Commercially available printers

Baseline tests were done under nominal operating conditions to evaluate the fundamental printing performance of different printing technologies. A baseline evaluation of the printing technology was done using premium paper on commercially available office printers where possible. Fifty images were produced with each printer.

2.3.2 Laboratory printer

Where a commercial printer was not available, the printing technology was put into a laboratory fixture that moved a white paper tape past the printhead under well-controlled conditions to generate baseline performance data.

2.3.3 Mailing machine transport

To generate the print samples on a variety of commercially available envelope stocks, a specially adapted Pitney Bowes mailing machine, which provided the ability to vary important printing parameters, was used. With this printer platform, fifty test images were generated on the envelope test set with each of the printing technologies of interest. The print quality and machine-readability measurements were taken from these specimens.

2.4 Substrate (envelope) materials

Due to large variations in ink and paper properties, it is extremely difficult to categorize properties important to digital franking. The main consideration in the selection of envelopes was to replicate, as nearly as possible, those envelopes that were most prevalent in the U.S. metered mail stream. Test envelopes were purchased from popular stationary outlets. In addition, the test set has been updated. Recycled materials and speckled envelopes were added to satisfy latest market developments.

3. MEASURING READABILITY

3.1 Scanners

Scanners differ in their spectral response, scanning resolution, optical systems, decode algorithms, and lighting uniformity. Because neither hand-held scanners used in automatic data collection systems nor in-line postal scanners has been specially designed for postal indicia reading, we must evaluate the impact of their specifications on the range of ink and paper optical characteristics found in the mailing environment. Printing parameters therefore have to be designed to optimize readability. Following is a description of the scanners used to measure bar code readability.

3.1.1 Hand-held scanners

Scanners usually decode the symbol and output the data, but offer nothing else in terms of symbol quality parameters. This limitation may be overcome by modifying the scanner software to allow for the capture of used ECC (UECC) for each scan. Such a modification provides more information than simply decoding the symbol. We therefore had a commercial scanner modified to support the evaluation process. The hand-held scanners were operated from a test stand to minimize operator error. In addition, the scanning height was controlled to maintain a camera resolution of 5-7 pixels per element size. The scanner was operated under normal lighting conditions (mixture of daylight and fluorescent illumination). In some instances, depending upon the light, the scanner used assisted LED illumination in the 660 nm range.

3.1.2 Verifier

Verification is an essential component of any automatic identification application, ensuring quality symbol generation and, as a result, down-line scannability. We used a commercial verification system that provides both verification reliability and 100% compliance with the AIM

Uniform Symbology Specification. The verifier identifies an overall symbol grade, reference decode, symbol contrast, print growth, axial non-uniformity, and the number of unused ECC.

3.1.3 Imaging system
A machine vision system was used for measurement and inspection of bar codes. The system allows for precise control of the illumination of the test samples and choice of optics. The control of illumination and optics provides a standard against which all hand-held scanners may be compared.

3.2 ECC statistical method for determining readability

This next step in the evaluation process measures the readability of 2D barcodes. Samples (typically fifty) of bar-coded envelopes were read many times to generate the statistics of UECC. UECC compensate for imaging defects that affect the readability of particular codewords. Several types of printing defects can require the use of ECC to decode a symbol and are defined here to clarify and classify defect types.

3.2.1 Systematic printing defects
Systematic defects can be continuous or intermittent; they are generally macroscopic, repeatable, and detectable with the naked eye. Examples of these defects are missing or misdirected nozzles and smears or smudges. Systematic defects always occur at the same position on a series of symbols. Systematic defects are not the subject of this study.

3.2.2 Random printing defects
Random defects are generally microscopic. They are due to ink/paper interactions and can be more specifically categorized as soft and hard defects.

3.2.2.1 Soft Defects
Random soft printing defects are distributed and occur randomly when a given symbol is scanned more than once. They can be subtle defects that occasionally cause a codeword to become unreadable. In addition to printing defects, system variables that can contribute to making a codeword unreadable include light non-uniformity, scanner noise, and the software decoding algorithms. Random soft printing defects are not severe enough to prevent decoding of a particular codeword every time it is attempted. When other system variables combine unfavorably, the codeword becomes unreadable. These defects are not predictable or repeatable.

3.2.2.2 Hard Defects
Random hard printing defects are localized and always seen at the same position each time a particular symbol is scanned. They occur randomly from symbol to symbol. These defects are not repeatable, and the printing defect in the symbol is severe enough that the scanning system cannot ever decode it. This is true even when other system variables are favorable.

3.2.3 UECC measurements and data analysis
Data from measurements of the number of ECC used to decode a symbol are graphically displayed as figures 2–5.

Figure 2 shows the results of fifty scans on each of fifty different symbols, for three different envelope types (producing three levels of relative print quality). These histograms provide insight into the distribution of the data as well as how close a particular symbol is to consuming all available ECC to decode the barcode. Examination of the individual histograms corresponding to fifty scans of each symbol shows that the minimum ECC used is always zero. These histograms contain exclusively soft errors.

Figure 3 gives the overall results of fifty scans of fifty different symbols for the same three envelope types in figure 2 but printed at lower print quality. Compared to the first series, there is a drastic change in the shape and position of these histograms on the x-axis. These histograms contain both soft and hard errors.

Figure 4.1, which corresponds to fifty scans of a single barcode, shows that the minimum of ECC used to decode the barcode is greater than zero. Some codewords cannot be read even after attempting fifty times. This minimum value represents what we refer to as the hard errors. Figure 4.2 shows the distribution of 2500 scans for fifty envelopes. Figure 4.3 represents the hard failures for the envelope set in figure 4.2 (the minimum UECC required to decode each of fifty symbols).

3.2.4 Postal readability rate predictor
One objective of this study is to define a test procedure for estimating the probability of achieving a required readability rate with a given envelope set, scanner, and printer. This procedure must estimate the effect of print quality on readability rate while eliminating operator and environmental effects. The scanner must output the number of UECC for each scan. The scanning algorithm should be commercially available, preferably the reference decode algorithm (3), modified to output UECC.

A statistical test is performed on a sample of N same-type envelopes printed under nominal conditions. The samples must pass a visual test with no obvious defects such as missing nozzles, smudges, or smears (systematic macroscopic printing defects). We start with N = 50, but one goal is to determine a minimum effective value for N. Each envelope is scanned many time (S ~ 50). The measured sample is S x N (S scans per envelope on N envelopes).

When UECCmin is greater than zero, it corresponds to the number of hard errors—codewords that cannot be, or most probably will not be, decoded by a given system (scanner, software, and filter). The random variable for this procedure is UECCmin. The larger the number of scans S for each envelope, the more accurate and more certain the estimate of UECCmin. The value of S needs to be consistent with actual operating conditions. A statistical analysis of the data shows that the histograms of UECCmin are, in general, reasonably well represented by a binomial distribution (hard defects), while the soft defects are represented by a Cauchy distribution. We are currently developing a procedure to grade the readability and predict (within a certain confidence level, typically 95%) the minimum guaranteed readability rate.

4. PRINT QUALITY MEASUREMENTS AND CORRELATION TO READABILITY

4.1 Print quality measurements
The print quality tests described in this paper were carried out in compliance with the methods and procedures described in the International Print Quality Standard ISO 13660. The measurements are of two basic types: solid-area measurements including reflectance, mottle, and graininess; and line measurements including line width, edge raggedness, edge blur, contrast, and fill.

4.2 Correlation to readability
The final objective of this study is to develop and understand the correlation between print quality and machine readability. We analyzed ink jet printing on envelopes and reduced the parameters controlling print quality for machine readability to the most significant. We plotted the number of UECC as a function of these print quality attributes (fig. 5–8). It can be seen that over some range of print quality attribute values (i.e., contrast, raggedness, blur, and line width growth) there is no relationship to the number of UECC required to decode a barcode (no UECC are required). However, the UECC required to decode a symbol begins to rise very dramatically at a threshold value of print quality. It is these values we are most interested in identifying. We need to do a statistical evaluation to say that with a high degree of certainty we can achieve 99.5% readability as long as the print quality is within specified limits.

Once this correlation is understood, it will be possible to make print quality measurements of a particular printing technology and determine if it will meet the application requirement for a mailing machine.

4.2.1 Used ECC versus print contrast
The relationship of UECC versus print contrast is plotted in figure 5. Clearly, most images with lower average contrast require more error correction codewords to decode the bar code. There are a few exceptions where symbols with low contrast were decoded using no ECC, but these are print samples made with high optical density inks on dark envelopes (R(ink) = 0.12, R(paper) = 0.33). These envelopes are not of practical interest in mailing applications because they do not meet the USPS minimum reflectance standards: 45% in the green portion of the optical spectrum and 50% in the red portion (5). However, these data points are interesting in that they help us to understand the importance of ink optical density and in creating symbols to be machine-read. If the ink is of sufficient darkness and other print quality attributes are within certain limits, even low-contrast symbols can be decoded. Sufficient print contrast would appear to have the ability to mask other print quality defects.

4.2.2 Used ECC versus line growth
Figure 6 shows a plot of line growth versus UECC. This particular plot is not very insightful in that there are data points with low print growth that use a large amount of ECC and values with high print growth requiring little UECC. Upon closer examination of the data it can be seen that these points correspond to print samples generated with a printer that used low optical density ink (relatively high reflectance). When this data is removed and the remaining data plotted on a different scale, it can be seen that a bar code reader is much more likely to require more UECC to decode a symbol when values get above 25%. The 25% maximum value corresponds to a C grade according to the AIM specification. Our expectation is that, as we generate more print samples with line growth above 25%, the correlation to an increased consumption of ECC will be more evident.

4.2.3 Used ECC versus blur
Figure 7 depicts the relationship between blur and UECC. The amount of UECC increases for blur values above 200 μm. Once again, if low-contrast print samples are removed, the blur value where UECC increases dramatically can be shifted to 300 μm. We have selected 300 μm as the appropriate maximum value for blur with the caveat that the optical density of the ink should be relatively high (print contrast ratios above 70%).

4.2.4 Used ECC versus raggedness
The effect of raggedness on UECC is shown in figure 8. The use of ECC begins to increase significantly at 30 μm. Our recommended limit is set at 35 μm. Again, this is done with the provision that ink yielding print contrast ratios of 70% or greater should be used when raggedness values approach this limit.

5. SUMMARY OBSERVATIONS

It becomes apparent that one can trade off gains in one area of print quality for degradation of another, perhaps less dominant print quality attribute. For example, it may be desirable to run a mailing machine at higher speed to obtain the required performance and suffer increased blur and raggedness as long as print contrast and line growth remain within acceptable limits. The reason is print quality attributes are interdependent and not linearly correlated.

Since many print quality attributes make up a printed sample, it is difficult to boil the print quality requirements down to a simple few. That is what this project attempts to do. Based on our data and experience to date, we would propose (Table 1) the following print quality attributes and tolerance limits as a guideline to achieve print quality consistent with the high readability required UPU (1) and USPS (2). As previously stated, our results must be considered preliminary and subject to change when a model is developed and the correlation between attributes is better understood.

Table 1

PRINT QUALITY ATTRIBUTE	VALUE
Print contrast ratio (PCR)	70% minimum
Line growth	25% maximum
Print reflectance difference (PRD)	0.40 minimum
Edge blur (one side)	60% of line width
Edge raggedness	7% of line width

6. CONCLUSION

In this paper we have demonstrated that the statistics of UECC permit a measure of bar code print quality and a distinction between two types of printing defects. Furthermore, the series of tests reported in this study illustrates the effects of print quality attributes (growth, contrast, blur, raggedness) upon the consumption of ECC. The methodology is applicable to any symbology using redundant codewords for error correction.

The results of this study, after some expansion, will be used to set the tolerance limits of the print quality attributes described in standards under development and to develop a method that will provide the Postal Services, manufacturers of postage meters, and providers and customers of PC-based postage-payment systems with a practical and objective way to communicate and evaluate basic print quality parameters of machine-readable postal indicia.

7. REFERENCES

1. UPU Technical Standards Manual S28-1, "Communication of Postal Information using Two-Dimensional Symbols," 15 July 1998

2. Information-Based Indicia Program (IBIP), "Performance Criteria for Information-Based Indicia and Security Architecture for Closed IBI Postage Metering Systems (PCIBI-C)," USPS, January 12,1999.

3. ANSI/AIM BC11-1997, "International Symbology Specification—Data Matrix."

4. C. Zeller, J. Auslander, W. Brosseau: "Print Quality Parameters and Decodability of PDF 417 Two-Dimensional Symbology, Information Based Indicia Program," *Technology Symposium Proceedings*, McLean, Virginia , November 25-26, 1996

5. USPS Publications, "Designing Letter Mail," *Publication 25*, August 1995

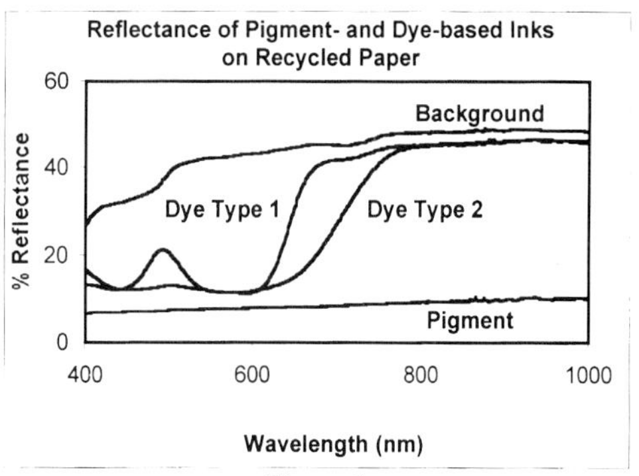

Figure 1

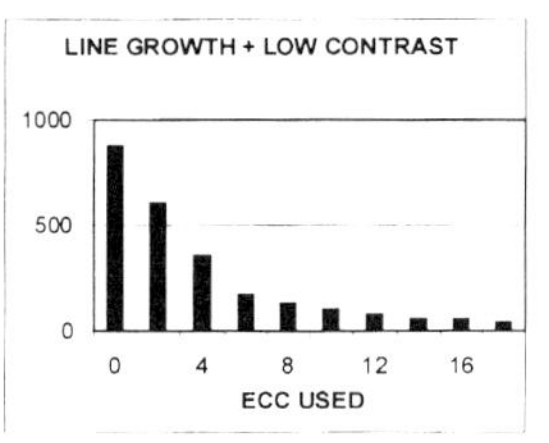

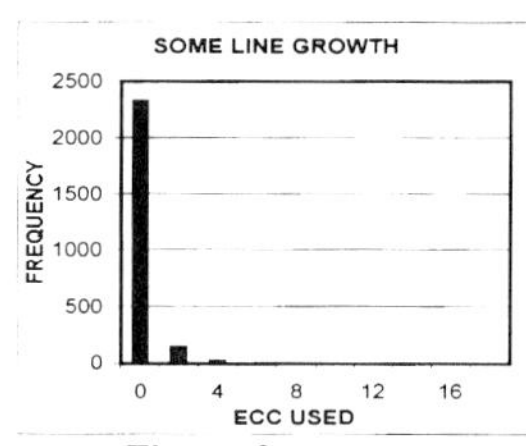

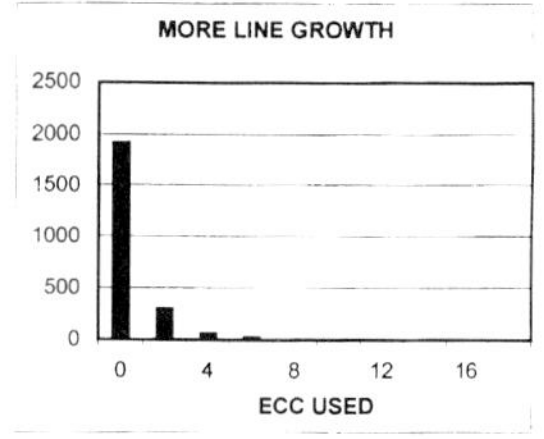

Figure 2

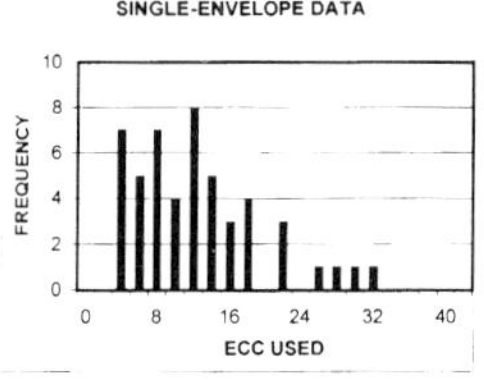

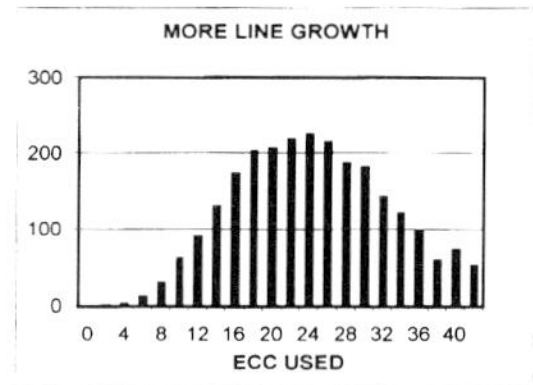

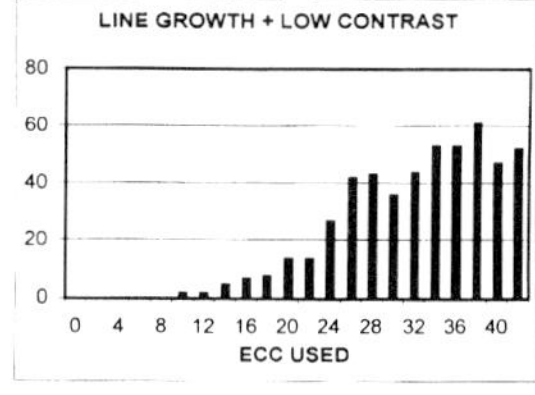

Figure 3

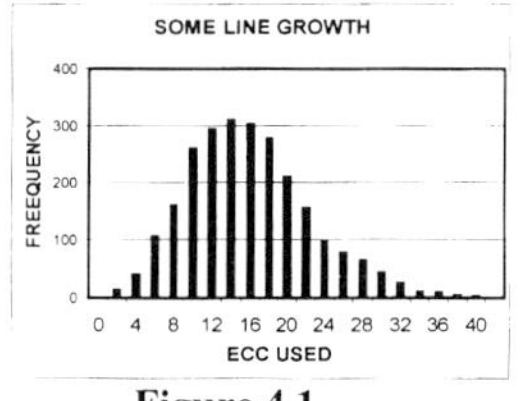

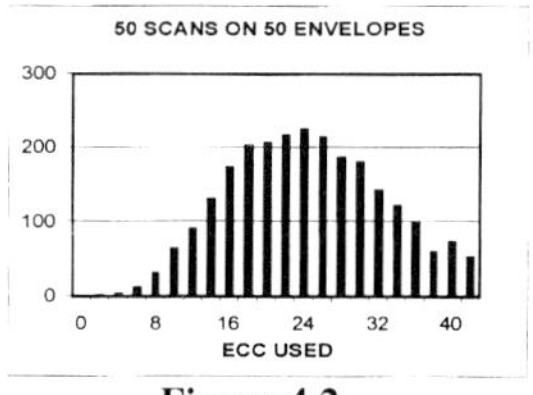

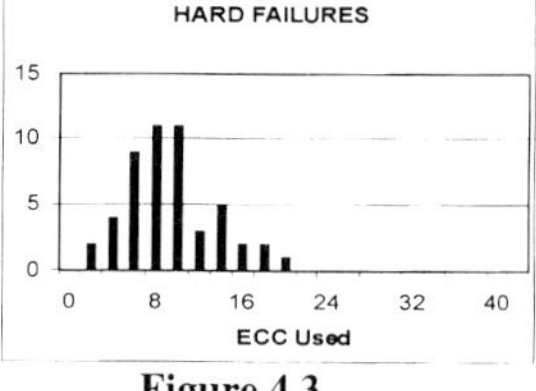

Figure 4.1   Figure 4.2   Figure 4.3

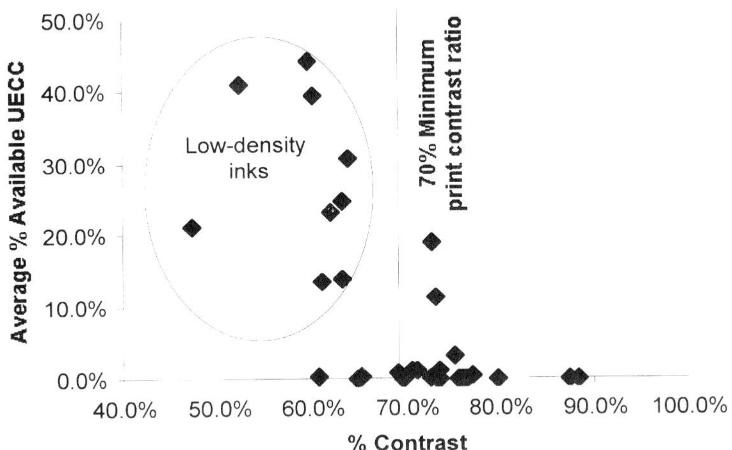

Figure 5

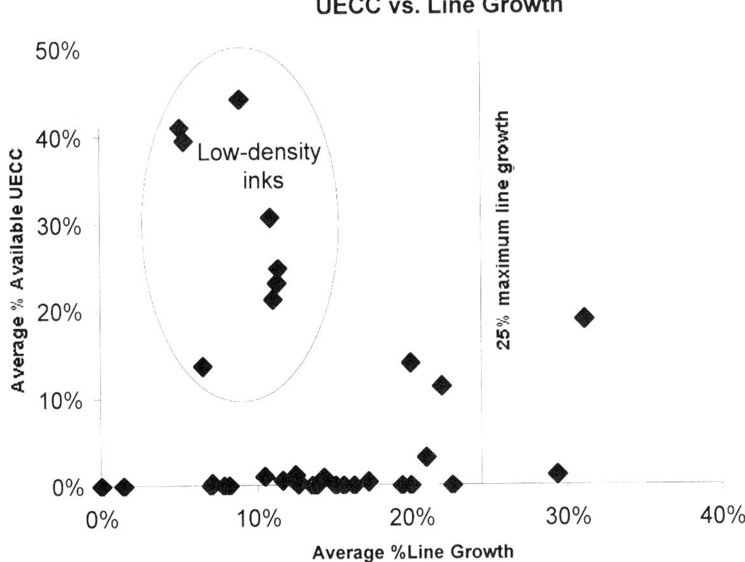

Figure 6

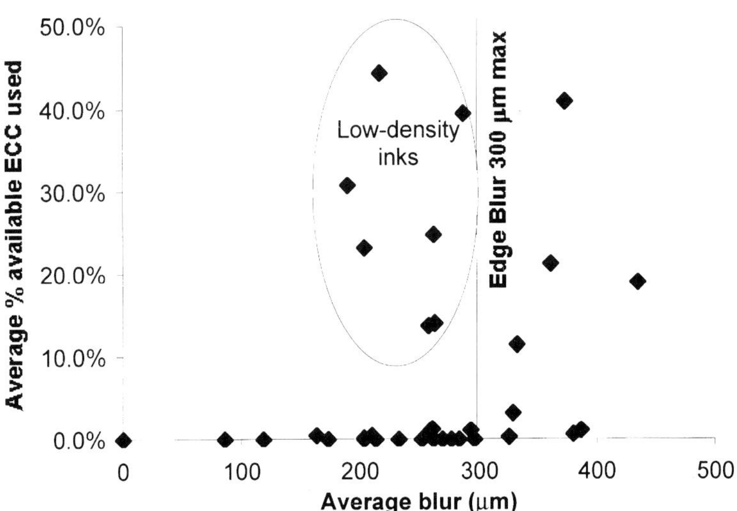

Figure 7

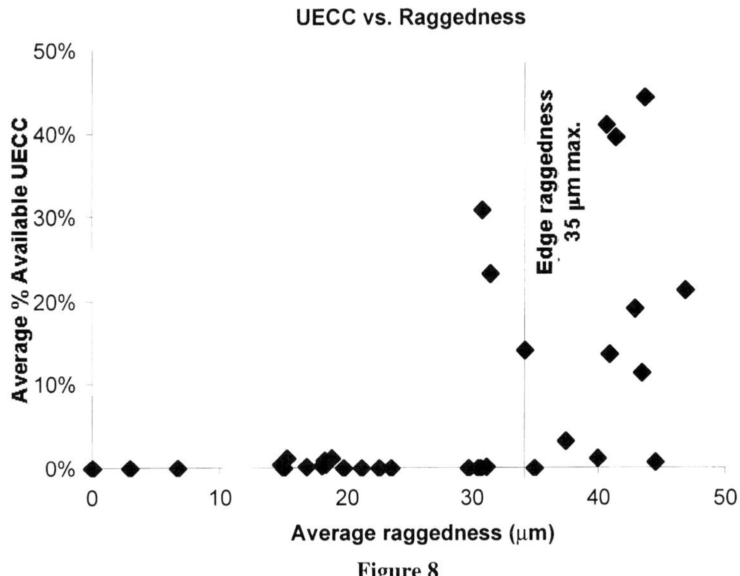

Figure 8

C568/003/99

Weighing – an essential element of mail technology

M J TUMELTY
Product and Technologies, Salter Weigh-Tronix, West Bromwich, UK

Weighing, in many postal applications, is now only one element of the complex systems used to manage the delivery process. The information needed for the task of mail-stream management includes class of service, destination, time of entry, value, method of transport and a myriad of other items, including weight. As the systems become more and more refined, the usage of this information becomes more critical. Properly integrated weighing is almost invisible to the user and operators of today's systems. ease of use, fully computerised integration with information systems, and improved weighing accuracy all add to efficiency improvements in the postal workplace.

The use of weighing in these systems is focused in three distinct areas;
A)   The retail postal counter (International)
B)   Work load assessment (Royal Mail)
C)   Manual scan-where-you-band section of mail routing (USPS)

As you listen to the following application descriptions we would challenge you, to imagine how your applications could benefit from the full integration of weighing and other functions to realise greater performance in moving the mail.

1. POSTAL CUSTOMER COUNTERS (INTERNATIONAL)

The most obvious and frequent points of interface between a postal organisation and its customers are twofold:
1.   The daily delivery of mail and packages to the customer's local address.
2.   The personal contact that occurs whenever the customer visits their local
        post office.
In the former case, the daily mail delivery often occurs unseen and unacknowledged by the customer. Unfortunately for the Post, this opportunity for positive contact with each customer is generally unrealised as regular and reliable postal delivery is frequently taken for granted.

On the other hand, each customer visit to a post office presents a definite and tangible interface opportunity. It is natural and inevitable that following each post office visit, the customer will depart with a personal impression, positive or negative, derived from their experience. For individual customers the physical point of contact with their post will almost exclusively be the customer counter.

The postal customer counter is a retail operation and as such, whether consciously or unconsciously, will be held in comparison with the myriad of other retail operations with which the customer comes in daily contact. A positive customer experience requires a number of co-operative factors including an attractive physical environment, a professional attitude evidenced by the postal clerk and the proper retail equipment for functional efficiency.

The postal counter scale is an often-overlooked piece of equipment that can contribute in many ways to the postal customer counter. Properly designed, the postal scale will enable efficient handling and maximise throughput to the limits permitted by the balance of the retail point of service equipment. The postal counter scale, whether a stand-alone piece of equipment or a vital component of an integrated retail automation system, can provide the postal customer with the confidence that is imperative in any retail operation which has weight-factored pricing. Salter Weigh-Tronix QPS Series Postal Scales use well-established, patented Quartzell technology to contribute most effectively toward achieving a positive customer experience at the postal retail counter.

The postal scale application, with its combination of letter mail and parcels, has historically offered one of the greatest challenges to the weighing industry. Letter mail tariff structures are generally based on weights beginning at 10g to about 2kg, while parcels weights range up to 30kg or higher. Due to the nature of postal customer counters as a consumer-trade application within the regulatory oversight of Weights & Measures authorities, there is an almost universal requirement for all postal counter scales to have legal-for-trade approval. Historically, this has often meant the use of two scales, one for letter mail (1gram resolution) and another for parcels (5gram resolution), an expensive and logistically undesirable situation.

Quartz Digital Technology (QDT) provided the first truly affordable, single-scale solution for international postal counter applications. Offering a 1 gram resolution for letter mail and a 30kg capacity for parcels in each scale model, Weightronix postal scales with Quartzell transducers provide the ability to weigh all letter mail and parcels typically accepted over he postal customer counter.

This minimises the equipment usually located between the customer and the postal clerk, and increases efficiency. Where parcels are being weighed at a scale in a different location than the clerk's window, there is no longer a need to lift and carry parcels back and forth to a larger scale. The postal clerk can conduct all weighing without leaving their station. Face-to-face customer contact is maintained.
Additionally the QPS range of Postal Scales with its 9.5mm high customer dot matrix display has the ability to show weight, price to pay and all services selected. Improving efficiency and customer service.

2. WHAT IS QUARTZELL?

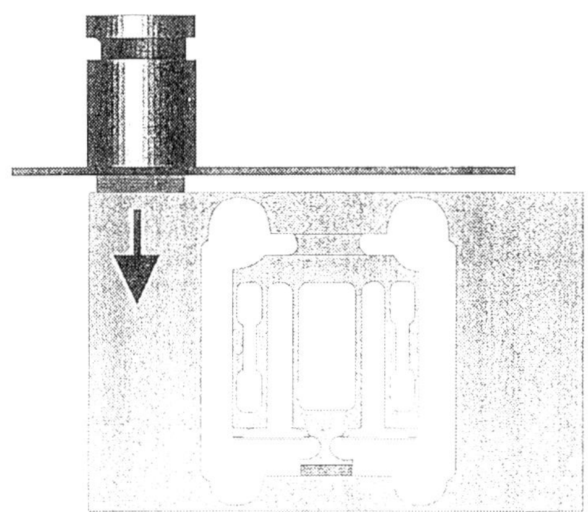

Quartz Digital Technology (QDT) is a method of force sensing that provides a high level of resolution at relatively low cost through a simple, interchangeable load cell style of weight sensor. Using a patented, force-sensing load cell arrangement, the Quartzell transducer compares the frequencies of two quartz crystals to determine weight. A major difference between this and any other weighing system is a digital signal that can be used immediately by the cell processor. The design does not use traditional analogue components like amplifiers, sensing circuits, and analogue-to-digital converters.

Two double-ended tuning fork quartz crystals are mounted in the cell within a three-stage interactive force division mechanism, which provides several unique advantages. The cell configuration may easily be viewed as two separate weighing devices; one with an ascending signal and the other with a descending signal. Common mode influences, such as temperature change, are easily rejected in this system because both weighing signals would move in the same direction. Such a change in frequency then would be recognised not as weigh, but as a common mode influence and the scale would reject this signal change as invalid.

The end result of this application of Quartzell technology to weight measurement is a very linear, high-resolution, and extremely repeatable signal that is unaffected by common mode influences across a broad temperature range.

The QPS Series scales offer a number of additional customer service benefits:
* Speed - the inherently digital signal of the Quartzell transducer allows the scale to weigh and display faster, especially in the weight ranges that are critical for flats and small packages. This saves time, improves throughput, and minimises customer queuing.

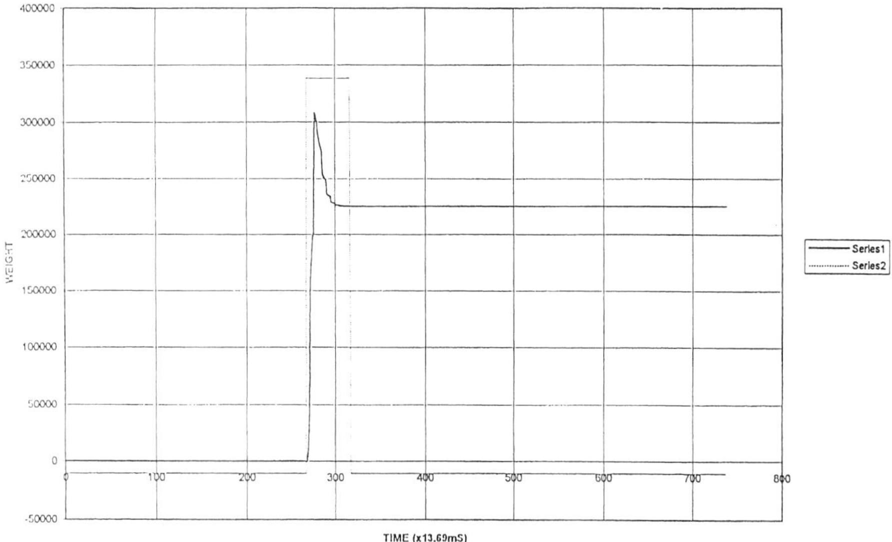

- Simplicity - QDT scales have only a few working parts, and a proven track record for functional reliability. This virtually eliminates a customer counter window being closed due to scale failure, especially critical in smaller rural or neighbourhood post offices having only a few windows.

- Reliability - the use of two quartz crystals in the Quartzell transducer eliminates external influences, such as temperature, radio frequency interference, electromagnetic interference or other environmental factors. The consistent, repeatable weight output increases customer confidence.

- Cost effectiveness - the many cost benefits of Salter Weigh-Tronix postal scales using Quartzell transducers help control the hardware and support costs of retail operations, directly benefiting postal customers. These include:

    1. The combination of quartz-crystal technology and modern manufacturing makes the cost of the scales comparable to type-approved analogue scales of inferior accuracy.

    2. Use of a single scale for letters and parcels increases a postal clerk's efficiency and frees up counter space.

    3. An adjustable gravity-constant value simplifies installation by eliminating a service call to calibrate the scale on site.

4. Fewer scales, and their very low power consumption, make it easier to interface with minimal impact into the complex counter systems that comprise the postal workstation, and also reduce servicing and maintenance requirements.

5. If service is required, the scale's relatively small size (350mm x 310mm x 100mm) and lightweight make them easier and cheaper to ship than heavier weighing devices.

6. Compliance to International Standards.

Weigh-Tronix postal scales with Quartzell transducers offer the opportunity for significant improvements in space utilisation, simple operation and serviceability for special customer service counter installations, together with greater resolution and faster speed at comparable cost to lesser technologies.   The NCI 7600 Series of postal bench scales offer weight-only models designed for use either stand-alone or within integrated postal retail automation systems.

The QPS scale of a modular postal rate computing design integrated by and available through Salter Weigh-Tronix. This is a design that can be readily modified for the rate structures, functions and language required by international postal authorities. Postal customer counter weighing and Quartzell are becoming synonymous.
For the benefit of postal customers, they should become inseparable.

3. SALTER WEIGH-TRONIX, WORK LOAD ASSESSMENT (WLA) SYSTEM (ROYAL MAIL)

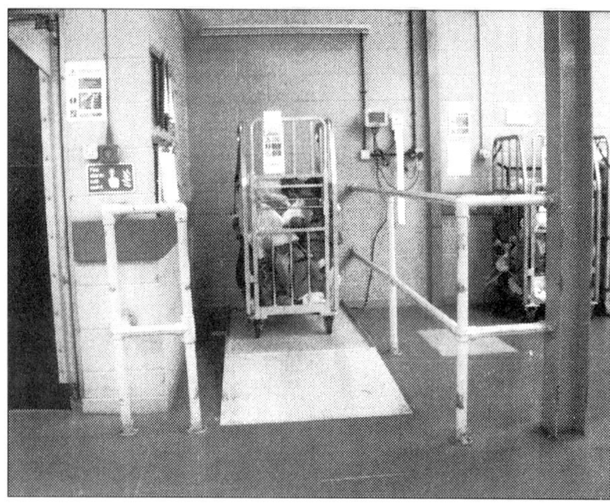

Salter Weigh-Tronix, is helping to deliver a better postal service by supplying weighing machines to over 1500 Royal Mail delivery offices across the UK.

The new equipment was installed under Royal Mail's continuing drive for further efficiency gains, leading to an even higher quality of service for customers and is being used as part of Royal Mail's work load assessment program. This process, which monitors and records the 164 million items handled every day by delivery offices, calculates the volume of letters being handled more quickly and simply, enabling Royal Mail to ensure that postal traffic is accurately measured.

Mixed unsorted mail, when entering the sorting office, is weighed and a printed record made. Each record includes the weight, a sequence number, and the time and date. At the end of each shift the information from the printed record is entered into a computer program by Royal Mail enabling the 'items per kg' and the 'total volume of traffic' to be calculated. This gives Royal Mail much needed management information for "Work Load Assessment".

The WLA implementation teams and Royal Mail Delivery Office Managers have a choice of platform design and size to enable them to select the platform which best suits the traffic volume and handling methods for an individual office. Once the equipment is selected, the Salter Weigh-Tronix team handles installation and training on location.

The most popular option has been an SLD LoDeck electronic platform scale utilising 4 Weightronix Weighbars for durability and accuracy combined with a WI-160 weight indicator and a dot matrix printer. The SLD is available in four standard platform sizes, again selected to suit the needs of the particular office. All SLD's are 600kg capacity. The design configuration has been adapted to meet the specific requirements of Royal Mail, with all scale platforms painted yellow for safety purposes, and featuring a run-up ramp with three side frames to prevent the cages from falling off.

The second option, primarily selected for smaller offices where space is at a premium, is the Salter SLH platform with a platform size of 600mm x 600mm and a weighing capacity of 100kg. Again, this is supplied with the WI-160 weight indicator and dot matrix printer.

All equipment is approved and stamped by the Department of Trade prior to dispatch and it is backed up by the professionalism of a known quality supplier with ISO 9001 accreditation. Additionally, all equipment had to be tested by the Royal Mail's own research and development team to ensure compliance with its safety end environmental guidelines.

4. MANUAL SCAN-WHERE-YOU-BAND (USPS)

A program used by the United States Postal Service, Manual Scan-Where-You- Band (Manual SWYB) is an integral part of STARSHIP (Strategic Traffic Analysis and Receipt of Shipments). STARSHIP allows for the automation of dispatching and receiving processed mail by automatically assigning mail pieces to airline flights and road transportation. The system serves to control the outbound and inbound mail availability and capture logistics management information, which in turn allows reduced handling.

Manual SWYB reads a bar coded distribution label, weighs the mail piece, assigns it to transportation and places a machine-readable (bar coded) Dispatch & Routing (D&R) label on it. The D&R Tag can be used to sort inbound mail at the destination Airport Mail Facility (AMF) and bypass the acceptance operation at the AMF. A Manual SWYB system is normally installed at a Processing and Distribution Centre (P&DC), Bulk Mail Centre (BMC), or International Mail Facility (IMF). The system accounts for the time delay to transport the mail from the mail facility to the airport.

4.1 Major System Components
A Manual Scan-Where-You-band site includes the following major components:

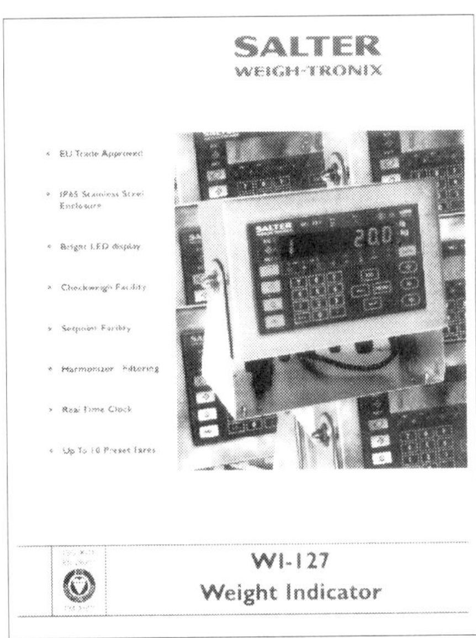

1. Manual SWYB System
   a. Static Scale: The scale is used to weigh the mail pieces. The weight information
      is supplied to the Personal Computer (PC) through a serial port and is also
      displayed on an indicator. The indicator, a WI-127 (see picture) is an NTEP
      certified unit capable of either static or in-motion weighing.
   b. Scanner and Interface Controller: The Scanner and Interface Controller are
      used to scan the bar-coded label on the mail piece. The information on this bar
      coded label is provided to the PC.

   c. Thermal Label Printer: The printer prints the D&R labels that are applied to
      the mail pieces. The printing of labels is controlled by the PC.
      A future upgrade will include using the scanner to scan the D&R label for
      verification.

   d. Personal Computer (PC): The PC receives information from the Scanner and
      Static Scale. This information is used to control the Thermal Label Printer to
      print the D&R label for that particular mail piece. The PC also uses an
      expansion card (ARCNET interface) that provides a link to the Transaction
      Concentrator.

2. Transaction Concentrator (TC): The TC is a Personal Computer programmed to serve
   as a bridge to link the Manual SWYB system(s) to the ADP. The ADP provides tables
   and files necessary to assign air or surface transportation information to the mail pieces
   processed with a Manual SWYB system.

3. ARCNET HUB: This device provides a connection from the Manual SWYB
   system(s) to the TC. The Manual SWYB system(s) and TC are cabled to the
   ARCNET HUB with coaxial cable.

This system provides an easily processable item for both the sending and receiving station.
Logistical information can be readily captured at the Manual SWYB system and applied as
necessary to further improve mail handling. Although weighing appears as a small portion of
the data involved, it is essential to the understanding of both air and surface transport.

SUMMARY

These three instances, the postal counter, work load assessment and Scan-Where -You-Band
all use weighing as an integral part of an efficient mail stream.
Weigh-Tronix, as a full-line international weighing supplier, is part of the continuing effort of
improving postal deliveries. At Salter Weigh-Tronix we see ourselves as a partner of postal
authorities in providing the best solution for the postal customer. Cost effective, efficiency
improving solutions can be the result of joint efforts between postal officials and weighing
equipment suppliers like Salter Weigh-Tronix.

Systems of the future will enable reduced labour and faster throughput via the use of better
weighing and more computer power on the counter. Higher levels of integration in the back
room between weighing and data systems and better controls for mail entry by using the

integration of scales and computers in all entry locations will also reduce labour and further refine these operations.

Weigh-Tronix will continue to provide improved weighing technology that easily integrates with other equipment in all of these postal applications. Quartzell for high accuracy and the Weigh Bar for high capacity and durability are two examples of the innovative solutions now being applied by Weigh-Tronix in the postal workplace.

Machines and the computer network at the Post Office

D HEITZ
La Post – Technical Research Centre, France

The French Post Office installed over 6000 machines in its post offices (Stamp machines, self-service franking machines and change machines). It wants to offer the public a group of machines in good working order and propose payment by banker's card on its postal machines. In 1996, a study was initiated by the Post Office's technical research department to analyse the requirement and propose a connection architecture. The first machines begin operating in Post offices and the initial results are encouraging.

INTRODUCTION

Between 1990 and 1998 the French Post Office installed over 6000 machines in its branch offices. These postal machines were autonomous. They include stamp machines, self-service franking machines and change machines. This document does not cover bank note cash machines that have a direct link with the Post Office's private network. The postal machines are mainly installed in the areas open to the public. The Post Office's wish to offer the public a group of machines in good working order, the high cost of processing coins and the large number of banker's card holders in France have led to the Post Office introducing payment by banker's card on its postal machines.
Payment by banker's card requires a connection with the bank servers. The Post office therefore used this connection to study the supervision and processing of its machines' accounts.

This is why, in 1996, a study was initiated by the Post Office's technical research department, in collaboration with the post offices network management and the relevant computer departments, to analyse the requirement and propose a connection architecture. It is this architecture and the organisation of this network which I intend to present to you below.

1 METHODOLOGY

The study was carried out in several stages. The first resulted in a dossier of choices of architecture. After addressing the field of study, the expression of requirements and a study of the existing architecture, this dossier proposed scenarios for the connection of machines inside post offices and scenarios for those machines outside.
Evaluations of the costs of development and deployment in the network were made in both cases.

The Post Office gave its opinion on the choice of scenario and detailed specifications were drawn up:

> - network architecture specification
> - the connection interface for machines on the post offices computer network

These two documents were attached to the consultation on the machines so that the manufacturers could provide for this type of connection on the machines proposed for the Post Office.

> - the development specifications for remote surveillance, remote operation and downloading
> - the development specifications for payment by banker's card.

The latter two documents constituted the general development specifications for the Post Office software development services.

2 CONNECTION ARCHITECTURE CHOSEN.

This architecture was to be based on the architecture already existing in post offices and introduce the least possible amount of additional equipment.

Migration should take place progressively and for this 4 stages have been fixed:

> - Connection to an existing computer terminal in the post office when the first machine with autonomous payment by banker's card is introduced.
> - The office server processes the connection, but payment remains autonomous
> - The office server provides the concentration of supervision and routes banking transactions
> - A server is set up specific to the Post Office for remote collection and authorisation of payments.

Progressive migration has led us to ask the machine manufacturer to supply a PC simulator for each stage to test the corresponding configuration.

And each configuration has been implemented in the field, except for the last one, which is currently being developed. The first type of connection was limited to connecting the post office cashier terminal to a machine. The cashier thus had information available on the supervision and accounting of the machine connected.

A first development led to connecting the post office server. The cashier still had available the supervision and accounting applications; but the machine alarms were sent to several terminals in the office.

Finally an external configuration, in which the cashier could obtain accounting information through a telephone call to a machine. The machine, in its external configuration, has a modem that allows the data to be routed on the telephone line. Exchanges with the banking world (requests for authorisation and remote collection) are also effected through the same telephone line.
These different pilot sites have enabled all the operating configurations to be validated.

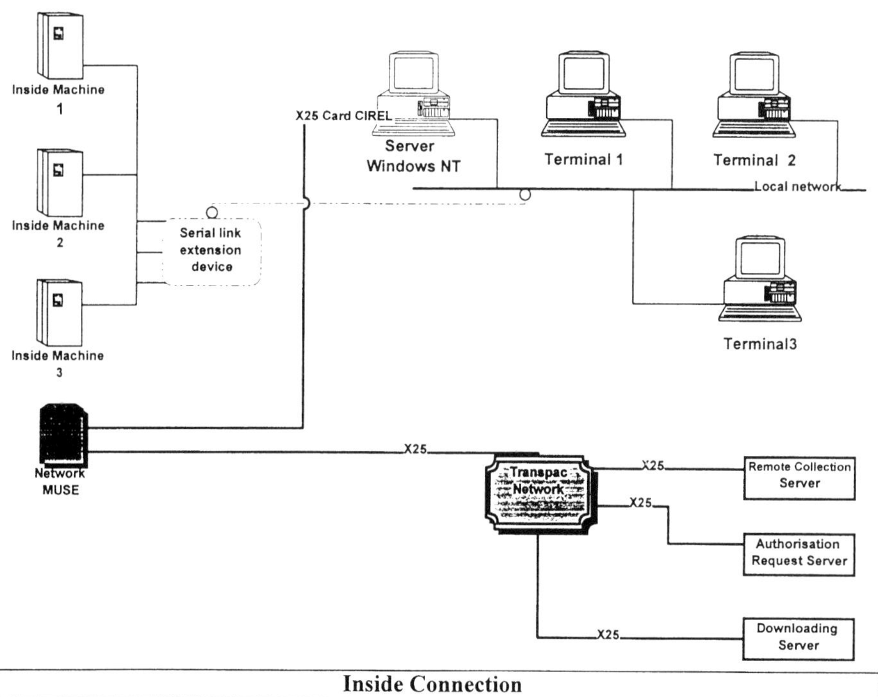

Inside Connection

3 THE STAGES IN DEVELOPMENT

In 1997 the first prototypes of machines distributing packets of envelopes were produced on the basis of this architecture by the company FORENIA. Validation was performed and four machines are being tested in Dijon, Paris Madeleine, Boulogne Hôtel de Ville and in the Saint Lazare train station in Paris. The latter prototype was set up to validate the connection of machines installed in sites outside the post office. The use of these machines does not pose any technical problems. But the marketing departments consider that these first prototypes do not offer customers enough products. This is why a new ready-for-posting machine was produced in 1998. This was based on a VENDO food vending machine and was integrated and connected by FORENIA. This choice enabled the French Post Office to capitalise on developments achieved with the first machine.

Today four machines are being used, two in the Paris region and two in Lille, and the plan is to equip the largest post offices within three years.

Finally, a new generation of self-service franking machines available in the year 2000 should replace the first generation and will be equipped with the same type of connection.

Mono product machine	Multi products machine

4 GENERAL PRINCIPLES OF EXCHANGE

For most exchanges, communication is initiated by the post office's computer. The only instances where the machine initiates the exchange concern processing linked to the banker's card module (Requests for remote collection and requests for authorisation). In the new version of the payment by machine manual, MPA Version 3, this will also be the case for remote configuration. The message between the supervision terminal and the machine is constituted by:

        - The machine's number
        - The command code which identifies the type of exchange
        - The data, whose content depends on the type of exchange.

5 EXCHANGES COMMON TO ALL MACHINES

The machine is monitored continuously. The supervising PC transmits a surveillance request towards the machine. The latter has to return its status, functioning correctly or error. The error message specifies the severity level of the error.

To facilitate the operation of the machines, the supervising PC can discover simply by reading data:

        - The status of sales which specifies the quantity of products sold and the method of payment, banker's card or coins.

        - The cashbox statements, showing the contents of the cashbox.

        - The remote collection reports, showing the transactions performed by banker's card. The remote collection report structure corresponds to that specified in the payment by machine manual. This ensures that remote collection is performed under the right conditions.

        - Product and cash supplies.

For each entity containing a product identified by its family code and its product code, it is possible to know the quantity added or removed and the date of these operations. It is possible to take the same reading for cash supply.

Finally, after consultation, purging the activities enables the information on supplies, sales, cash and remote collection to be removed from the machine's memory. The machine or the payment module must be able to retain the record of these elements for a minimum of five days, before the machine switches out of service.

It is also possible to load object programs into the machines to process developments. A control key is used to check that the object program conforms.

6 SPECIFIC CHANGES TO EXTERNAL MACHINES.

The connection of machines outside post offices is performed via the switched telephone network. But to avoid fraudulent calls, when the call is made, the machine must identify the calling supervisory terminal.

The supervision rate of these machines is also much lower to avoid penalising the operating costs of this type of machine.

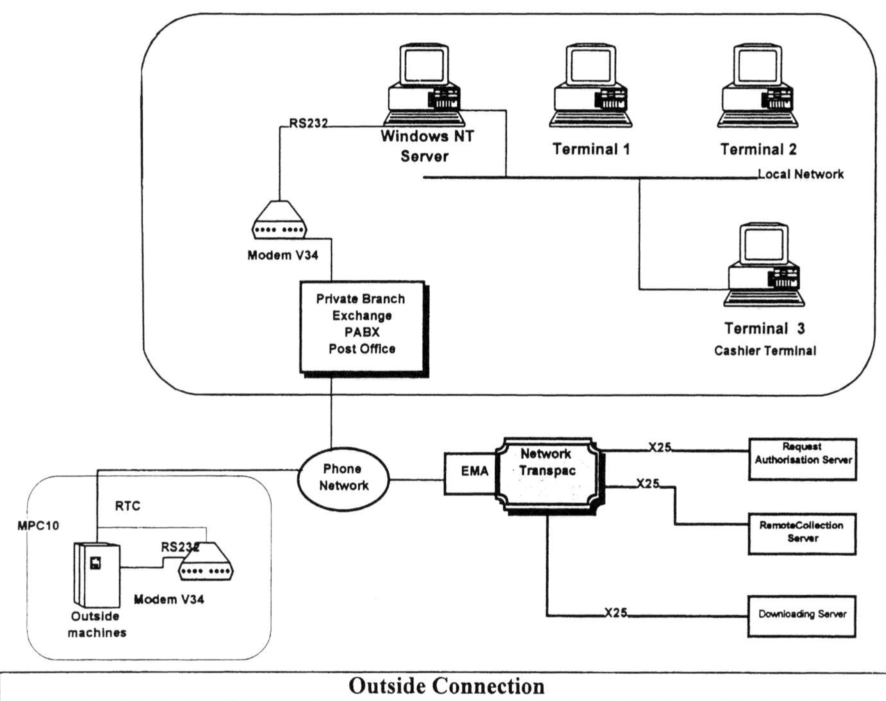

Outside Connection

7 INTEGRATION OF THE PAYMENT MODULE.

The payment by banker's card module integrated into the Post Office machines is constituted by a central unit to which are connected:

    - A screen keyboard for entering information coming from the client (confidential code, etc.)
    - A modem which allows connection to the bank servers. For security reasons, this modem is dedicated to bank exchanges. It cannot be used to upload machine supervision information. A modem must be provided to be dedicated to non-bank exchanges.
    - A connection to the host machine. This interface is currently an RS 232.

A specific protocol enables a dialogue to be established between the payment module and the machine. Printable information from the banker's card module is transmitted to the machine's control unit. The latter than takes charge of managing the printer and supplies the receipt to the customer. Dialogue messages can be routed to the main screen of the machine.

8 ADVANTAGES IN CONNECTING THE MACHINES

The accounts processing process is lightened.

Until now, as the machines were autonomous, at each close of accounts, that is every day, the officer responsible for the machines, generally the cashier of the post office, had to read the sales counters and calculate the turnover for each product. He checked the contents of the cashbox, emptied it and entered all the information in the post office accounting application. With the connection, the data are automatically uploaded and loaded into the accounts.

In relation to banker's card processing.
The most optimistic predictions have been exceeded. On the first machines installed, customers mainly use banker's cards.

A reduction in operating costs
In the area of maintenance, savings are based on two levels. The indication of the fault severity level must allow repairs to be grouped together. There are about one or two changes of software per year. In particular self-service franking may be enriched with new services which would lead to modifications to the machine software. Until now these changes were made by changing the memory (PROM) and required the attendance of a technician on each site. Remote downloading reduces the costs of these updates.

In machine operation, the cost of processing cash is high. It takes into account the cash statement, counting, reloading the dispensing trays and the accounts processing of the coins. Automatic processing of payment by banker's card constitutes a large saving.
Before initiating studies and developments, the Post Office produced profitability simulations, assuming that 10% of operations performed on the self-service franking machine and 25% of operations performed on the stamp machine are paid by banker's card.

For the supervision part, management costs for all the current machines and updates of the software were used to calculate depreciation. The conclusions of this simulation showed the depreciation from studies and investment in additional material (payment module and supervision) over 6 years for the supervision part and 3 years for the payment part. However the life span of machines is currently above 5 years.

9 REMAINING DIFFICULTIES

The interfaces are specific to the Post Office machines. The information uploading interfaces are not standardised and require developments in relation to the new machines and the Post Office network. Standardisation does not seem to be forthcoming in this area. Operating modes vary greatly between the different machine operators. The most sophisticated sweet vending machines have a collection terminal which allows accounting information to be read each time they are restocked. But it is not possible in this case to supervise the network of machines. An agent is required to attend regularly.

The second specificity is due to the payment module interface. It also depends on the payment module manufacturer. Each constructor must set up a physical interface for the module and develop the dialogue on its own machine. For our postal machines, this interface is an RS 232. The dialogue is a Monetel specific dialogue.

Finally, bank processing is specific to the French interbanking network. It is specified in the payment by machine manual (MPA Version 3). The machine or the payment module must receive the certification of the French banker's card economic interest group (GIE CB). This body is the product of French banking principles. Either the payment module integrated into the machine is approved, in this case a reduced procedure approval is possible; or the module has not been approved and the machine must be approved in its entirety by the GIE CB. For the moment the Post Office has opted for the first solution, using the MPC10 module from Monetel which already has approval.

The Post Office would like to see standardisation with regard to machine supervision exchanges, but also between the machine's central unit and the sub-assemblies constituting the machine (change maker, payment by card module, distribution cups, user interface, etc.). The MDB standard seems, in the change taker maker field, to be a more widespread standard.

This lack of standardisation leads, when a new machine is installed, to developments in relation to the machine manufacturers and computer applications for supervising these machines.

Seeking standardisation can only reduce the price of machines and improve the service offered to users.

10 SOFTWARE DEVELOPMENT

Developments must be separated, as far as possible, from modifications due to the introduction of a new machine into the network.

Only the final version of the software architecture will be presented below. Intermediary validation stages have been necessary to reach this configuration.

All the machines are connected to the NT server of the post office. Each type of machine has a driver on the server. Call processing varies between the internal and external machines (calls can be configured 2 to 3 times
per day, etc.). So for each type of machine there is an internal machine driver and an external machine driver.

Each driver makes data (DLL) available to the application in an interface common to all the machines. The application can thus be developed exclusively. Introducing a new machine necessitates developing exclusively the internal and external drivers only.

The server also supports call processing towards the remote collection, remote configuration and authorisation bank servers. These calls are supported by the private X25 network of the Post Office for the machines in the offices. In machines outside, the machine uses the switched telephone network to access the various bank servers. In this case the payment module behaves like a traditional payment terminal.

Finally the server assures the distribution, via the local network, of warning messages towards the workstations in the post office.

CONCLUSION

It appears to me that the key to success in reducing operating costs for a dense machine network lies in standardisation. But this standardisation remains embryonic and the operators remain fixed in specific configurations. It is within the framework of consultation between machine operating services like the Post Office and manufacturers in this field that there is a real hope of seeing the development of these modern distribution methods.

Capture of information on the use of postal services

D T GILHAM
Neopost Limited, Romford, UK

CAPTURE OF INFORMATION ON THE USE OF POSTAL SERVICES

Abstract:

The aim of this paper is to demonstrate that it is possible to allocate a new role to postal mailing systems used by business mailers to capture information on the postal services used. To present a cost-effective solution to this previously unsatisfied need shared by Postal Administrations world-wide.

To describe an effective solution that provides a new level of user convenience and is flexibly enough to survive a dynamic postal service environment that is subject to increasing competition and change.

1. MAILING SYSTEMS

Used extensively by business mailers to conveniently purchase and access postal services. More than 3 million systems are in use world-wide collecting one third to one half of the postal revenue in the major markets.

Mailing systems for the Postal Administration fulfils two basic functions:
- To print a postage mark which can be validated and having a value appropriate to the postal service purchased, and
- To accurately record in non-volatile registers the postage purchased and the postage used

These functions are performed in a totally secure environment. Additional functions are also fulfilled for business mailers to satisfy their productivity and added value requirements

2. TECHNOLOGICAL EVOLUTION'S

First mailing mechanical machines were introduced in the mid 1920's. Microprocessors were introduced in the late 1970's and online payment transactions in the late 1980's. Systems are now digital and increasingly becoming software based products.

During the evolution to digital new functions were added, in particular:
- Storage and processing capabilities, and
- Communications:
    - To a postal service computing weighscale
    - To a printer
    - To a PC
    - To a remote computer system or server

3. DATA CAPTURE NEEDS

3.1 Postal Administration

Postal Administrations world-wide are asking a basic question - Do we know about our customer? This question is not unique to our industry. A recent survey across blue-chip companies showed that 90% of companies do not really know about their customers. Yet at the same time companies believe that IT solutions to target customers more effectively would help them to achieve revenue growth.

An IT marketing tool that puts customer purchasing information of postal services into the hands of Postal Administrations would be an ideal solution. Apart from the obvious security benefits, this would provide postal management with information to help them "enrich customer services".

As Postal Administrations embrace technology and prepare to face a more deregulated market the need to know customers better could not be greater. Pushing into new areas requires the strategic pursuit of innovation to monitor customer needs and the delivery of new and differentiated products and services. Our industry can no longer afford to offer the same package and not everyone would want the same package if it did.

As new postal services are introduced then the need to quickly know whether these products are a good answer to what customers are waiting for, are they used and by whom are essential marketing issues.

3.2 User Needs

Business mailers wish to better manage their communication expenses including post-mail, especially the more significant volume users. They wish to allocate expenses to departments with more detailed information on the service products used for better management and future planning purposes.

3.3 Postal Service Data

Postal products are categorised by item, service mode, complimentary services, for example registered, insurance, oversize…, also by weight and destination.

The service is known once the mail piece is prepared at the customers office location, first by the identification of the item (letter, printed matter, parcel…), then by destination (domestic, EEC, Europe outside EEC, International…) and mode (surface, air…) and then weight and complementary services. Clearly the number of possible service selections is high and the change in service specifications occurs regularly. This requires for a flexible solution and ideally one that is transparent to the user since the selection process does not want to be a complex matter.

4 EFFICIENT DATA CAPTURE

Most users today work with a few basic service postage values, frequently relying on memory, used for the majority of their normal requirements. This is a disadvantage in several respects. Errors in applying correct postage, both at the expense of the user and the Postal Administration. Lost useful information. Lack of process and cross training of staff. And lack of appreciation that suitable services are provided by the Postal Administration that can compete successfully with private operators.

Clearly this approach is often used because to process by service selection can be too complex although this would be an answer to these problems. New digital mailing systems can now integrate a postal service rate calculator, see example in figure 1. This can be combined with a choice of optional weighing systems, for example static, dynamic or automatic, with different weight capacities and oversize measurement possibilities. The user interface can include quick select keys that allow frequently used selections to be made with a single key press. Defaults can be pre-set to save the user time in selecting most common requirements. In this way the user can work faster and more systematically. At the same time allowing postal service data to be captured in a completely transparent fashion by the mailing system.

5 EFFICIENT ORGANISATION AND STORAGE OF DATA

The number of possible service selections is high; 256 would not be unusual and the trend is to offer more services or complementary services, for example, proof of posting, delivery confirmation, notification of undeliverable mail, time and proof of delivery, date specified delivery... The memory capacity of mailing systems is ultimately a function of cost; small systems for low volume applications are more sensitive to price than for high volume applications. It is also desirable to store information for two time periods so that previous data can be recovered if necessary. The service product code and cumulative weight, quantity and postage value for each service will need to be stored for the total quantity of items processed.

There are two basic approaches, fixed service code selection or a flexible and smarter selection. The former requires a Postal Administration to make a pre-selection of the product services to be captured on the assumption that it will not be practical to have sufficient capacity to capture everything for everyone, big and small customers alike. Plus pre-selection suggests a degree of inflexibility that is unlikely to pass the test of change over time. This method is not recommended.

Live tests typically show that users purchase up to ten service products over any one month period; often the quantity is smaller; four or five. However, users purchase different products regardless of company size or industry classification. Also in general high volume users purchase a wider selection of services. This leads us to the conclusion that a flexible system is preferred and a scaled capacity to store service information according to the volume of mail processed. Then the system capacity can be typically an order of magnitude smaller than the total number of possible service selections. Flexibility means that each mailing system will capture during any period, usually one month, the services used and in the order they are purchased up to the capacity number of the system, for example 40 for high volume systems. Any services used beyond the capacity level will be captured in a "other services" register within the memory storage of the mailing system. For each period the sequence is repeated and consequently the profile of usage captured can be different from one period to another. This provides for a flexible solution, one that can evolve with changes and one that is cost effective to implement due to the reduced memory capacity required.

Service selection and the corresponding service coding are managed through the rate generator, which can be an integral part of an optional connected weighscale or an integral part of the mailing system software. It is common practice to update the rate generator whenever service rates and/or service structures change. This is generally achieved by an exchangeable memory pack or by an online transaction.

6 TRANSMISSION OF CAPTURED DATA

Mailing systems vary according to market and technologies evolve over time. The choice of system architecture must address these dynamics. Implementation of such a data system in France used a Provider Frontal to capture first level data direct from the users mailing systems, see figure 2. Transmissions are initiated by the user system either on a time or function basis, for example, when downloading new postage payment. Transmissions, therefore, can be spread over each time period for manageable loading. A transmission will transfer for each postal service code used during the preceding period the accumulated weight (if available), total postage and quantity of pieces, see typical data structure for content transmitted in figure 3. A successful transmission will release the second period set of registers to capture the next set of data; note that the preceding set is still available for one period in the first set of registers, should a subsequent enquiry be necessary.

The Provider's Frontal can facilitate any difference in level of mailing system specification based on an identity that is transmitted at the start of any transaction. Clearly mailing systems specifications will evolve over time with technology and so different standards will need to be managed by the frontal. The Frontal consolidates the data captured and reformats the information according to the Postal Administrations requirements. These reports are transmitted to the Postal Server each day or each week according to requirements, see figure 4.

Transmissions from the Mailing System to the Frontal are made via a standard modem and the public telephone network, for example using a V32 connection and X28 protocol. The information transmitted can be protected for confidentiality by encryption, if appropriate. The transmission between the Provider's Frontal and the Postal Server can be made over a standard quality data line with access protection, such as, over a X25 Transpac network, see figure 5.

7 APPLICATIONS

Once data has been received by the Postal Server in a standard transfer format from providers the Postal Administration now has the possibility to make different analyses according to specified marketing and security applications. For example, the complete purchase profiles of customers with automatic and regular updates can be generated. Trends can be followed, market segments can be analysed and reserved and non-reserved service accounts can be separated. Data can be merged with other customer profile information, such as SIC references; then new prospects can be identified and new services targeted more effectively. Image and customer loyalty can be enhanced. Profile data can be used to support customer relationship programs and the operations of call management centres. And much more …

Figures 6 – 9 inclusive, show typical profiles for two customers; profiles by postal service products during a specified period for both volume and postage revenue. Previously, both these actual customers could have looked very similar to the Postal Administration but in fact the product profile shows there are significant differences. These differences can mean different needs, which should be understood and satisfied. As the industry moves forward with increasing deregulation and commercialisation of services, knowing customers better becomes an increasingly important requirement.

Figures 10 &11, show the consolidated profiles for a population or segment of customers by postal service product during a specified period for both volume and postage revenue.

8 ALTERNATIVES

What are the alternatives? Sample monitoring of mail streams is expensive, reports arrive late and they do not provide one to one marketing data. Although this method is commonly used today by Postal Administrations to prepare mail flow reports it is recognised that it is not a satisfactory solution.

In the future it is anticipated that more information will be captured from a digital postmark (2D machine-readable) to enhance service capability, see figure 12. Online data capture is compatible with developments in digital postmarks. It provides efficient on-line EDI files, which are complete, and error free. These files are solid references in which to correlate data captured from digital postmarks. Online capture also provides a closed loop system of control plus quality customer information.

SUMMARY

Neopost undertook a project "Credinet" to test the solution described above. This live test captured transparently daily information about the quantity, the postage and the services purchased by business mailers. Quick service select features where designed for business users to achieve high productivity processing on modern mailing systems. About 50 high volume users tested this data capture solution over a six-month period. The results were very successful and now this project will enter the first phase of a national roll out in France. A survey of users was made as part of the market test and responses most often mentioned the value of detailed information and modern technology as the most important reasons for interest. 95% of other potential customers surveyed in this market segment said they would like to have this capability. It should be noted, however, that in France the Postal Administration operates a monthly billing payment arrangement. Consequently, the online data capture system has additional benefits in automating this process for the Postal Administration and the customer. Several other Postal Administrations have now indicated interest in an online solution for capturing the customer purchase profiles for their postal service products. The importance of a system design that gives sufficient flexibility and that can keep up to date with rate and service structure changes is confirmed.

Figures

Figure 1: Mailing system with an integrated postal services rate calculator and quick select features plus modem for online communications

Figure 2: The provider's frontal interfaces with the population of mailing systems over the public telephone network. Mailing systems may be at different technology levels. The frontal standardises and consolidates the data received and creates an online report for the Postal Administration.

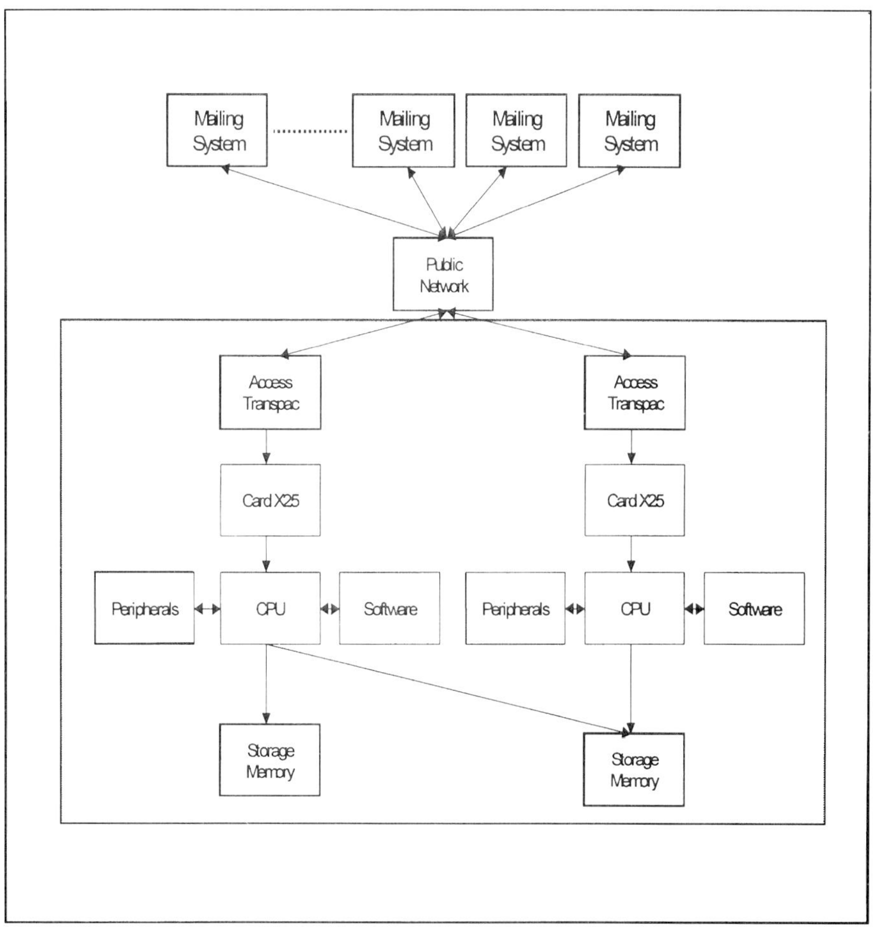

Mailing System
Mailing System
Mailing System
Mailing System
Public Network
Access Transpac
Access Transpac
Card X25
Card X25
Peripherals
CPU
Software
Peripherals
CPU
Software
Storage Memory
Storage Memory

Figure 3: Structure of a message for a single frame and including a frame check sum.

FRAME START FLAG	I D	TYPE	FRAME LENGTH	ENCRYPTION FLAG	MESSAGE TYPE	MESSAGE LENGTH	MESSAGE DATA	FRAME END FLAG	F C S

Figure 4: Typical data content in a transmission from the frontal to the postal server

NEOPOST-VER.1.00	14/0001116489
SK0000	15/0001116899
01/START OF PERIOD	16/0001117123
00/AUTOMATIC	17/0001117569
28-03-1998/00:00:00/0001111156	18/0001117569
29-04-1998/23:59:59/0001120359	19/0001117569
28/0001111156	20/0001117780
29/0001111156	21/0001117820
30/0001114519	22/0001117950
31/0001114789	23/0001118920
01/0001114894	24/0001119823
02/0001114923	25/0001119823
03/0001115111	26/0001119823
04/0001115111	27/0001119956
05/0001115111	28/0001120156
06/0001115213	29/0001120359
07/0001115289	00/COHERENT
08/0001115302	033/000128/0000003000/00001055
09/0001115325	574/000200/0000006203/00005600
10/0001115999	F020029
11/0001115999	01/END OF PERIOD
12/0001115999	
13/0001116356	

Figure 5: Interface between the frontal and the postal server

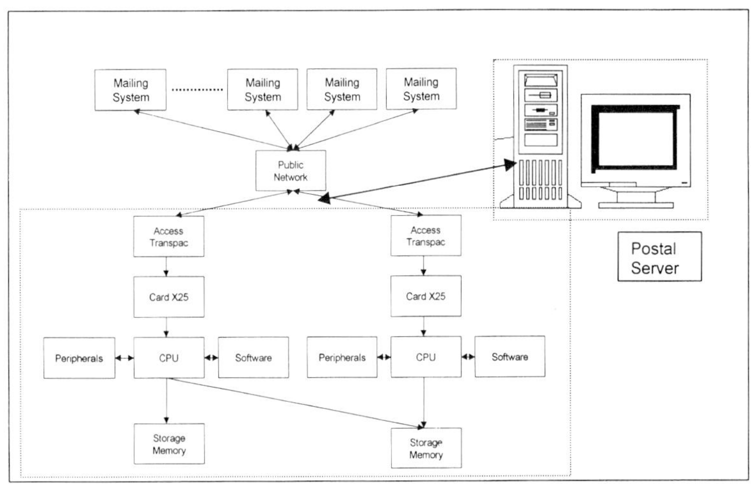

Figures 6 – 11: Purchase profiles of two customers by volume and value over a defined period of time and by postal service product.

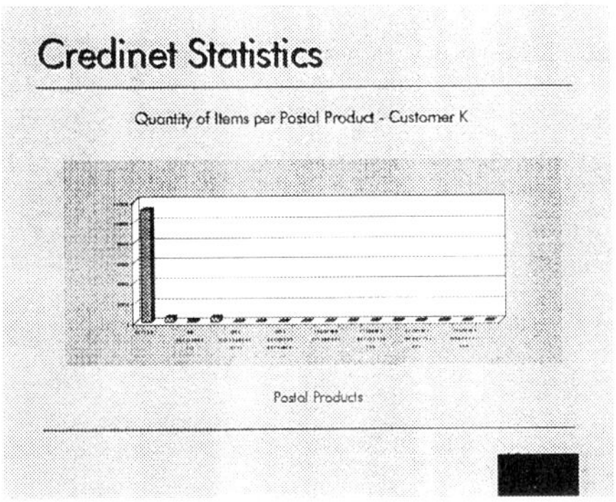

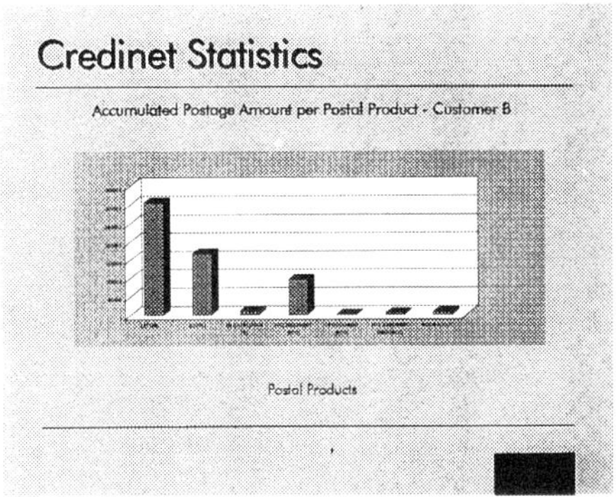

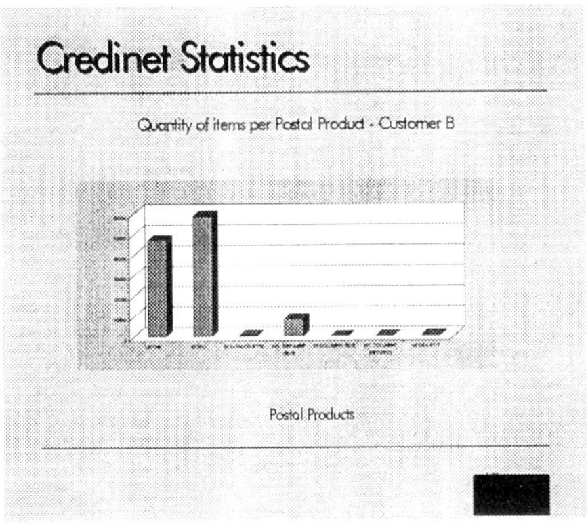

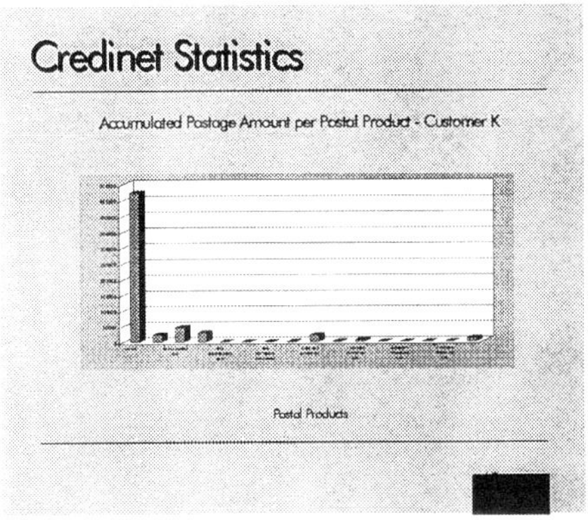

Figures 10 & 11: Consolidated results for number of customers; population or segment

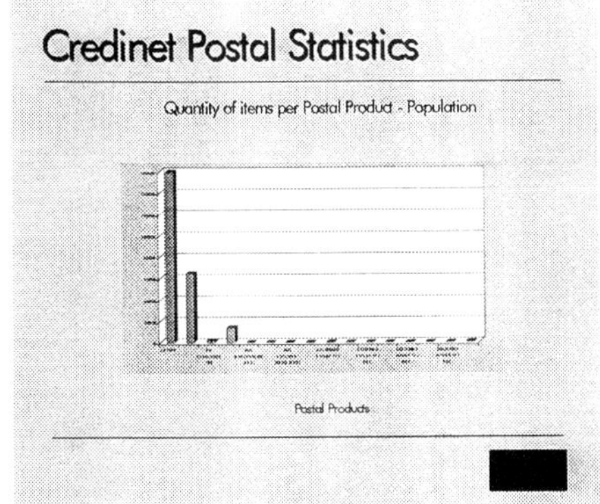

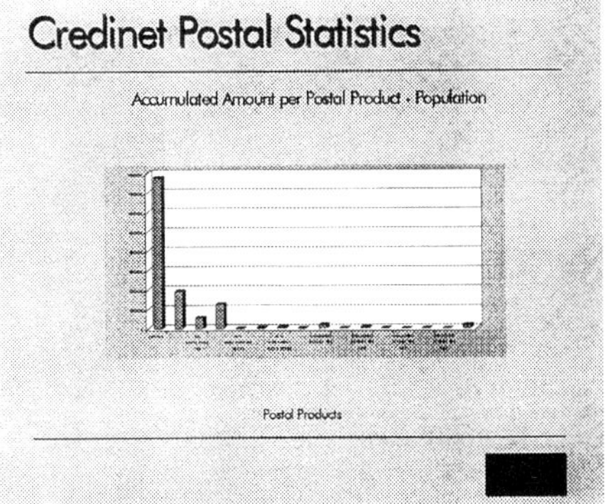

Figure 12: 2D digital postmarks containing machine readable data for postage evidencing and service enhancement applications. The first example uses PDF417 2D symbol and the second example uses Data Matrix 2D symbol.

LOS ANGELES HOLLYWOOD   RATE
MAR 18 98   9876543210 $0.000
METER   NNJ6UUUU
U S POSTAGE

RATE
$0.000
9876543210
NNJ6UUUU
MAR 18 98
LOS ANGELES
HOLLYWOOD

****

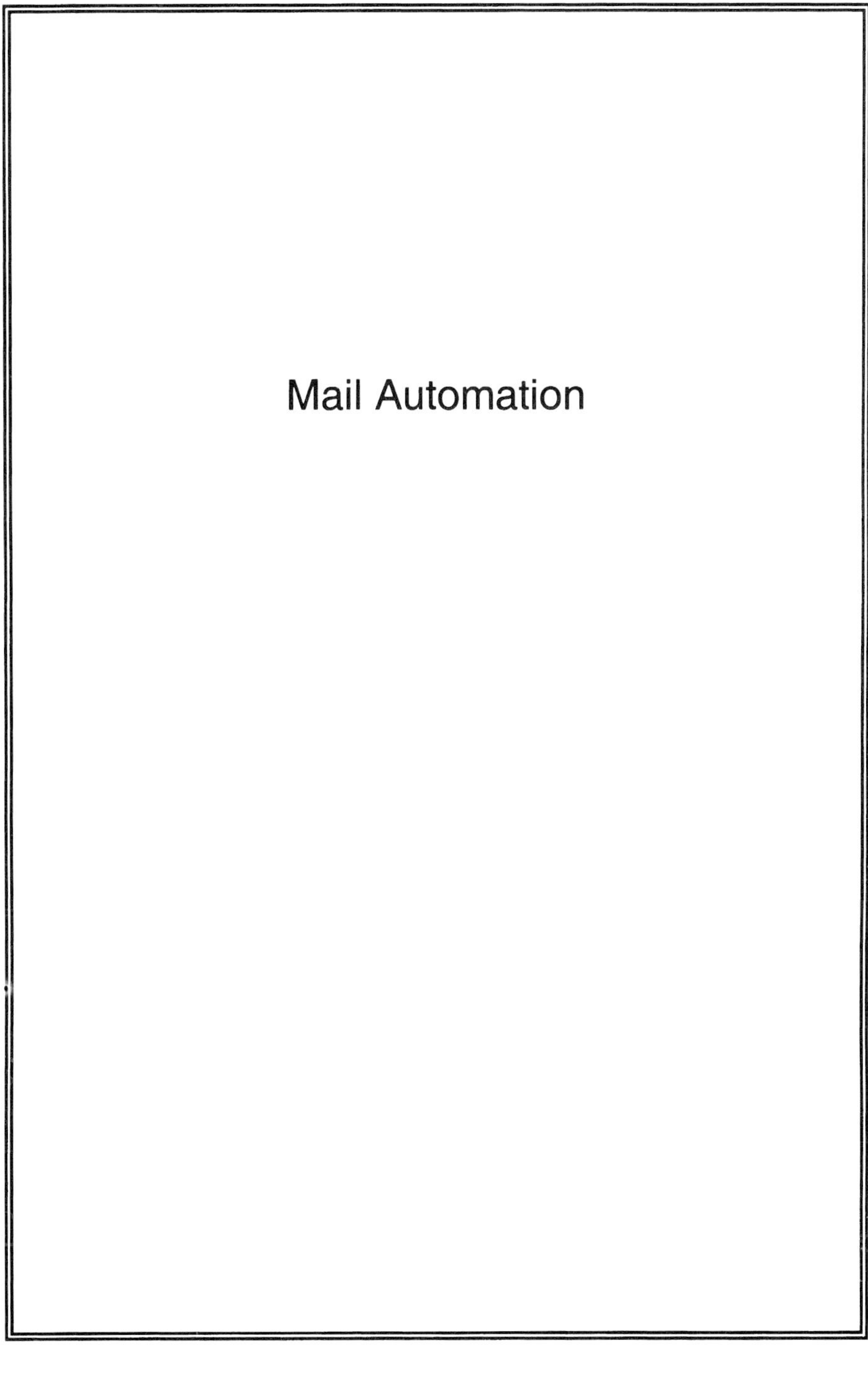

Mail Automation

MARS – the global branch office solution

J SCHOFIELD and **F GILLET**
Mannesmann Dematic Postal Automation – Business Development, France

1   ABSTRACT

Currently, most Postal Operators have chosen to adopt a centralised structure for their Mail Processing Network. This approach is based on a number of interlinked Main Processing Facilities (MPF), each relying on a cluster of satellite Branch Offices. In this scenario the mail processing equipment is centralised in the MPF whereas the Branch Offices rely solely on manual sorting. This scenario is able to process today's mail, however in the future, given the growing volumes and a global desire to increase the automation of the sort process through to delivery point sortation, it will no longer be the optimal solution. For this reason Mannesmann Dematic Postal Automation has designed the Mail Automated Reordering Sorter (MARS). This system is designed to provide the Postal Authorities with a cost effective sorting and mail processing solution adapted to the Branch Offices. MARS will revolutionise the way in which postal administrations operate by passing from a centralised approach to a decentralised approach, whilst spreading the workload over a longer period of time. The new functionalities of the machine will also lead to its utilisation for a longer period rather than the classic two-hour sequencing phase.

2   REDUCING A SIGNIFICANT COST ITEM

Improved automation and significant investment on behalf of the major Postal Authorities has led to the optimisation and effective utilisation of the resources within the Main Processing Facilities. This is reflected in Figure 1 below, which presents the current breakdown of the processing cost of a mail piece.

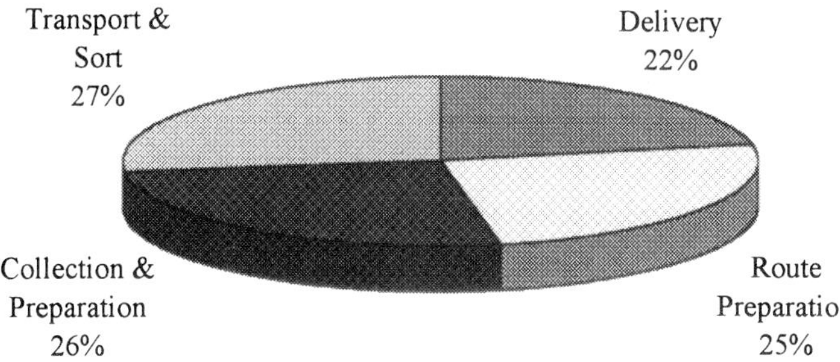

Figure 1 - Mail Processing Cost Breakdown

Of this global processing cost less than 27% of the cost can be input on the MPF, whereas almost three quarters are generated in the Branch Offices (Collection & Preparation 26% + Route Preparation 25% + Delivery + 22%). Now that the MPFs have reached a high level of automation, Mannesmann Dematic Postal Automation now believes it is time to address the Branch Offices. It is obvious that certain tasks cannot be automated today, for example the physical delivery, however, others can be quite simply implemented and have a significant impact on this pie chart. In fact the Branch Office is one of the few remaining areas where the sortation of the mail pieces is systematically, but for a few exceptions, carried out manually. For this reason MARS was developed.

3   WHAT IS MARS?

MARS is a compact sorting machine designed for use in Branch Offices. Two configurations are available with either 16 or 20 large volume stackers (400 mm).

Figure 2 - Mars Machine Overview

The system is based around an ergonomic design, which places the conveyor in front of the output stackers. The machine in its basic configuration is 2050 mm wide and 6635 mm in length without the add-on module and 7930 mm with. With an input capacity of up to 6000 mm and a throughput of 40 000 mailpieces. MARS is also equipped with fluorescent and black & white bar code readers in addition to a number of optional features. Mars uses the high speed conveying technology previously incorporated within the other products included in the Mannesmann product range. Indeed the technology used is based on conveying the mail piece in a landscape mode, between two pinch belts at up to 4 m/sec.

Figure 3 - Mars 400 mm Capacity Stackers

To avoid double feeds the feeder uses controlled air valves behind perforated belts in order to singulate the mailpieces. Furthermore, the inclination of the feeder magazine aids the singulating process and further reduces the number of potential double feeds. Lastly, an additional device installed at the feeder level halts, should a double feed occur, the second object.

One of the design prerequisites was that the machine requires a single operator. For this reason the ergonomic design, whereby the mail feeder is integrated within the machine body, was chosen. In addition the large capacity feeder magazine provides the operator with sufficient time to carry out other tasks in parallel.

4   DECENTRALISED SEQUENCING – THE MAIN FUNCTION

Few will disagree that in the modern Postal Services the idea of automating Carrier Route Sequencing is at the forefront. By automating the labour intensive task of carrier walk sequencing it is possible to obtain a saving of up to 16% of the overall processing cost described in Figure 1 - Page 2. However, once a Postal Service has decided that indeed it wishes to benefit from the savings described, it is faced with another dilemma. What approach should be adopted, a centralised approach using the Letter Sorting Machines (LSM) installed in the Main Processing Facilities, or to install smaller machines in the local Branch Offices. Following an analysis of the latter solution, MDPA came to the conclusion that for a significant number of Postal Authorities the decentralised approach seemed justified and could provide a gain over the centralised approach of approximately 40 % in terms of initial capital outlay.

It is for such a purpose that MARS was designed. The major constraint in a Branch Office often being the machine footprint, special attention was paid to this area and the overall footprint is now under 8.5 square metres for the 16 stacker configuration.

In the sequencing mode the recirculation of mail pieces can be a complex procedure which if incorrectly carried out will hinder the process. If the operator were to empty the feeders and invert the order then the mail contained within, the inverted stackers would be rejected as the machine would no longer recognise the order from the previous pass. By incorporating the feeder conveyor in front of the stackers, the recirculation phase is simplified thus avoiding such errors, as shown in Figure 4.

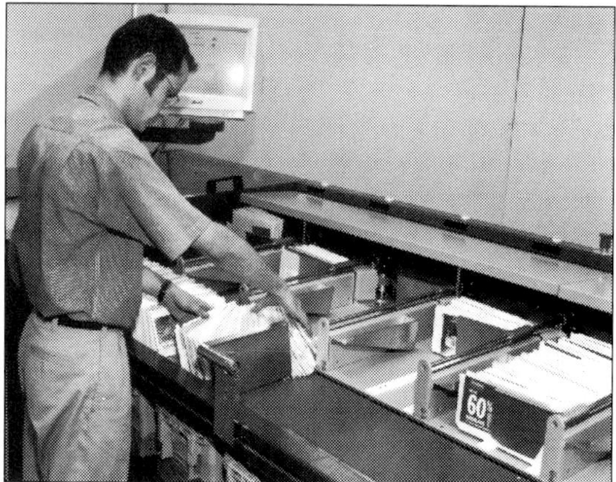

Figure 4 – Ergonomic Mail Recirculation

In order to fully optimise the operation of the sequencing sorter it can be of interest to artificially combine several carrier routes and to sequence them as a single virtual carrier route.

At the end of the entire process this single route is then broken back down into the separate actual routes. This is achieved by adding colour-coded separators to the mail to be processed. These show where one route ends and the next begins, as shown in Figure 5.

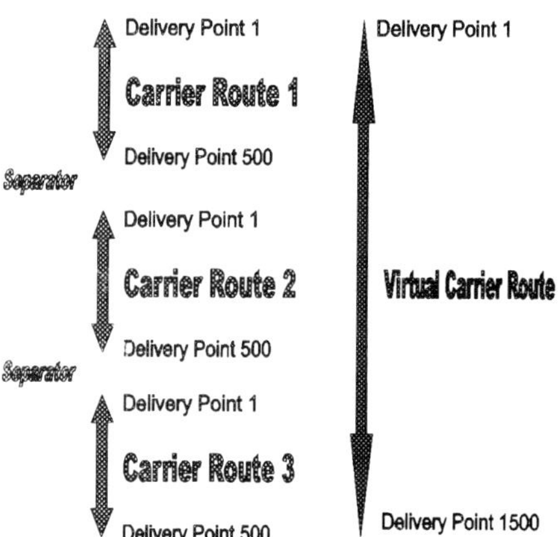

Figure 5 - Merging Carrier Routes for the Sequencing Phase

By adding these separators it is possible to sort 10 to 12 mail carrier routes simultaneously.

Furthermore these separators can also be used in order to "overlap" the different sort passes carried out on the machine. Mars uses a three-pass approach implying that the mail is passed

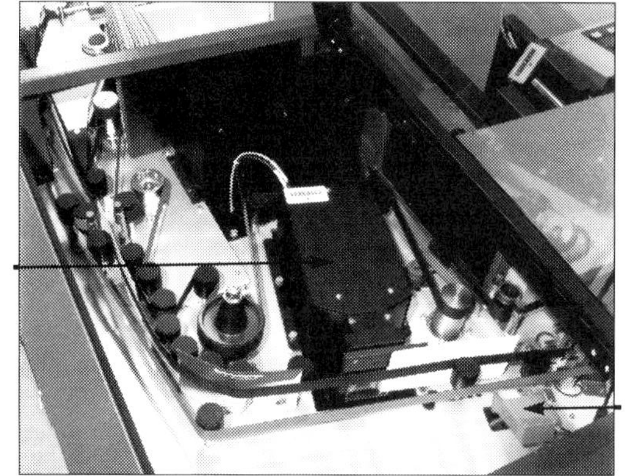

Figure 7 – MARS Mail Path

When the second optional module is added to the system the following components are incorporated within the module. A phosphorescent detection unit and a serration detection unit for stamp detection linked to the cancelling unit, an image acquisition unit, and two ID-Tag Printers, one for the UPU code on the reverse face of the mailpiece and one for the front. The module can be added as an optional item after the machine has been installed and is in operation.

5.1.1   Facing & Cancelling

MARS has been designed in order to be able to add, simply and in a modular nature, a stamp detection and cancelling unit. By incorporating a stamp location module it is possible to locate and cancel mail using MARS. The process is relatively simple, but remains efficient and provides an additional service. The mail passes through the machine. If the stamp detector (which can be a phosphorescent or a stamp serration detector) locates a stamp in the correct position it is cancelled and sent to a designated stacker. The operator then transfers all the non-faced and cancelled mail from the stacker, turns it around and it is once again processed through the machine. Such an approach should enable the Branch Office to face, orient and cancel up to 15 000 letters per hour with a single operator. This will help avoid large quantities of mail needing to be faced and cancelled in the evening, when the workload is at its highest, on the Culler-Facer-Cancellers in the Processing Facilities. In addition, fewer CFCs will be required by the Postal network for the same number of mail pieces processed. By adding this function the Postal Operator will also improve the Quality of Service provided as the cancelling will take place in the Branch Office and will better reflect the true time at which the mail was collected.

The CFC function will enable the detection of phosphorescent and fluorescent stamps. In addition, by adding the image acquisition unit to the system MARS will be able to detect stamps by comparing the image of the stamp acquired with those in the database. This will be of increasing interest as the number of "collection" stamps increases and the use of fluorescence and phosphorescence decreases.

5.1.2   Image Lift

By fitting an image acquisition head and an Id-Tag printer, the system can start to process the mail for the outgoing sort. When the mail passes through MARS an image of the letter is acquired and an Id-Tag printed. This then enables the images to be processed by a local or three times through the machine. "Overlapping" implies starting the second and third passes before fully emptying the outputs from the previous pass. This is done by adding the separators at the end of the mail to be processed. The identifying bar code printed on the separators is read by the Bar Code Reader shown in Figure 6 and Figure 7. The separators are then fed, one to each stacker, at the end of the pass. The operator can then use these as references when emptying the stackers. By "overlapping" the passes the machine's performances can be increased by up to 30%.

5   COLLECTION MAIL AUTOMATIC PREPARATION

In addition to the sequencing role the MARS design enables a number of additional manual functions to be automated. These include Facing, Cancelling, Customer Bar Code and ID-Tag Reading and Printing, Image Lift for local or remote OCR/VCS, etc. These additional uses result in a machine which will operate for a longer period of time than just the sequencing phase and therefore generate a more rapid ROI.

The new functionalities listed are included into an optional add-on compact module placed at the feeder end of the machine, as shown in Figure 6.

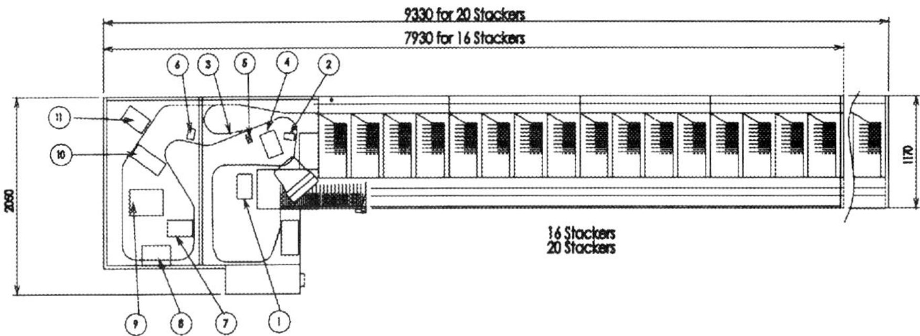

Figure 6 - Mars Additional Functions Module

1. Fluorescent Bar Code Reader (Face)
2. Separator Card Reader (Reverse)
3. Switching
4. UPU Fluorescent Bar Code Reader (Reverse)
5. Customer Bar Code Reader (Face)
6. Serration Stamp Detection Unit (Face)
7. Image Acquisition Unit (3.2 m/sec)
8. Phosphorescent Stamp Detection Unit (Face)
9. Cancelling Unit (Face)
10. Id-Tag Printer (Face)
11. UPU Id-Tag Printer (Reverse)

In the sequencing mode the mail passes through the first module (containing units 1-5). This includes the feeder section and the delivery point identifier (fluorescent or black code located on the front or on the rear side of the envelope), as shown in Figure 7.

centralised OCR/VCS whilst the mail is physically being transported from the Branch Office to the Sorting Centre.

The advantage of having an image lift and local OCR in the Branch Office is that mail for the local area does not need to be transported from the Branch Office to the MPF and back again for.

In the case of a local OCR it is also possible to start a pre-sort during which the 16 or 20 stackers can be used to sort the mail to the large destinations or to keep local mail within the Branch Office. Again this process enables the mail flow to be smoothed over the entire day rather than having a peak in the evening.

Easy access around the machine has been ensured in order to provide this option and so that the utilisation of the machine is as ergonomic and user friendly as possible.

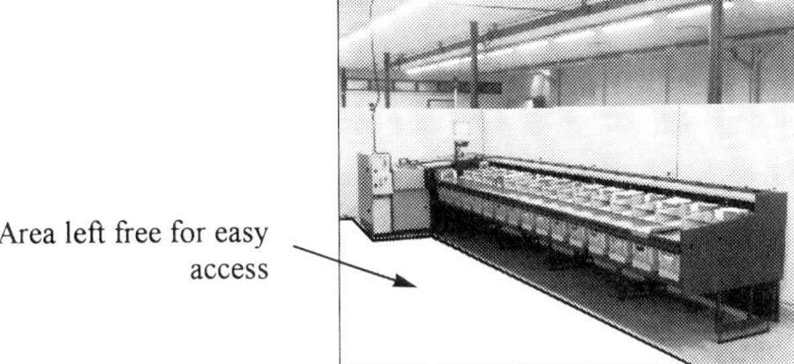

Figure 8 - Mars Easy Access

In addition to being used for Optical Character Recognition and Video Coding System purposes the Image lift function can also be used to detect the mail class and the mail to be processed first. The use of a Digital Post Mark (Franking Mark) is also of interest as it enables the Postal Authority to control both the invoicing and tariff for the mail piece.

5.1.3 Forwarding

If we assume that MARS is fitted with the Image Lift described above, it is possible to carry out the forwarding (redirection) of mail at the beginning of the mail processing chain rather than at the end. Obviously this will require that a number of fundamental features be implemented for example a National Change Of Address (NCOA) file containing the addresses and names of the persons having moved and wishing to have their mail forwarded. MARS processes the mailpiece, the image is acquired and an Id-Tag printed. The image is then sent to an OCR/VCS and should the addressee be identified as potentially requiring forwarding the image is double checked visually by a dedicated VCS operator and the mail piece is separated from the normal flow during the outgoing sort in the Processing Facility. Today, a human intervention will always be required, as it is necessary to establish that the mail piece to be forwarded is indeed for the person having moved. For example;

Mr J Smith
12 Birch Avenue
Northampton

has moved and wants his mail forwarded, however his brother Mr S Smith of the same address has not. An operator is therefore required to differentiate between these two persons.

Capturing the mail to be forwarded at the earliest possible stage avoids a great deal of additional processing and transport costs and thus provides significant savings. This function seems to be very difficult, if not impossible, to implement in a centralised approach due to the time constraints.

5.2   MARS - The Global Branch Office Solution

To summarise, MARS and the decentralised approach have a number of significant advantages:

1. The capital outlay and the operating cost are less.
2. Constraints, which exist in terms of organisation and logistics for transporting the mail once sequenced, are no longer required.
3. The Postal Organisation has a number of backups available should a problem arise.
4. The Quality of Service is improved through more accurate cancelling times and expedition locations.
5. Capturing, at an early stage, mail, which needs to be forwarded, becomes possible.
6. The Main Processing Facilities' workload is reduced, and smoothed over a longer period and the peaks in terms of CFC, OCR and VCS functions, which occur in the Main Processing Facilities in the evening, are minimised. Furthermore the machines are used longer and labour costs can be cut if so desired.

This new approach of decentralised mail processing using MARS, will revolutionise the Postal sector, not only through the restructuring the networks, but also through a more adapted and improved utilisation of the equipment in the MPF and Branch Offices.

Blueprint for the second wave of flats automation

G SCHINDLER
Siemens ElectroCom GmbH, Konstanz, Germany

The automation of the flat mail sorting process has proceeded beyond infancy. Siemens ElectroCom has intensively investigated the flats handling process within the complete mail handling continuum from collection to delivery. It proposes a new functional blueprint for the continuing automation of the flat and regular mail streams that will not only fully integrate automation from the outset, but also reduce the residual manual labour requirement for flat mail handling to desireable levels.

1. INTRODUCTION

With several hundred Flat Mail Sorting Systems in operation worldwide, the industry has progressed beyond that phase where we discuss *visions*. We – suppliers and users – have reached the point at which, with a wealth of experience to support our efforts, we can intelligently propose a plan, a blueprint, for the next wave of flat automation technology. In this discussion, I would like to propose the automation products that will extend the domain of flat automation from its current, sorting centre-bound, inward/outward mechanisation focus to encompass the entire flat handling operation.

From the Siemens ElectroCom perspective, this "2nd Wave of Flat Automation" will also be the actual catalyst that transforms the much maligned postal "islands of automation" into an end-to-end automated workflow landscape. Accordingly, although the focus of our blueprint is on the mechanisation of flat sorting, we will present the reduction of the *total residual manual labour* as a by-product of a correctly automated flats mail stream. After all, manual labour is still the glue the operational sorting and consolidation continuum that is the backbone of the postal process:

Collection → Cull/Face/Cancel → Outward/Inward sorting → Carrier Route Preparation

Presently, when one includes flat mail in this sequence of postal operations, we find, to our surprise, that as much as 80% of the process remains manual! Given the high level of regular letter mail automation, it is clear that non-automated flat processes are affecting all of the other mail handling that must progress in step with flat mail processing to achieve a desirable end product.

Thus, tying the integration of automation islands and flat handling becomes less nebulous. Indeed, it takes on significant scope within our blueprint discussion when we realise that most of the current gaps in flats automation correspond to the well-known operational bottlenecks presented by collection, culling/facing/canceling and delivery preparation where flats and regular mail are intermingled.

Against this backdrop, we can define the necessary automation bridges that need to be built and the bottlenecks that will be addressed as part of our blueprint for the 2nd wave of flat automation. The building blocks of this blueprint are the new processing paradigms through which we will achieve the automation goals discussed.

2. BLUEPRINT FOR THE 2ND GENERATION OF FLATS AUTOMATION

Our initial focus in our search for a rationale and corresponding blueprint for the 2nd wave of flats automation is on understanding those attributes of flat mail that make automation significantly different and operational aspects distinct from the experience we have gained by automating regular letter mail processing. Before proceeding with our discussion, let us clarify the physical differences between regular and flat mail, and highlight the business and logistical environment that makes flat mail processing a unique challenge.

2.1   The initial approach

First of all, we should start with the issue, "Is there really a problem we need to solve here?"

After all, postal services have well established operational principles and proven models for mechanisation of regular letter – nominally postcards through C5 envelopes. Referring back to our blueprint's objective of automating the flat processing continuum of

Collection → Cull/Face/Cancel → Outward/Inward sorting → Carrier Route Preparation

we find successful, proven solutions, both mechanical and in OCR technology, dating back to the 1970's. These automate regular mail sorting, almost flawlessly cull, face and cancel envelopes and, using bar codes, sort mail to an almost infinite number directions. As a case in point, route preparation carrier walk sequencing of regular mail can be performed within a multi-pass bar code sorting operation using any of several well known carrier sequence sorting multi-pass paradigms.

So the question becomes, more precisely, "Given that we have existing, proven methods of postal processing automation that cover collection to delivery end-to-end, why do we need a new blueprint, new automation techniques and revised principles of operation?"

But this question is based on the premise that current automation principles and, more specifically, state-of-the-art regular mail automation technology, can be carried over one-to-one to flat mail. You might call the underlying premise of this approach to flat handling technology "the Darwinian model of flat automation."

Basically, the Darwinian concept of flat automation assumes that our letter mail automation hardware adapts proportionately to become flat mail systems and follows the same processing paradigms. At first glance, this may seem reasonable, since one could point to the current generation of flat sorter OCR's (FS/OCR), which do look like large-dimensioned letter mail OCR's. Closer examination of the FS/OCR, however, proves that it is far from the being a mutant letter Integrated Reader Video system or other letter sorting OCR configuration.

Essentially, it is fair to say that FS/OCR's did evolve from their smaller cousin's mold but, as we in the industry have seen, the FS/OCR's quickly developed their own mechanical character. Unfortunately, the developments were not all "natural," but technology or project-driven compromises, even plain old fixes. However, because of the changed operational environment, particularly the physical character of the goods we are handling, our flats systems have evolved to accommodate that environment: trays have replaced bins and stackers; fully automatic sweeping has taken the place of manual sweeping of bins and stackers; and, most recently, automated assists for the feeder operator are being implemented to replace the purely "flip the tray" approach from letter sorter OCR's.

Thinking about it, we can only conclude that these developments were the clear response to the problem of *size*, by which we are literally referring to the sheer bigness of flat mail volumes compared with letter mail. Take a normal sorting centre with a flat mail volume of 20,000 flats per hour. At 33 items per tray, we are talking about handling 600 trays, a stack about 150 meters tall, *every hour*. In fact, the trays aren't just twice as big, they are twice as heavy and, more to the point, they are extremely unwieldy. So, staying within the Darwinian model, we evolved – practically mutated – responses that allowed our evolving creature to survive. And, as in nature, time will tell whether these are positive, negative or neutral mutations.

Returning to our letter OCR to flats OCR devolution however, we find that, once under the machine covers and into the intelligent machine processing, flat and letter OCR's reflect very different image processing strategies and scenarios. Flat mail OCR, for example, has a technological sophistication for the handling of clutter associated with "region of interest" location that is orders of magnitude beyond letter mail OCR technology. And these differences continue as we move along to the intricacies of FS/OCR video coding relative to the same operation in a letter sorting OCR.

We need only to look a bit more carefully at our Darwinian model to find support for the idea of a new automation blueprint and a new generation of flats automation equipment. Letter and flats are, frankly, birds of a different feather. We posit that fundamentally different physical and technical solutions will be necessary to achieve a truly effective postal processing continuum for flat mail.

2.2 Reassessing flat mail solutions

To get a better feeling for the real solution requirements that will underlie and affect each step of flat mail handling in seeking our goal of end-to-end automation, we need to reassess flat mail – bottom-up, as a mail stream distinct from and equal to regular mail.

Before getting on to how the flat mail automation blueprint meets next generation requirements, some additional remarks on the impact to date of having started the first generation of flat mail automation with the Darwinian model, i.e. letter OCR as the direct precursor of flat OCR.

By taking the approach that seemed obvious to all of us, we came up with flat mail automation solutions that are, in important ways, the same islands of automation that their letter cousins were. We missed the opportunity to go the additional length to achieve the seamless connection to the collection/pre-processing processes taking place upstream and have even less integration with the delivery preparation processes downstream in the delivery units.

Now we, the suppliers, have our machines, which reflect the highly centralised processing strategies of our customer postal services. We anticipate that the future, however, holds the need for both centralised and decentralised solutions. But the key to that decision will lie in the details of a processing concept the fully integrates the postal process continuum

3. A CUSTOMER-CENTRIC VIEW OF FLATS

If suppliers were to adopt a more customer-centric point of view, it would be clear that regular mail and flat mail processing are more different than they are alike. This customer-centric viewpoint will be a major facet of the blueprint that follows. From our extensive customer contacts, we at Siemens ElectroCom have seen the interesting perception that
(1) Flats are a unique postal media
(2) Flats are, if anything, more related to parcels than to regular mail!

Taking these observations as our starting point, let us now assess some of the unique characteristics of flats when viewed from a customer's standpoint. Using their opinions and perceptions regarding the "business opportunities and threats of flat mail," we will proceed to derive a blueprint for the 2nd Wave of Flats Automation.

3.1   Dimensions
Although the definition of a flat seems well specified, with nominal dimensions of between 140 x 90 mm and 400 x 300 mm, thickness of 0.5 - 30 mm, and allowable weights of 10 - 1500 grams, we find that no two major posts have the exact same mail specification! In addition, many of the largest handlers of flats (e.g. FedEx, UPS, DHL, etc.) initially receive their flat and small parcel mail intermixed from the customer. This causes us to *operationally* view everything from large, stiff letter sorter rejects to small parcels as potential flats. One might conclude that the specifications above are more a reflection of the suppliers' and engineers' desire than a customer-perceived category.

3.2   Wrapping
A unique customer-driven characteristic of flat mail not seen in regular mail is the wide variety of enveloping material, which, particularly in the case of plastic, creates entire classes of difficult-to-machine flat mail pieces. It is noteworthy that, because plastics are frequently the wrapping of choice for bound printed matter, this category of flats is increasing in volume and cannot be simply "bumped" to the next processing level. In letter mail processing, non-machinable pieces are often simply redefined as flat mail and sent to another section, but bumping a flat over to parcel processing causes other problems. That was fine earlier, but today the serious postal services are benchmarking their degree of automation at 100%, so this option does not get us very far.

3.3   Origin
In terms of volume, flat mail originates almost exclusively from business and professional sources.   Unlike regular mail, flat mail hardly ever originates in private households. Accordingly, the potential for cooperative large volume mailer preparation is a real opportunity to be exploited in an end-to-end automation solution blueprint. This includes services such as pre-printing of bar codes or walk sequence bundling by the carrier.

3.4   A mail stream within a mail stream
The postal market for flats in the form of overnight express letters and small packages is dominated by well-organised, dynamic, profitable private sector competitors. They normally have a much lower degree of automation than their governmental counterparts. Postal entities

such as ChronoPost in France are successfully entering this segment and finding that the "arbitrary" flat mail standards set by automation planners are at best irrelevant but more likely merely complicate fast, smooth processing.

Generally, private express operators, newly initiated postal express operations, as well as smaller postal services worldwide have only a blurred processing distinction between flats and small parcels of up to around 3 kg. Not only customer shipping but also operational economies of scale dictate that flat mail and small parcels be handled within the same automated process.

In any case, customers cannot be expected to accept an envelope size (i.e. C5, B5, etc.) as a flat specification.

3.5 Higher monetary value

The monetary value of flat mail items runs the gamut from nominally valued flyers to high-value, parcel-like contents. Across the board, however, flat mail value tends toward a much higher unit value than regular mail. Thus, we have a higher proportion of flat mail requiring tracking and tracing as part of standard service levels. This necessitates a simple user option for initiating track and trace *without* making a trip to a post office counter.

3.6 Impact of *technological substitutes* for regular mail on flat mail volume

Whilst regular mail volumes are being eroded by fax and e-mail coupled with high quality printing, flat mail volumes (not to mention parcel volume) are, in contrast, being enhanced by these competitors. We all now know that Internet (e-commerce) and fax requests generate enveloped flat mail responses and/or small-enveloped purchases.

But if regular mail volumes begin deteriorating in absolute terms in the next ten to fifteen years, we need to consider how the mail stream that takes on the priority role is going to look.

Another example of the positive correlation between flat mail and the technology substitutes for regular mail is the recognisible trend away from folded inserts in regular mail and toward very high quality printed information sent as flat mail. We find that the rapid technology and product growth of high speed, high quality printers also indirectly creates potential for a whole new flat mail stream, which we have dubbed *aesthetic hybrid flats*. By *aesthetic* we mean

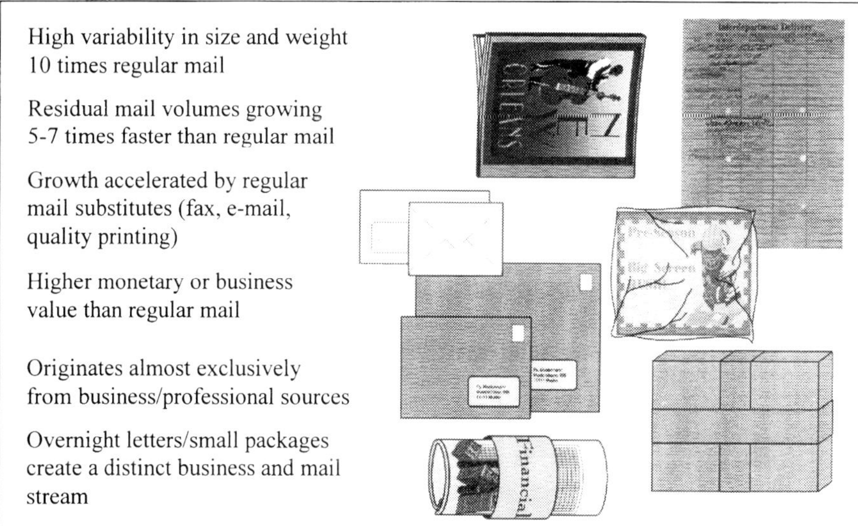

Fig. 1: Major factors underlying the customer-centric views of flat and regular mail

hybrid-created flat mail that is indistinguishable from corporate mailroom mailings that go to great length to convey an image of high-quality, personalised correspondence. We see this as a small but important mail stream because of the high value placed on it by the sender.

4. OPERATIONAL FACTORS DRIVING THE FLATS AUTOMATION BLUEPRINT

We have discussed those aspects of flat mail that make it unique or, at least, more like parcel mail for most postal services. Now, let us examine the effect this has on the paradigm on which we want to base our blueprint.

4.1 Manual handling inefficiency
This is a key driver for our flats automation blueprint.

The extreme variability in weight and dimensions of flat mail severely limit and essentially preclude automation processes predicated on repetitive loading and unloading of containers by an operator. As mentioned above, the sheer unwieldiness of our goods make it impossible for operators to exercise the dexterity, productivity and reliability necessary to make our recirculating bar code sorting paradigms work efficiently for flat mail inward sorting or, most critically, for delivery walk sequencing.

Equally critical, the high degree of variability of flat mail dimensions compromises operating paradigms that depend on gravity-assisted stacking or recirculation.

4.2 Growing prominence of flats that cannot be automatically in-fed
In recent years, we increasingly have do deal with a subclass of non-machinable mail items referred to in the industry as RESMail (i.e. Residual Mail). Clearly, by its very nature, we

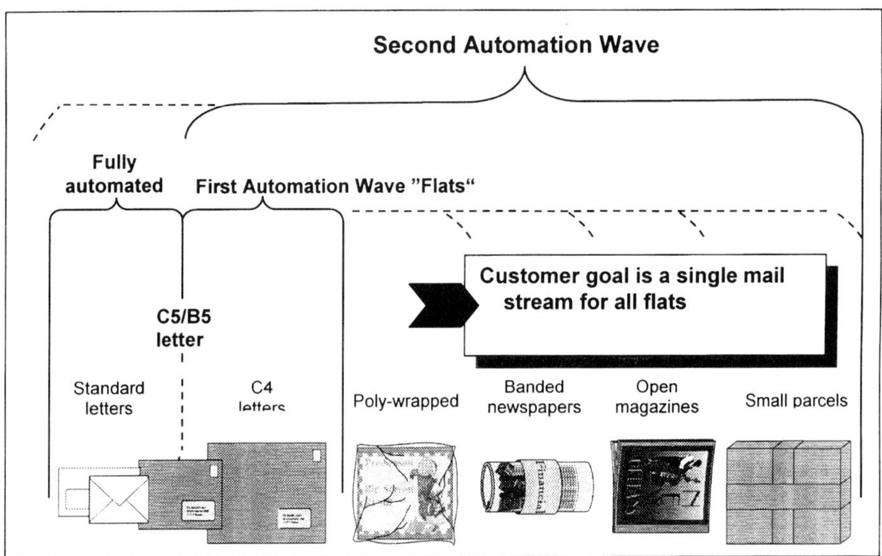

Fig. 2: Extending the range of flat mail that can be automatically handled is key to further progress in automation.

cannot specify RESMail by dimensions or weight since we also find perfectly normal looking flats denoted as RESMail due to the texture of their plastic envelope – which unfortunately inhibits automatic feeding. A common distinction for RESMail is to classify it as:

- Those items that cannot be automatically fed, but which can be manually in-fed and sorted.
- Items that are neither automatically nor manually feedable[1]

Our main concern is that RESMail is a growing segment of flat mail. Regular mail in advanced countries normally contains less than 2% non-machinable items whilst flat mail contains over 10%, with a tendency toward 15%.

4.3   Impact on delivery unit operations

In some U.S. Postal Service delivery units, flat mail is reaching 50% of total delivered volume. At this magnitude, flats create in essence an additional pass in carrier delivery preparation:

→ Merger for delivery of sorted flats/bundled flats/regular mail

When flat mail must be manually walk sequenced using pigeon hole casings, and then merged before departure with regular mail, the level of labour in the delivery unit remains virtually unchanged even when delivery point sorted[2] regular mail is introduced. Hence, the route preparation activities are undoubtedly the most labour-intensive part of flat mail processing and need to be automated with the same intensity and focus that we have applied to sorting centres until now.

5. THE BLUEPRINT

Based on the preceding discussion, we approach the flat mail automation blueprint with some generalised, high-level prerequisites. We know that the blueprint must:

1. Describe a process that avoids the pitfalls already recognized in regular mail automation. We should examine whether the existing delineation between regular and flat mail needs to be fundamentally redefined.
2. Structure operator tasks so that mail handling dexterity is not taxed or on the critical path.
3. Include all process steps up to route preparation.
4. Intensively review the automation requirements of flat mail and regular mail at those points where they must be consolidated or in which they are intermingled.

Our purpose here is not to write volumes on methodology and design alternatives that can accomplish the objectives set. Instead, we will outline and explain a functional blueprint that we consider essential to the achievement of effective end-to-end automation of flat mail from its collection through to its delivery, which is consistent with the major points of our preceding discussion.

---

[1] This latter category is RESMail flat is, on occasion, *promoted* to parcel and handled accordingly. This is less than desirable since parcel handling may be too rough and parcels are often handled in different facilities, so that the transfer causes additional delays.
[2] Automatic Carrier Walk Sequenced

5.1   Culling/facing/canceling

In line with the goal of achieving a fully automated process, the CFC handles regular mail, flat mail and RESMail as peer mail streams. The specification for the automatic feeder of the CFC will need to accommodate everything from regular mail to flat mail, with manual feeding and scanning of RESMail.

The RESMail will be image-lifted and replaced in the stream by "surrogates" that fit seamlessly with machineable flats. These surrogates allow us to delay handling of the difficult pieces until we actually reach the delivery phase. Techniques have been developed that manage the separation of logical processing from the physical flow, which are detailed in associated patent filings.

5.2   Outward and Inward Sorting

In our new model, a key change lies in the redefinition of the flats sorter's role. It is no longer a multi-pass sorting system, but rather a purely mechanical sorting device. The role of the flat sorter is significantly reduced because the volume is off-loaded to bundle collation (via cooperative large volume mailer programmes) and generation of aesthetic hybrid flats, which will become important parts of the delivery function.

The remaining flat mail is handled by a flat sorter whose role is limited to outward sorting a limited number of directions (schemes) to each sister centre. In this marginalized role, inward flats sorting is just a single-pass operation in which the sorting granularity corresponds to large consignments of flats sorted by carrier groups. For example, a delivery unit with 20 carriers could be covered by two sorter outputs (ten carriers per output), each nominally batching 2,000 flats (200 per carrier) which are then consigned to the delivery unit for final sorting.

These relatively large consignments per output mean that the number of *necessary outputs per sorter are roughly half* of what is now being installed in sorting centre operations.

5.32   Route Preparation in the Delivery Unit

For the first time, the delivery unit becomes an integral part of the automation process. In fact, we view the delivery unit as a manufacturing process that combines Just-in-time and

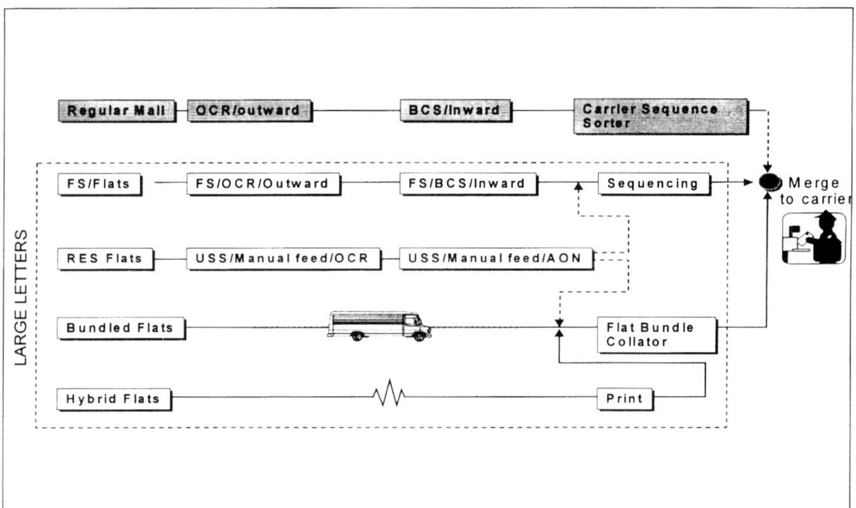

Fig. 3: The organisation of the main mail streams within the functional blueprint of the 2nd Wave of Automation

                                                                                                  C568/048 © IMechE 1999

continuous process flows. The key to success lies in the successful transfer of the control over the process to the delivery unit, which is now confronted with the organisation of a highly complex task.

In the delivery unit, the mail consignments from the sorting centre are first carrier walk sequenced. Sequencing of the large flat sorter consignments is done using closed cycle sorting techniques. Here, the operator interacts only on a limited basis with the mail as it circulates and is operated upon until delivery sequence is achieved.

Thereafter, the four sequenced mail streams – regular mail, sorted flat mail, collated flat bundles and hybrid – are merged by walk. The final merging is completely automatic.

6. SUMMARY

Automation of flat mail processing has progressed beyond infancy. It is time that we revisit our initial efforts at flats automation, our long history of regular mail automation and also consider the developments in regular and flat mail processing in other industry sectors. Using our insights, we must develop products that extend the scope of flats automation across the complete mail processing continuum and finally reduce residual manual labour.

Distancing ourselves from existing specifications, we have attempted to describe a customer-centric view of flat mail. In doing so, we find that flat mail is seen as either a unique media or, at least, more like parcel mail than regular mail in critical ways. Reviewing these critical aspects, we find that three factors will drive the next wave of flats automation:

- Dealing with the variability, weight and unwieldiness of flats
- Handling the growing amount of non-machinable flat mail
- Alleviating the impact of growing flat mail volume on delivery units

As a result of this effort, we have established a functional blueprint for the 2nd Wave of Flats Automation. After showing that three streams of flat mail -- large consignment flats, bundle collated flats and hybrid flats – have crystallised, the blueprint outlines the impact of our changing mail structures on the culling/facing/canceling function, on outward/inward sorting functions, and on the delivery function. It indicates the areas in which we see opportunities to streamline and facilitate not just flat mail processing, but also regular mail processing.

There are still a number of open issues in our blueprint that we have not addressed in this paper. Siemens ElectroCom invite discussion of these issues and possible concrete solutions with you, the postal service, the owners of the mail processing problem.

DUAL – a double feed detector for flat sorting machines

D THIERIOT and **D AUDRAN**
Service de Recherche Technique de la Poste, Nantes, France

SYNOPSIS

Every day, vast quantities of letters, forms, cheques, advertising, etc... are processed on production lines which have become more and more automated. The French Post Office has successfully automated the sorting of letters, and hundreds of millions of letters are sorted automatically every year.

However, flats (postal items ranging in size from C5 to 260 x 380 mm) are much more difficult to process automatically : the form of such objects varies tremendously, not only in size but also in thickness, rigidity and type of envelope material (paper or plastic). As a result, de-stacking or separating them is a particularly tricky operation.

Because it is difficult to improve the destacking process at the sorter input, SRTP has concentrated on a method of detecting unseparated objects and rejecting them. In this way, there is little risk of objects being dispatched in the wrong direction, which is a high cost factor for La Poste.

1. THE SOLUTION : DUAL, DOUBLE FEED DETECTOR

DUAL double-feed detector is designed for use with the Mannesmann sorting machine. The device is installed under the conveyor, before the first output box, so that all the objects processed by the machine are scanned. When DUAL detects a double-feed, the unseparated objects are rejected systematically and returned to the input station.

The operating speed of DUAL is six images per second. It has a detection success rate better than 80 %, with only 0.5 % hyperdetection (false detection of double-feed).
Double-feeds account for about 1% of all flats processed by machine.

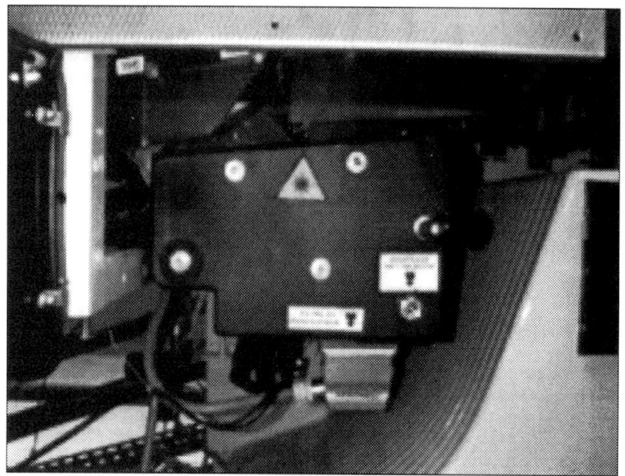

View of DUAL mechanical unit

2. DESIGN

2.1.Mechanical unit

The unit is installed just before the first output bin, under the sorting carousel.

Two objects which have been fed into a single box are first separated by two belts which run through the notches in the bottom of the boxes (there are 4 openings in all: the belts run in the first and third notches from the outer side of the sorter). The speed of the belts is slightly higher than the conveyor speed.

When the mailpieces arrive on the unit, the belts raise them by about 20 mm and their speed is increased slightly. This tends to separate the objects, improving the detection of double-feed objects.

The unit can be uncoupled rapidly in case of a jam or for maintenance.

Uncoupled unit

2.2. Image capture

The camera and necessary lighting are housed inside the unit.
The black-and-white matrix camera is operated every time a box passes it, by means of an inductive sensor which detects each box. The image is captured at the second notch in the box, i.e. between the two separating belts.
An oblique light beam enables double-feed objects to be detected: the space between the objects shows as a dark area..

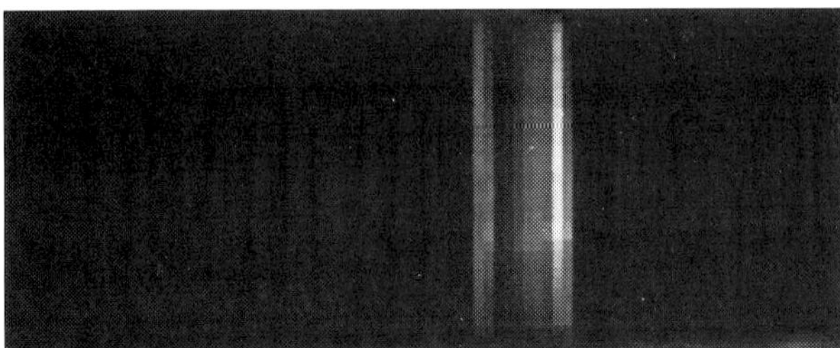

2 paper envelope objects

Another light source, consisting of a laser diode, detects polythene envelopes, which tend to create hyperdetection errors if they contain several documents.

1 polythene envelope object with 1 paper envelope object

2.3. Computerised processing and control

A PC Pentium II (300 MHz) controls the whole operation (at a rate of 6 objects per second) and processes the double-feed detection.
The processing time is about 20 ms.
When DUAL has detected a double feed, an electrical signal is sent to the sorter, which rejects the objects.

The image-processing is a sequential algorithm composed of 4 main steps :

- Determining if a box is empty or contains a mailpiece

- If the box is not empty, the profile of the image captured is analysed to detect any troughs (which would indicate a gap between two objects).If no trough is found, the process stops (the box contains only one object)

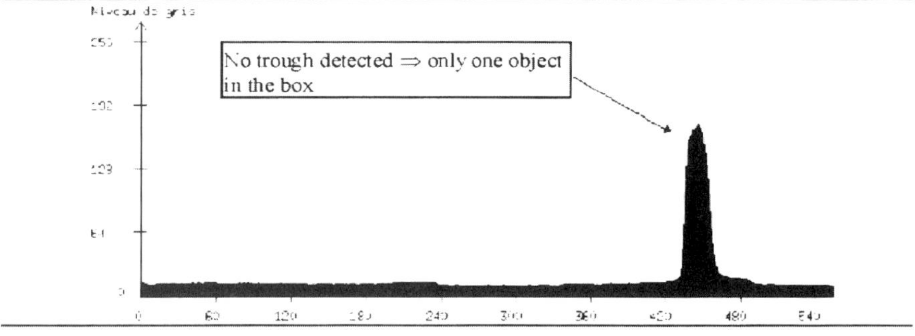

One object profile

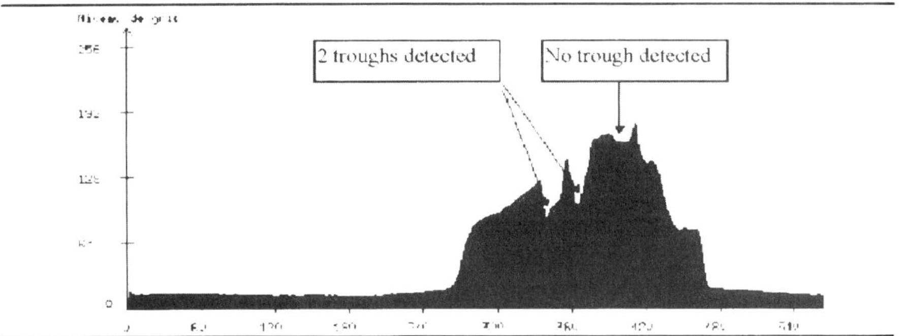

Multiple object profile

- To signify the presence of two objects, a trough must cover most of the height of the profils.

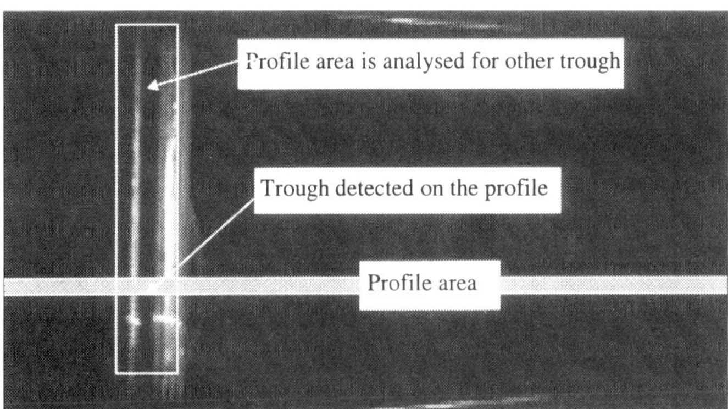

Example with 2 objects

Sometimes, particularly with polythene envelopes , one object will give a profile with a significant trough.For this reason, the laser light bean is also analysed. If the bean is detected in the trough, the trough does not indicate a gap between two objects.

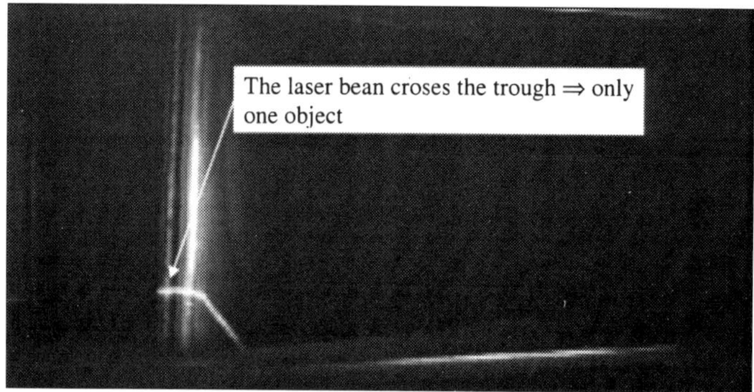

Example of one polythene object

2.4. Electrical cabinet

The cabinet is installed together with the PC and a voltage regulator at the centre of the flat sorter, and runs off a 220 volt electrical supply from the power cabinet of the sorter.
The PC screen and keyboard are fitted on the side of the sorter, near the mechanical unit.

3. PERFORMANCE

In actual operating conditions, Dual's detection rate is over 80%.

Hyperdetection errors are most common with expanding envelopes, which are in any case classed as non automatable by La Poste since they are a source of numerous multiple feed errors in the destacking process. Objects wrapped in plastic film are also problematic: they are the cause of most of the detection errors made by the image processing software.
Overall, the hyperdetection rate is no more than 0.5% of total automated mail flow.

4. CONCLUSIONS

SRTP has developed a system which meets perfectly the needs of the French Post Office, and La Poste aims to install a total of 40 DUAL detectors on sorting machines by the year 2001.

The resulting increase in efficiency means a better quality of service and a decrease in the work of processing wrongly-routed mail.

If the cost of a badly-routed mailpiece is estimated at 0.30 Euros, and the mechanised mailflow is 250,000 objects per day for 300 days a year, the annual saving that can be made by using DUAL is approx. 180,000 Euros per flat sorting machine.

To instal DUAL on to other types of flat sorting machine, it is necessary to make notches in the bottom of the carousel boxes. The necessary.adaptations to the mechanical and electrical parts and the software can be carried out by the installer, using the SRTP plans.

C568/001/99

Parcelforce Worldwide national and international hubs

L BROOKS and **D SIMMONDS**
Parcelforce Worldwide, Milton Keynes, UK

Parcelforce Worldwide is currently restructuring its national and international parcel distribution networks. The paper covers the background to the project, infrastructure changes, and two new sorting centres.

INTRODUCTION

As early as 1991 it was recognised that due to changes in the parcel market the distribution network of Parcelforce Worldwide would need to be reorganised. Standard traffic was declining and there had been strong demand for express services. This was the starting point for what would lead to the biggest project in Parcelforce's history — Project Gemini.

At the heart of the final strategy are two major sorting centres — one for inland distribution and one for international — the kernel of two 'hub and spoke' networks.

Running to over £100 million for the buildings, sorting equipment, computer systems and infrastructure changes to support them, the project is large and covers many different disciplines.

During 1996 the hunt started for a suitable site. It had to be large enough to accommodate the two centres, provide easy access to the motorway and rail networks and in close proximity to an airport. Coventry Airport was eventually selected.

Work started on site in July 1998.

THE DESIGN BRIEFS

The National Hub
The design brief for the National Hub was:

- provide a centre capable of handling 40,000 parcels per hour varying from a match box to a 500 kg pallet in any shape or size!
- no damage to items
- maximum use of automation to minimise staffing costs
- 100 outfeed docks
- the number of infeed docks to be defined by the supplier
- equipment and systems must meet the arrival and despatch profile
- be resilient and avoid single points of failure
- extensive 'recovery' facilities to cope with unreadable bar codes or missing pre-advice (parcel sorting data)
- provide inputs, operation and outputs for handling unitary loading devices (the containers loaded into planes)
- provision of integrated maintenance and asset management systems
- provide a vehicle control system capable of managing 2,000 vehicles a day in to, out of, and between the two centres

The International Hub

The International Hub is a smaller but more complicated operation.

The design brief for the import operation was:
- a processing system capable of handling 2,500 items per hour, either all bags, all loose items or any ratio of the two
- 4 infeeds, 3 outfeeds
- Customs facilities

And for the export operation:
- 8,000 items per hour real-time — both bags and loose items
- 12 infeeds and 18 outfeeds
- security screening in the form of x-ray and related air security equipment
- conforming to Universal Postal Union (UPU) requirements for despatch, labelling and pre-advice data

In addition there had to be extensive parcel tracking and bar code scanning, unitary loading device operations and 'recovery' facilities for unreadable bar codes in both import and export operations.

TENDERING AND ADJUDICATION

A major part of the project was identifying user requirements. Under PRINCE (Projects IN a Controlled Environment, the project methodology used by the Post Office) users and customers must be involved at all stages.

Traditionally Parcelforce systems were specified in considerable detail. While there are benefits to this approach it has drawbacks. In focusing on the solution it is easy to overlook business needs. Is it cost effective? Does it fit with business strategy? Is it flexible?

A methodology now favoured by the Post Office is the functional approach — the aim is to specify **what** is needed, it is for suppliers to decide **how** it is achieved.

It takes considerable confidence to adopt this approach. Control and responsibility for the physical solution no longer rests wholly within the organisation. The main benefit is that the specification process focuses on the really important issue — does the proposed system meet business needs? — while harnessing the innovation, expertise, resources and experience of the market place.

Those of you who have been involved with large public funded projects will be well aware of the need to conform to GATT/WTO (General Agreement on Trade and Tariffs/World Trade Organisation) and European Union procurement regulations. The regulations seek to provide a fair and level playing field for potential suppliers. This process was followed rigorously, requiring extensive contact with suppliers, bench-marking and precise planning and co-ordination.

Because of the size and complexity of the project the National and International Hubs were tendered and adjudicated separately.

The Parcelforce board determined that there would be a single contract to cover all aspects of the project, with the builder taking the prime contractor role and the automation sub-contracts being integrated under the main contract. JCT 81 was selected as the contract conditions.

Parcelforce elected for this route because of experience in attempting to manage and integrate large building/automation projects in-house. The new credo is stick to one's area of core competency — distributing parcels — and buy in the additional expertise.

THE SUPPLIERS

The functional approach is not favoured by everyone. Those schooled in more traditional approaches are horrified by the lack of detail and the opportunity, at later stages, for misunderstanding and confusion.

However, the tender returns for the project were a striking testament to the companies who responded. They had captured the essential elements of performance, design constraints, cost and requirements of the operation. It was fascinating to see how different companies approached the same design brief.

Some 'threw' vast quantities of equipment at it, others were innovative with new technology.

Choosing between design solutions is extremely difficult as anyone who has attempted a major tender adjudication will know. It is a balance between cost, quality and performance. The cheapest solution is often not the best value for money.

Sir Robert McAlpine were chosen as the prime contractor. They are one of the largest building companies in the UK. They have been involved in several major projects and in partnership with Laings are constructing the millennium dome.

Alstom Automation Ltd, formerly GEC Alsthom, were selected to provide the automation and IT systems for the International Hub. Alstom is a massive company with engineering interests spanning the globe, from power stations, to baggage handling and railways.

Crisplant are a Danish company who have specialised in tilt tray sorting systems for many years. To some they may be better known for gas bottling systems. They have been major suppliers for airport luggage handling systems, most notably the baggage systems in the new Hong Kong airport. They were selected to provide the automation and data systems within the National Hub. Additionally they were contracted to produce a vehicle control system for the whole site.

INFRASTRUCTURE CHANGES

The hubs themselves are but one part of the total system that has to be implemented.

To ensure that the new hubs work at maximum efficiency, bar codes are applied to all items. This is essential for both national and international operations. They allow items to be sorted automatically and provide the means by which they can be tracked.

Parcelforce bar codes are simply unique numbers and have no meaning in themselves. Without additional electronic information — 'pre-advice' (provided in advance of processing, hence the term pre-advice) — the bar code would be of little real use.

Pre-advice is provided principally by Parcelforce local depots — this is where items are accepted into the two networks. Operators input relevant information to a central system called PMS — the Parcel Management System. This information is made available to other functions and processes within Parcelforce, for example for billing and customer enquiries.

The National and International Hubs clearly represent two significant new clients for PMS.

Bar codes and pre-advice are essential for the efficient handling of items at the hubs. Just as important though is getting the items there in the first place. To enable the project to work, the whole transport network had to be revised to ensure items were provided to the hubs at the right time, and could be distributed in line with service requirements.

The further away the items are collected from, or delivered to, the more time critical their transportation becomes. Express items from Scotland and Northern Ireland going internationally must be flown into Coventry Airport to ensure they meet service targets.

THE NATIONAL HUB

Overview

The National Hub will be a slick, fast operation, designed to process 40,000 express parcels per hour (next day and within 48 hours). There is no storage. All items processed in a shift are despatched for delivery. The cycle time for the hub is around fifteen minutes.

Building

The building is 360m long, 50m wide along its core and 80m square at the ends. Its construction is basically steel structure with cladding.

The electrical power for the building is provided by a combined heat and power plant, making the site the biggest gas user in the Coventry area. The excess power is sold back to the National Grid. In the result of a failure with the combined heat and power plant the hub's electrical power will be provided by the National Grid, making the supply of electrical power to the hubs extremely reliable.

The National Hub — the central core

Mechanical Handling

The input profile for the National Hub is approximately 85% suitable for full automatic processing. A further 12% require some manual intervention but suitable for low technology automation, with the remaining 3% having to be handled via a completely manual process. For these reasons there are three distinct processing systems.

Regularly shaped and sized items can be fully processed automatically with the destination decision being achieved by dynamic reading of bar codes. Four Crisplant tilt tray sorters, each approximately 1 kilometre long, provide a theoretical sorting capacity of 54,000 items per hour, although only 34,000 items per hour initially is required to meet functionality.

There are 48 infeeds, 24 at each end of the building, split into banks of six. As a vehicle is unloaded the operator decides which of the three systems is appropriate. Items suitable for the fully automatic system are placed on an extending boom conveyor and are then conveyed to a tilt tray sorting machine induction. Each sorter feeds all outfeeds.

Virtual reality can help bring a project to life

Items which cannot be handled on the tilt tray sorters (eg tubes, spherical, light, heavy, long, unsymmetrical items) are passed to a separate semi-automatic system positioned beneath the tilt tray sorters. Items for this system are placed on the infeed boom conveyor but are diverted via a vertical switch to a mezzanine area.

Here the bar codes are hand scanned and a highly visible numerical identifier is placed on the product. The item is then conveyed to a carousel where the operator looks for the identifier appropriate to his area. Diversion is achieved manually — staff push the item onto a roller glacis (a chute with a holding area at the bottom). From there the item is placed in one of 25 tow line carts. When the cart is full it is hooked up to the Egemin in-floor tow line system and transferred to the outfeed area.

The tow line system was not the first choice for transferring items to outfeeds. There was a concern the system could prove inflexible. As it turned out, the level of flexibility required was quite limited and the tow line was by far the most efficient, cost effective and safest means of internal transportation. Moreover they have been used in many environments and the technology is robust and durable.

One hundred outfeeds are arranged along the central core of the building. The tow line carts are automatically diverted into one of 14 sidings. Each cart contains items for 4 outfeeds and they are manually sorted to their appropriate selection.

Boom conveyors are provided by Caljan and the conveyors by Van Riet.

To further raise sorting capacity, and to provide greater flexibility, cross-over systems are provided, giving the capacity of 8 sorters.

Each sorter is equipped with omni-directional overhead bar code scanners. Once the bar code has been read the system identifies the shortest route to the appropriate outfeed. If necessary the item is transferred from one sorter to the another via the crossovers. By selecting the shortest route the sorters are empty at each end of the building, thus achieving true 'double ending' and doubling capacity.

'Recovery' areas are located at either end of the building to handle bar code read fails and other problems. Items are manually processed to generate pre-advice, rectify read fails or undertake other short duration recovery processes. In the event that the bar code has become detached or requires more involved recovery, it is directed to a secondary recovery area.

Items which cannot be handled by either of the other two systems are handled totally manually. If suitable, these items are placed on carts for transportation on the tow line system. Pallets can be transported on the tow line by the use of palletisers fitted with a tow line safety tow pin.

Maintenance management systems

In order to improve the availability of the sorting systems supplied and prevent unpredicted stoppages a maintenance management system has been provided. The main purpose of the system is to:

- plan preventive maintenance
- supervise of the sorting system
- deal with system-identified and user-identified corrective maintenance
- allow stock control
- generate statistical information
- document the system supplied

This sophisticated and comprehensive system has been extended to the International Hub and uses remote pagers for key alarms and bar code technology for the rapid collection of preventative and corrective data.

National Hub data systems

The operation requires the extensive use of data systems, both for operational purposes and to allow information collected and generated in the hub to be made available to other Parcelforce systems.

Most of the technology is standard PCs. A number of high powered servers provide the processing muscle allied to a massive communications infrastructure in the form of internal LANs.

The main interface to the outside world is what is known as the Front End Processor (FEP). All information relating directly to the operation pass through the FEP. This isolates the hub from the rest of Parcelforce both logically and contractually.

THE INTERNATIONAL HUB

Overview of the operation

The international operation is considerably more complex and 'bitty' than the inland one.

In addition to sorting items, the outputs of the operation must conform to UPU specifications, e.g. extensive documentation and pre-advice to foreign postal administrations and partners.

Customs and air security clearance facilities are statutory requirements.

The main drive behind the development of the international operation has been to raise quality through the use of extensive tracking of items at all stages of processing.

The heart of the import operation is what is known as IIG — International Import Gateway — a process of 'converting' foreign products to Parcelforce inland product. A new label and bar code are applied and the address and product information captured electronically for later use.

Import items are not sorted — they are passed to the National Hub which is responsible for inland distribution.

With over 400 potential destinations and many different products, the export operation is complicated and time critical. Some items must be processed in as little as 45 minutes, others may not be despatched for some time, typically by sea.

Items are scanned every time they are handled. A detailed log is built up of how they have been processed. This is a heavy overhead but ensures quality is implemented at every stage of processing.

The international operation has to facilitate the efficient and secure provision of items to and from HM Customs. Typically these are items from certain higher risk countries, items for which duty is payable (or reclaimable), or items requiring export licence checks.

Items flown internationally must be security screened, principally for explosive devices. A number of devices are provided, from x-ray machines of different sizes, to a particle scanner.

Building

The international operation is housed in a building 100m by 120m constructed of steel covered in cladding. Combined heat and power is included in the International Hub scope of supply.

The International Hub - Nearly ready to accept the all-important mechanical handling systems

Export mechanical handling at the International Hub

The principal sub contractor for the International Hub is Alstom Automation Ltd.

There are two distinct operations in the building.

The export operation comprised of a single Crisplant tilt tray sorting system, with split trays and an approximate length of 300m. There are 29 telescopic boom conveyors, 12 infeeds, 13 unscreened outfeeds and 4 screened outfeeds, manufactured by Caljan Ltd.

The operator places machinable items and bags onto the infeed boom conveyor which transports the product to the induction lane. The bar code is scanned on the induction lane prior to passing on to the sorter. In anticipation of a lower read rate for bagged items, operators man induction lanes hand scanning items which have not been read. Items are discharged via Safeglide™ chutes to outfeeds for distribution by road or air.

Items transported by road are fed by boom conveyors directly into the back of vehicles. Items to be flown on any part of their journey are x-rayed and then discharged onto carousels, from where they are manually sorted into unitary loading devices to be flown directly from Coventry or transferred to another airport.

Items unsuitable for the tilt tray sorting machine are handled manually.

Import mechanical handling at the International Hub

Import items are off-loaded from vehicles using 4 boom conveyors. Bags are transferred to one of three bag opening carousels via an overhead conveyor system supplied by Schierholtz. Loose items are transferred to the processing area on conveyors.

Import acceptance processes are undertaken at PC workstations. Customs cleared items are transferred via three booms conveyors to shuttle vehicles destined for the National Hub.

Items requiring secondary Customs inspection or specialist Parcelforce handling are taken by conveyor to a separate work area for processing.

International Hub data systems

The data systems are similar to the National Hub. All information relating to the international operation flows into and out of the hub via the FEP.

VEHICLE CONTROL

Managing up to 2,000 vehicle arrivals a day across a huge site with 3 main gate-houses is a complex operation. In addition to operational vehicles, there are customer vehicles and goods vehicles (e.g. food delivery).

All operational vehicles must be efficently managed on to and off infeeds, outfeeds and parking bays, and all relevant paperwork must accompany the driver as he leaves.

Clearly there is no point scheduling a vehicle on to a non-functioning infeed or outfeed, so the vehicle and sorting operations must be integrated.

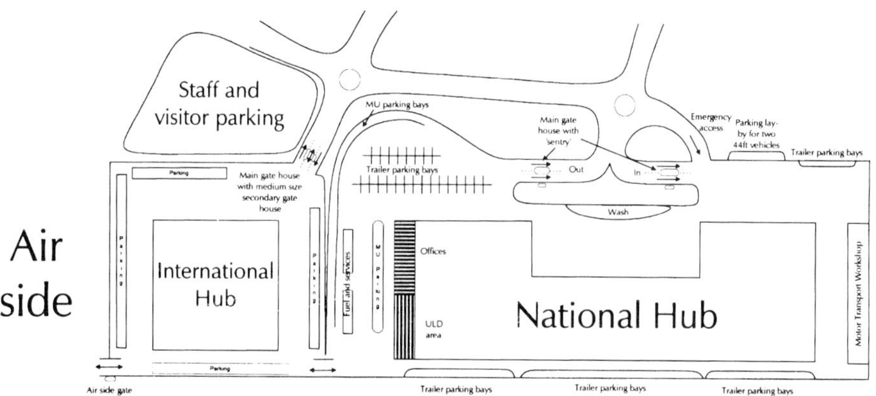

Graphic of the site for vehicle control

Crisplant, the automation and data systems provider for the National Hub, were commissioned to provide vehicle control for the whole campus.

One of our principal concerns was the efficient (automatic) registration of vehicles on and off site. Gatehouses are traditionally manned by staff looking at closed circuit screens and conversing to drivers via intercoms. It can work pretty well. But don't underestimate how awkward the task is on a wet, dark night with the wind howling and a back log of vehicles.

We wanted to avoid these problems by electronically tagging vehicles and having automatic sensing at the barrier. Pre-scheduled, pre-advised, vehicles are automatically registered and given access without gatehouse intervention.

On the exit side, automatic logging allows additional quality checking to be implemented. If the driver has not picked up his documentation, for example, the system can prevent him leaving site.

Underpinning the whole vehicle operation is the vehicle schedule, a pre-planned template of vehicle arrivals, movements on site and departures. It can be updated as the shift runs and there are extensive 'tools' to manage exceptions and variations.

CONCLUSION

Gemini has been a massive project by Parcelforce Worldwide standards. It is critical to its future success, and its ability to compete in a very competitive market.

We believe the combined operations will make Parcelforce a very competitive distribution far into the 21st century.

FURTHER INFORMATION

For further information regarding this document or the project, please contact Claire Lawrence, Communications Manager, Parcelforce Worldwide, Solaris Court, Davy Avenue, Milton Keynes, MK5 8PP, phone 01908 687388.

FACTS ABOUT PARCELFORCE

Parcelforce Worldwide is the largest parcel carrier in the UK. Its traditional market profile is parcels posted by members of the public, however, now the majority of items are business to business. Express parcels requiring timed next day or 48 hour delivery with signature and tracking.

Parcelforce employs over 12,000 staff and runs some 8,000 vehicles. Unlike its competitors, it is required by law to deliver to every UK address — over 23 million, from the Shetland Islands to the Channel Islands — at a fixed price.

The international operation covers 239 countries, 99.6% of the world's population.

Parcelforce carries 135 million parcel a year.

C568/026/99

The 'Bullfrog' automated small parcel sorting system

N TAN
United Parcel Service, Connecticut, USA

SYNOPSIS

This paper presents the high-level controls for the 'Bullfrog' automated small parcel sorting system that has been developed by United Parcel Service (UPS). It describes the impact of the Information revolution and the internal functional requirements that provide the business context for the development. Based on the vision to create a scalable standard high-level control system for all of UPS' automation needs starting with the 'Bullfrog', a component-based, multi-application client server architecture has been selected. With 19 systems already deployed successfully, the paper shares some of the key lessons learned as well as insights about the future direction.

1. INTRODUCTION

As the Information Technology revolution opens new markets and drives global business need, the process of delivering parcels and mail is bound to become more sophisticated and automated. To meet this challenge, the development and implementation of automation must increasingly add value to the business. The added value is not only in the form of improved production but will increasingly be derived from being an integral part of the distribution business process such as supply chain logistics. The high-level control systems that integrate the plant floor automation to the enterprise IT systems will play a crucial role.

United Parcel Service (UPS), the world's largest package distribution company, transports more than 3 billion parcels and documents annually (1). More than 1,700 facilities are in use to provide service to over 200 countries and territories, including every address in the United States. Over the past several years, UPS has been stepping up its program to automate its facilities.

The purpose of this paper is to present the technology development and operation of the high-level control system for one of UPS' automated sorting systems nicknamed the 'Bullfrog'. It also includes a review of some of the key lessons learned and a look at the next steps for the future.

2.   BUSINESS OVERVIEW

The technology development of the high-level control system for the Bullfrog automated sorting system is best understood in the context of its high-level functional requirements.

2.1 Functional requirements for machine

- Handle small sized packages - typically from 'flats' to about 20" (508 mm) wide x 16" (406 mm) long x 12" (305 mm) high and 6 lbs. (2.7 kg) weight
- Sort up to 400 destinations and 6,500 pieces per hour
- Multiple machines can be grouped together to form different sort configurations
- Low capital cost (modular construction, 'off-the-shelf' parts wherever possible)
- Easy to maintain (minimal moving parts and adjustments, simple parts change).

2.2 Functional requirements for high-level control system

- Process a package with a Maxicode (2) shipping label
    ◊ A Maxicode shipping label (figure 1) is applied either by the shipper or by UPS by means of key or voice encoding stations
- Sort the package according to a distribution sort plan
- Record the tracking number of the package and associate it with its transport container
- 'Remember' the tracking numbers of packages that are deliberately left in their containers between two consecutive sort operations
- Upload the package tracking records to the mainframe package tracking system
- Provide systems monitoring and performance statistics
- Provide high system reliability.

3.   MACHINE DESCRIPTION

The Bullfrog machine primarily consists of a series of sorting trays or cells on a linear conveyor (figure 2). Packages are manually inducted or placed onto the trays. It sorts the package by tilting the tray left or right with respect to the direction of travel. The package momentarily lands on a short inclined chute before it enters its transport container (a re-usable bag). A valve is used to trigger the tray's tilting action by allowing low-pressure air to inflate the bellows under the tray (hence the 'Bullfrog' nickname). A PLC (Programmable Logic Controller) provides the machine level controls including interfacing to the scanner/reader and the high-level control system. Although originally designed by UPS, the machine is now commercially available (3).

4.   SYSTEM OPERATION

4.1 Installed base

The first Bullfrog system was installed in a UPS facility in March 1996. Up to the end of 1998, a total of 19 systems have been installed in 8 facilities. All the systems are located in the USA and altogether they process just over 10 percent of UPS' domestic US volume for small sized packages.

4.2 General operating layout

Figure 3 shows a typical operating layout. There are four operator positions, viz. de-bagger, data acquisition keyer or voicer, inductor and bagger. The process flow of the Bullfrog sorting system and the role of each of these positions is as follows:
1. The de-baggers are the first to handle the packages that arrive in the Bullfrog area. They remove the packages from their transport containers that are either bags or totes. Packages without a Maxicode label are sent to the data acquisition section. Packages already with a Maxicode label are fed directly to the inductors.
2. At data acquisition, a keyer or voicer enters the address label information into the workstation, scans the tracking number, obtains a Maxicode label from a printer and applies it on the package, and places it on a conveyor that takes it to the inductors. The description of the key or voice encoding station is out of scope for this paper.
3. The inductor receives a package from the feed conveyor and places it onto one of the empty Bullfrog trays. His/her main job is to ensure that the package is placed centrally on the tray with the Maxicode label facing up towards the reader.
4. The tray transports the package under the reader where the Maxicode label is decoded. Based on the selected sort plan, the high-level control system in turn determines the destination chute where the package should be sorted. When the tray carrying the package reaches its desired destination chute, it tilts and discharges the package down that chute.
5. The package completes its Bullfrog journey when it slides down the chute into an open transport bag. When the bag is full, a bagger performs the following tasks:
    - Scans the chute position bar code
    - Removes the full bag and replaces it with an empty one
    - Closes the full bag
    - Scans and attaches a bag tag to the bag
    - Places the bag on a conveyor that takes it away from the Bullfrog area.
6. The high-level control system consolidates the required tracking information about the package and its transport container, and uploads it to the mainframe.

5.   CONTROL SYSTEM DESCRIPTION

5.1 Technical overview

To implement the functional requirements, a client server architecture has been adopted for the high-level control system. The selected computer hardware is Pentium class PC servers and workstations operating under Windows NT. The workstations are currently industrial grade PCs located on the shop floor next to the machine PLC (Programmable Logic Controller). Ethernet is extensively used for networking, except for the current use of serial communications between the machine PLC and workstations. However, the planned switch over to Ethernet should enable the eventual conversion to less expensive desktop workstations located in a separate control room from the shop floor.

5.2 Application architecture

Figure 4 shows the relationship architecture of all the applications that constitute the Bullfrog's high-level control system. The control system acts as the middle layer between the lower level machine controller (in this case a PLC) and the higher-level facility/enterprise I.S. systems. The following is a brief description of the applications grouped together by function.

Sort plan generation – The Sort Plan Editor is a server based application that is used to create or edit the data tables that determine the machine sort destinations for a given package address and service class. Periodic changes and updates to the corporate data tables are automatically downloaded from the enterprise Sort Planning System. The operational data tables are stored locally in an Oracle database. Prior to the operation of the Bullfrog, a sort plan is selected and downloaded to the Sort and Container controllers.

Sort manager human interface – This application acts as the central coordination for all the other applications. It collects and displays information from the applications, enables the human operator to set up and change the sort configuration, and provides synchronization between the applications. It is designed with the flexibility to operate locally or remotely through a network connection.

Sort control – The sort control function is provided by the Maxicode reader and the Sort Controller application. Both are connected to the machine PLC. The Maxicode reader is a commercially available CCD-based device that can decode both Maxicodes and barcodes. The Bullfrog generally uses two such readers for primary and back-up modes. Developed as a workstation application, the Sort Controller receives package information from the reader, determines the sort destination, directs the machine PLC to divert the package, receives the divert confirmation from the PLC, forwards the relevant package tracking information to the Package Tracking Data Store application, and collects operational statistics.

Transport container control – This function is provided by an application that consists of three major components, viz. a workstation software program, a hand mounted wearable scanner and a radio frequency (RF) network that provides mobile communication between the scanners and the workstation. It is used by the bagging operators to provide a) the link between a pre-printed barcode tag and its container (in this case a re-usable bag), and b) exception handling via manual scanning. The combination of the wearable scanner and the RF network was chosen to provide non-restricted, hands free movement to the bagging operators.

<u>Package Tracking Control</u> – Two server-based applications combine to store and upload tracking information about the processed packages and transport containers. The Data Store receives and temporarily holds the information. Synchronization between the applications ensures the logical association between the packages and containers. At specified time intervals, the Data Export application re-formats the information into files and uploads them to the mainframe.

<u>System monitoring and statistics reporting</u> – The server based application that performs this function is designed to report alarm conditions via SNMP 'traps' that have been instrumented in the other applications. Similarly, it polls the other applications at regular intervals (every 2 minutes in the Bullfrog's case) via a standard application interface specification. The statistics can be presented at on-line client workstations, in standard reports that are automatically generated at the end of the sort, or from queries of historical data. Recent historical data are stored in an Oracle database and older ones are stored in back-up tapes.

6.   CONTROL SYSTEM DEVELOPMENT APPROACH

The underlying strategy for the development of the Bullfrog control system is to be able to expand it to become the standard control system for all of UPS' automation needs. In order to make this a reality, the development approach has to meet the following goals:

- A scalable system for various automation types, sizes and operating scenarios
- Reduced IT development and support costs as well as deployment time
- Standard user interfaces so that automation will have a similar 'look and feel' throughout the operations
- Standard system interfaces for machine controls and enterprise I.S. systems for ease of system integration and project management
- Implementation of business rules consistently across all automation projects.

The technical path chosen to achieve these goals was client-server. Furthermore, a modular approach was adopted to develop and link the applications through the extensive use of communications networking and re-usable common components that represent business and technical 'building blocks.' Considering the importance of attaining control system reliability (4), this approach has its own element of complexity and risk that must be managed. This matter is covered in the next section.

Figure 5 shows the component-based development approach. In simplistic terms, it is not dissimilar to what the auto industry does or will do with car production. A car company develops and assembles a series of basic components (engine, transmission, chassis, etc.) as well as some other more product specific components to produce a base vehicle that can be tailored to changing market conditions or eventually be customized to order. In this case, the software common components are determined on the basis of their business and/or technical function. For example, the Communication Manager component is used by all applications to 'talk' to each other. It also handles all the communications protocols that support the specified standard interfaces with the machine PLC and the enterprise I.S. systems. These components typically take the form of a DLL (Dynamic Link Library), OCX or ActiveX object. Each application is built using these components as well as its own application specific software. The applications are typically developed using Visual C++ and Visual Basic. If required, an Oracle database is included. They are also typically multi-threaded operating under Windows NT. Several such applications are assembled or integrated to form the initial lab. system. An installation script (typically written in InstallShield) ultimately 'wraps' the entire system software into a form that can be released for production deployment after it has passed the usual rounds of testing.

7.   KEY LESSONS LEARNED

After about two years of development effort that has produced four versions of the control system software and a very successful installed base of 19 Bullfrogs to date, some key lessons have been learned on both the technical and non-technical fronts.

Lesson 1: Be prepared for the large scope of system integration.
Positioned between the machine PLC and the enterprise I.S. systems, the span of the integration that the high-level control system has to fulfill can be considerable. The issues and 'cultures' at two ends of the integration spectrum are very different and have to be effectively covered. This translates into a) a key people issue in terms of the skills and relationships that must be factored into the project, and b) requirements that must be well defined and understood.

Lesson 2: Make automation a business process agent.
Initially the project focused on making sure that the control system 'did its job' to operate reliably. However, early in the deployment phase it became apparent that the combination of package sorting and tracking performed by the Bullfrog sorting system streamlined certain steps but highlighted gaps in others. Such gaps could be technical, operation or business related. By performing an 'end-to-end' analysis of the system's performance against its desired objectives, all the stakeholders were able to focus their efforts to improve the effectiveness of the system.

Lesson 3: Make automation much easier to use.
This lesson is the logical extension to the previous one. Furthermore, based on UPS' experience with a large part-time and relatively high turnover workforce, the imperative to make automation much easier to use is a compelling one. The 'look and feel' of the control system and its user interfaces should be intuitive, meaningful and predictable. The support infrastructure should increasingly be more proactive and integrated with the operations compared with the more traditional Help Desk and third level technical support.

Lesson 4: Manage the risks - a successful, reliable system does not happen by chance
Our experience has shown that the component-based client server approach for the high-level control system is very powerful and works. However, there are real risks that must be addressed on both the technical and project management fronts. The major ones are:

- Assess the technology choice realistically based on a '360 degree' view of business, technical and political considerations
- Make sure that there are people who have the required technical skills and knowledge
- Employ sound design and development principles for real-time, multi-threaded applications that take time and effort to become robust, particularly when it is not often possible to fully replicate the automation environment in the lab. The following have proven to be very helpful:
    - ◊ Extensive logging capabilities built into the applications
    - ◊ Test plans that stress the system in terms of expected and unexpected volume and behavior responses
    - ◊ Adequate time for the alpha and beta tests to truly shake out the bugs
    - ◊ An internal disaster recovery plan that ensures there is sufficient isolation between applications (in certain cases, the system will continue to sort packages with an application down) and an easy restart that automatically picks up from the point of the application crash.

8.   FUTURE DIRECTION

8.1 Internal plan

The next development phase for the Bullfrog control system is targeted at greater ease of use and alignment with corporate strategic processes. The significant development effort, however, is the expansion of the current Bullfrog control system to provide the high-level controls for the next UPS automated facility that is slated to open in June 1999. Instead of sorting small packages, it will be used to control machines to sort large sized packages in the automated facility. This is consistent with the vision to expand the Bullfrog control system to be the high-level control system for all of UPS' automation needs.

8.2 Long term perspective

The work of Michael Dertouzos (Director of MIT Laboratory for Computer Science) and his dream of the Information Marketplace (5) comes to mind. His ideas concerning the power, value and meaning of information, and the 'user friendliness' of Information Technology are very appropriate not only as we ponder the possibilities of the Information revolution but also how mail technology itself will be transformed.

If the Bullfrog control system could be viewed as an example of the impact of Information technology on mail automation, we can see it in the interplay between the physical movement of the package and the information that flows in and out of the automation control system. If it has not started already, some of that information will flow into the Information Marketplace as suppliers and consumers use it to give increased value to the physical package itself. The challenge for mail service providers is to become part of that Information Marketplace. For some, if not all, it includes harnessing the power of Information Technology for automation.

9.   ACKNOWLEDGEMENT

The author would like to gratefully acknowledge the contribution of his technical development team located at Danbury, Connecticut USA. Their ideas, skills, motivation and dedication have been paramount to the successful development of the system.

10.   REFERENCES

1.   United Parcel Service company worldwide web site – ups.com
2.   ANSI/AIM BCI-1997 International Symbology Specification - Maxicode
3.   'Pneumatic Tilt Tray Sorter Model 2475', Form 6050 2/98, Rapistan Systems, worldwide web site – rapistan.com
4.   'Industrial Computing', 12/98, pages 20-23 and 25-27, ISA Publication
5.   'What will be : how the new world of information will change our lives', Michael L. Dertouzos, published by HarperCollins, 1997, ISBN 0-06-251540-3.

Figure 1:   Example of shipping label with Maxicode

Figure 2: The Bullfrog sorting machine (viewed from induction end)

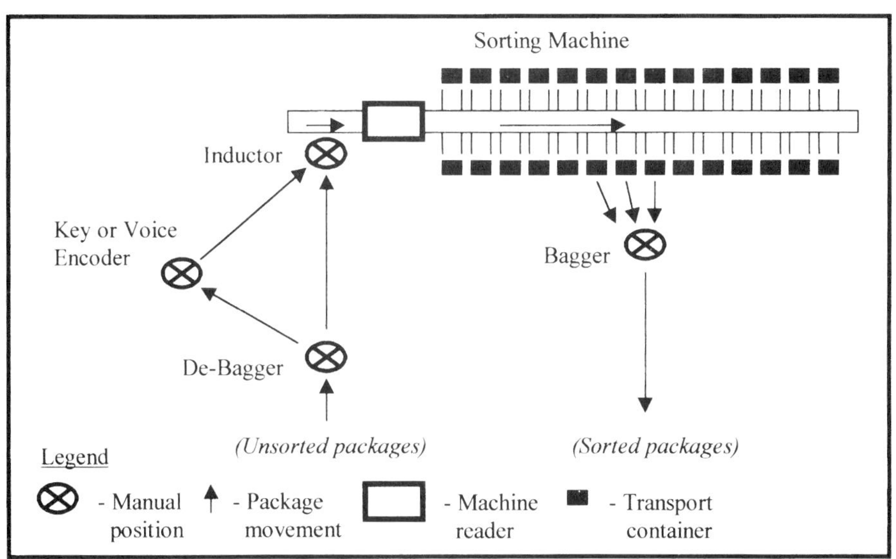

Figure 3: Operational layout of Bullfrog sorting system

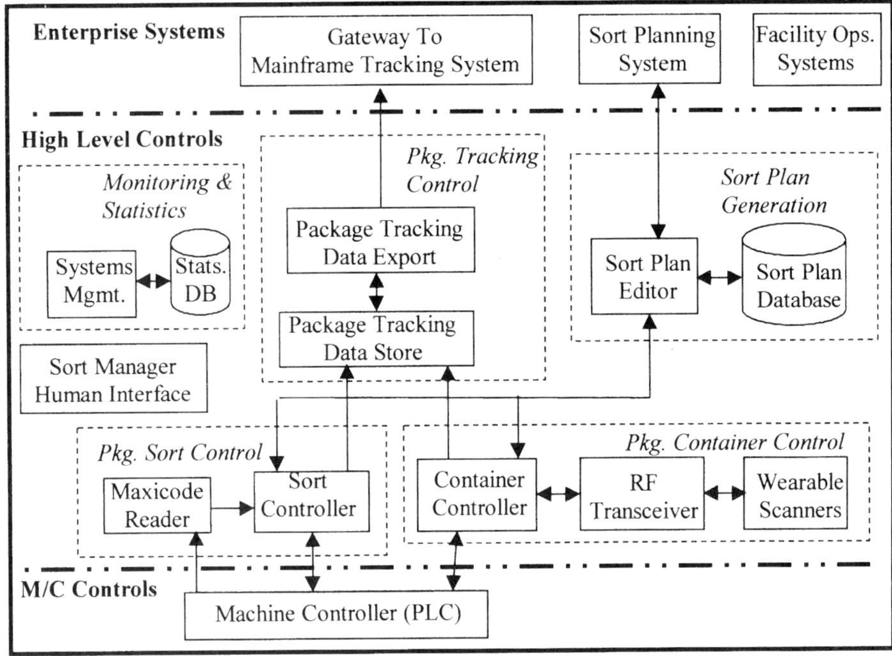

Figure 4: Overview of high-level control system

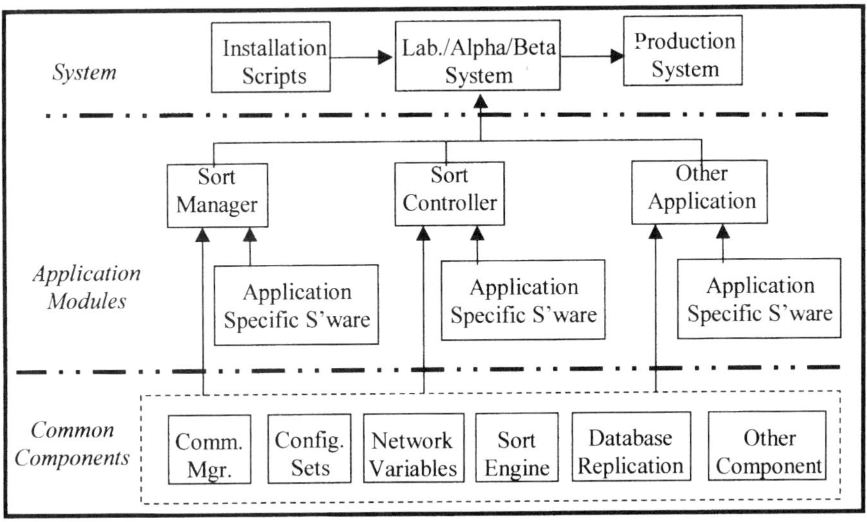

Figure 5: System development architecture

C568/043/99

Kinetic energy regulator – an alternative solution for the kinetic energy control

P FAUCHARD
Service de Recherche Technique de la Poste, France

SYNOPSIS

The parcel sorting techniques used involve the continuous processing of a combination of light and heavy parcels that have very different kinetic energy values.

In order to counteract this problem, SRTP has developed a device which produces a regulated kinetic energy at the end of the sorting process for all parcels, ranging from a few grammes to several kilos.

1. INTRODUCTION

In an effort to increase their market share and keep their customers, routing organisations (such as mail order firms) are having to process parcels which are ever more bulky and heavy. Handling equipment must keep pace with this development by being able to process such objects automatically.

The design of this new machinery has to take account of items of vastly differing weight, size and texture which must be processed at the same time and at ever-higher speeds.

Inevitably, this increase in the size and weight of objects means that the different kinetic energy values of objects being processed automatically will become more and more of a problem.

Solutions exist, but the machines are generally not cost-effective (the heaviest objects are often statistically exceptional), and inflexible (they operate at a constant speed).

The "DREC" kinetic energy regulator has been designed as an alternative, flexible solution to the problem.

2. DESIGN

2.1. Mathematical formula for kinetic energy
Kinetic energy (Ke) is calculated with the formula
$$Ke \text{ (joules)} = \tfrac{1}{2} M \text{ (kg) } V^2 \text{ (m}^2\text{/s}^2)$$
where $V^2 = 2 \gamma \text{ (m/s}^2) L \text{ (metres)} + Vo^2 \text{ (m}^2\text{/s}^2)$

2.2. Aim
The aim is to obtain a near-constant kinetic energy for objects of differing weight, size and texture at the end of their travel down an inclined plane. The device is to be self-regulating and easy to install, and should satisfy the following equation:

$$Ke = \tfrac{1}{2} Mi\ V^2 = K \text{ (target value)}$$
therefore, at the end of the process: $V(mi) = \sqrt{2K} / \sqrt{Mi}$

2.3. Given situation
Each object has an initial speed (V_0) and moves under the effect of gravity down an inclined plane. For a given friction coefficient, the acceleration and thus the speed of the objects are the same, irrespective of their weights ($\gamma = g\cos\alpha\ (tg\alpha\text{-}tg\varphi)$, and $V = V_0 + \gamma t$).
The unit mass of each object is a constant.

2.4. The principle
This consists of utilising the mass of the object in motion to activate a decelerating device. To achieve the aim we have set, the heavier the object is, the slower its speed must be at the end of the process.

2.5. Design

The design we have adopted consists of several overlying inclined planes which can pivot about a common fixed axis, situated at their base, to fit into one another. A return torque, generated by counterweights, keeps each separate plane in a predetermined high position limited by stops. Each plane has a different friction coefficient.

In its simplest form the machine must have two planes, a smooth upper plane and a rougher lower plane.

When an object of mass M starts travelling down the upper plane, it generates a torque acting in the opposite direction to the plane's return torque.

- If the object's torque is lower than the return torque, the object slides down the upper plane without pivoting it.

- If the object's torque is higher than the return torque, the plane is forced down and fits into the second plane, which has a high friction coefficient. As the object progresses down this second plane, the torque it initially generated will decrease progressively until it is lower than the return torque of the smoother upper plane, so that the upper plane returns to its high position and the object continues its movement down this plane. The heavier the object, the later the return action. The process is repeated on as many planes and as many times as necessary to achieve the desired speed at the end of the objects travel.

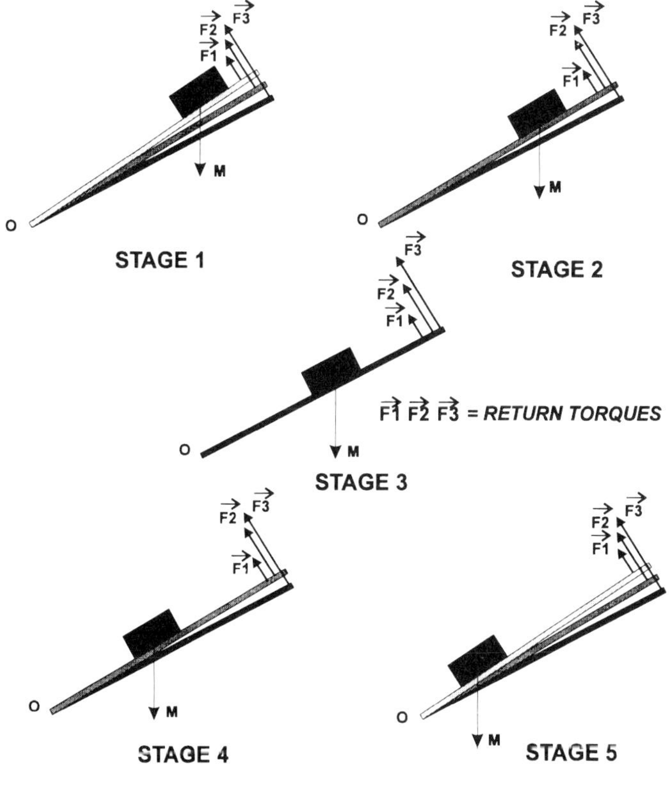

WORKING PRINCIPLE

3. PROTOTYPE

A prototype was developed in order to test the feasibility of this design in operating conditions. For practical reasons, the prototype had only two planes, a pivoting smooth plane and a fixed rough-surfaced plane. This simplified design was sufficient to confirm the device's feasibility.

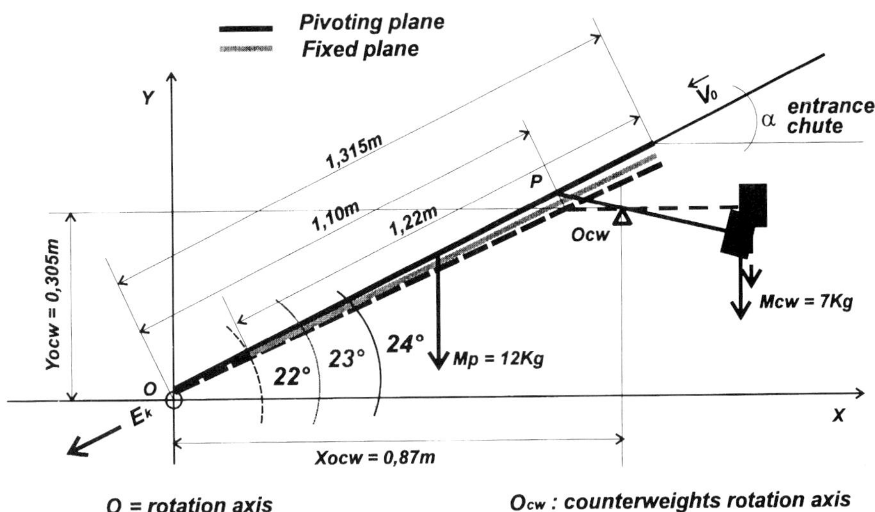

DREC PROTOTYPE PHYSICAL CHARACTERISTICS

3.1. Physical characteristics of the prototype

3.1.1. Pivoting plane
Length : 1m.315
Width : 0m.9
Material : Stainless steel
Weight : 12 Kg
Inclination: high : 24°; low : 22°. Friction : 16°. Structure : channelled along the whole length (7 channels). To avoid the risk of pinching, the channels are wider at the top than the bottom. Average width of channels: 50 mm.

3.1.2. Fixed plane
Length : 1m.22
Width : 0m.9
Friction material : Polychloroprene LSM in 28-mm wide strips. Friction 50° ±10°
Inclination : 23°

C568/043 © IMechE 1999

3.1.3. Return system
Mass of counterweights : 7 Kg
Total length of lever arm : 0m.455
Distance of fixing point (P) from axis of mobile plane : 1m.10
Co-ordinates of articulation point from axis of mobile plane : X-axis: 0m.87; Y-axis: 0m.305

3.2. Environment
The DREC was mounted on castors and set in place just downstream of one of the outlet chutes of the sorter. At this point the objects had a speed of 4.43 m/s, thus $V^2 = 19.6$ m²/s². The sorter outlet chute consisted of a wooden board inclined at 33°. The weight of the objects varied from a few grammes to 30 Kg. They were of differing size and texture.
At the DREC outlet, the objects were recovered manually on a roller table.
At the DREC entry a 30 Kg object had a kinetic energy of 15 Kg x 19.6 m²/s², or 294 j.

3.3. Measurement devices
The speed measuring device consisted of two optoelectronic direct-reflection sensors, placed 100 mm apart at the DREC outlet. The time taken by an object passing between these sensors was measured in milliseconds. The speed of the objects was calculated with the formula $V = 100$ / reading taken.
The dimensions and weight of the objects were determined using a tape measure and scales calibrated from 0 to 70 Kg, with a tolerance of +-50 g.
The installation is shown on the front picture.

3.4. Results
The following graph shows the results obtained from the test, which was carried out on some 400 objects. The theoretical kinetic energy curve is not a straight line due to the random movement of the masses on the X axis.
Initially the device which we tested was to be installed in place of the sorter outlet chute, so that V^2 entry = 6.25 m²/s², the aim being to obtain a kinetic energy close to 20 joules.

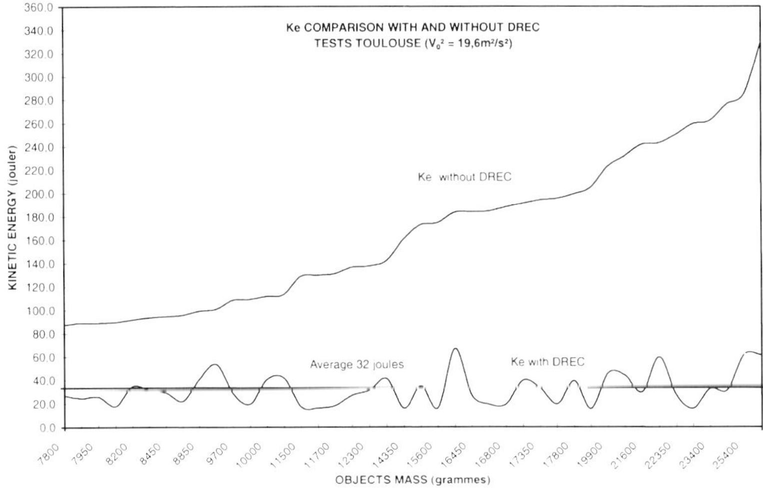

Analysis of the results shows that DREC acts as an efficient regulator of kinetic energy. The width of the band in which the kinetic energy varies depends for the most part on the relative friction coefficients of the rough-surfaced plane and the texture of the objects.

Thus, in order to determine the required characteristics for a given machine, certain external constraints must be taken into account : the range in weight of the objects to be processed, their texture, and so on. These would be input into a computerised mathematical model of the operation of the DREC.

4. MATHEMATICAL MODEL

A comparison between the theory and the practical results has enabled the mathematical model to be validated, a necessary step in the adaptation of the DREC to practical needs.

4.1. Model of the system

The methodology for the design of a DREC model is shown in the following diagram. The methodology uses object technology (C++).

MATHEMATICAL MODEL METHODOLOGY

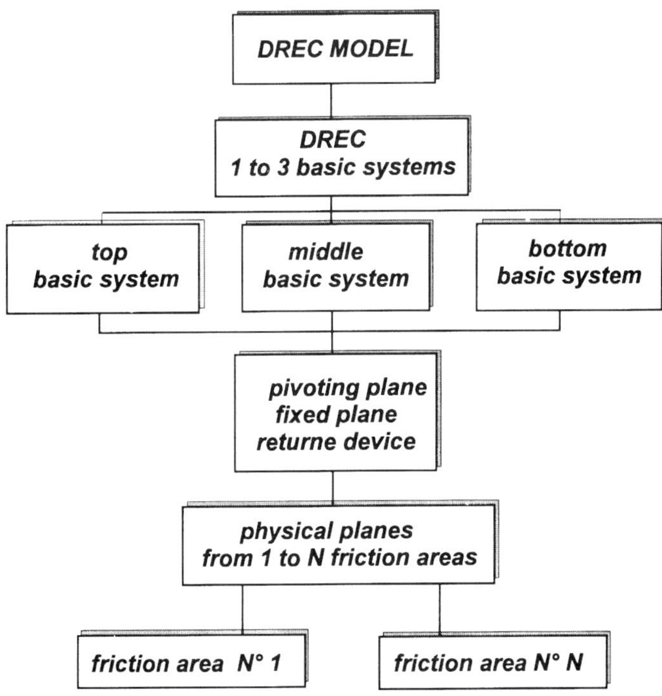

Physical planes are considered as surfaces which may be composed of several friction areas.
Pivoting and fixed planes are inclined physical planes
The mathematical model accepts a succession of three DREC systems.

     C568/043 © IMechE 1999

4.2. Theoretical formula

This uses Lagrange's theorem of virtual powers :

$$D(\delta T(\Sigma/R_0)/\delta(dq_i/dt))/dt - \delta T(\Sigma/R_0)/\delta q_i) = \{\Sigma->\otimes_j\}*\{d(\delta U(\otimes_j/R_0)/\delta(dq_i/dt))dt\} + \{\otimes k->\otimes j\}*\{d(\delta V(\otimes_j/\otimes_R)/\delta(dqi/dt))/dt \ , j<K$$

4.3. Numerical equation to be solved

$$\alpha_+{}^2 + P(X_+,X,X_-,\alpha,\alpha_-).\alpha_+ + Q(X_+,X,X_-,\varphi_{imp}) = 0$$
$$X_+ = R(X, X_-,\alpha_+,\alpha,\alpha_-)$$

MATHEMATICAL MODEL OF A BASIC SYSTEM

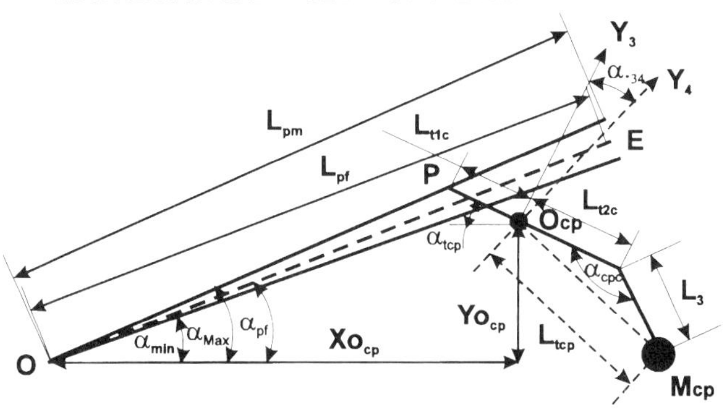

O : Pivoting plane rotation axis
Xocp, Yocp : Counter-weight rotation axis coordinates
P : Pivoting plane/counter-weight fixing point
Lpm : Pivoting plane length
Lpf : Fixed plane length
α min / α Max : Pivoting plane tilt angles mini and Maxi
α pf : fixed plane tilt angle
Mcp : counter-weight mass
E : system entrance point

PARCELS MATHEMATICAL MODEL

Gc : center of gravity
Xgc : center of gravity co-ordinate

5. PRACTICAL APPLICATION

Originally, it was envisaged for the DREC to be installed in place of the outlet chute of every parcel sorting machine in operation. However, this proved too costly to put into practice.
A practical application will soon be in use on outlet chutes of a luggages sorting machine with a trolley speed of between 1,6 and 1,8m/s.
The chutes are 4,5m long stainless steel planes inclined at 35°.
The weight of the objects varies from 3 to 60 Kg.
The following drawing gives the practical application proposed to be installed

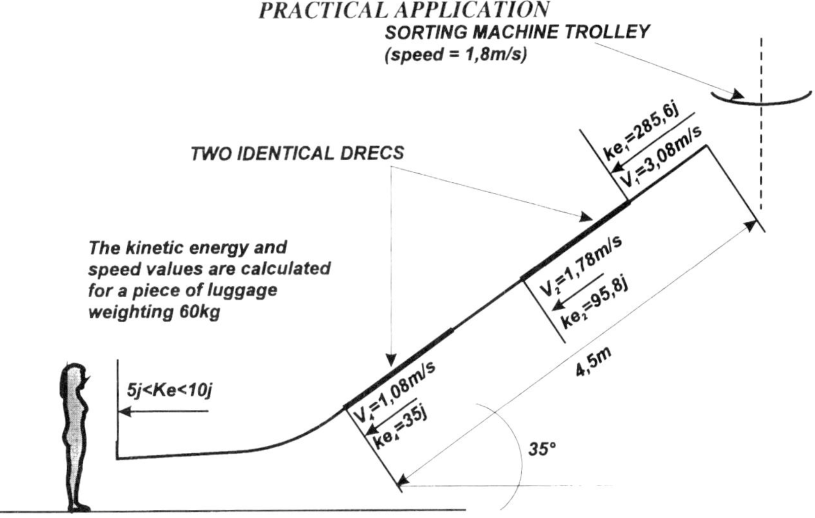

Apart from the speed at the entry point of the DREC, the mathematical model generates all the values.

5.1. Physical characteristics

5.1.1. Pivoting plane
Length : 1m60
Width : 1m
Material : stainless steel
Weight : 15 Kg
Inclination : high 35°, low 33°
Friction : 16°
Structure : Channelled along the whole length

5.1.2. Fixed plane
Length : 1m50
Width : 1m
Friction material : Polyurethane 50° average friction coefficient
Inclination : 34°

5.1.3. Return system

Top DREC : counter-weight mass 9,5kg
Bottom DREC : counter-weight mass 4,8kg
Total length of lever arm : 0m8
Distance of fixing point P from axis of mobile plane : 1m40
Co-ordinates of articulation point from axis of mobile plane : X-axis = 1m37, Y-axis = 0m75

5.2. Theoretical results

The following curves give the results of the mathematical model.
At the end of the process, the kinetic energy should be between 80 and 90joules to ensure that
the luggage travels smoothly to the end of the chute. Final adjustments will be made in situ.

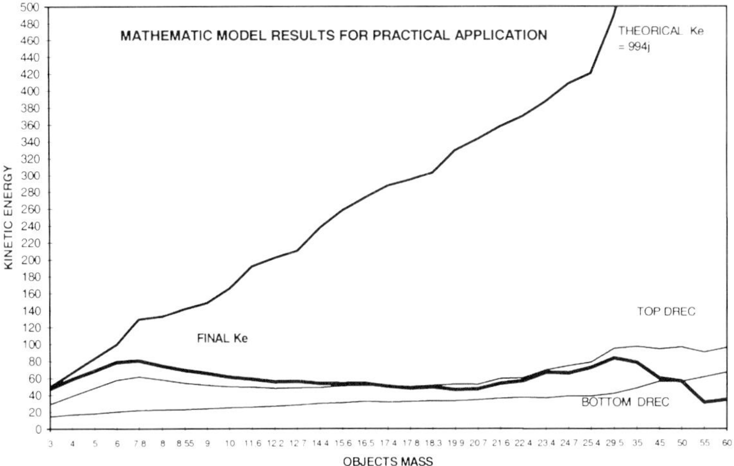

5. CONCLUSION

DREC is a viable alternative system for solving kinetic energy problems. It is entirely
mechanical, self-regulating and needs no external power source.
It can be designed either for the regulation or for the efficient reduction of kinetic energy,
wherever a change of physical level may result in damage to objects.

C568/044/99

TABOU – an innovative concept of mail sorting machine

J P DUGAST and **P DARCHIS**
La Poste Service de Recherche Technique, Nantes, France

SYNOPSIS
In the French Post Office (La Poste), mail sequencing has until now been carried out by hand by individual postmen in each delivery office in France. For the last two years, La Poste has been equipping its mail sorting centres with videocoding systems and remote-OCR systems. Now, this equipment is able to encode the handwritten delivery line of the address. So, in 1996, the Research Center of La Poste (SRTP) designed an entirely new mail sequencing machine for delivery offices, called TABOU. Two prototypes of TABOU have been trialled during the past year in the delivery office of Sartrouville near Paris.

1. MAIL PROCESSING IN FRANCE

1.1. Organisation of the Postal System in France

In France,the postal network is modelled on the country's administrative divisions.
France is divided into 95 «departments». Each of these is identified by a two-digit number (from 01 to 95). As a rule, each department has one sorting center. In some departments with a very high level of mail traffic there are several sorting centres (for example, Paris has 8 sorting centres). On the other hand, certain departments which receive very little mail have no sorting centre; their mail is processed by a neighbouring department.
In all, France, has 103 mail processing centres (CTCs). The size and number of automatic sorting machines in these centres depends on the amount of mail they sort each day.

1.2. The Postal address

The Postal address consists of three main pieces of information :

the Post Code :

Each delivery office in France has its own post code which is a 5-digit number :
- the first two digits are the number of the «department» in which the Office is situated,
- the last three digits identify the delivery office within the department.

the Routing Line :

This is the last line of the address.
It is made up of :
 Either the post code followed by the name of the delivery office to which it must be sent,
• Or a specific post code (different from that attributed to the Delivery Office) followed by the name of the delivery office and the word CEDEX (Courrier d'Entreprise à Distribution Exceptionnelle - special delivery for business users).

the Delivery Line

This includes :

• either the number, type and name of the street (delivery information),

• or special delivery indications (each acronym identifies a special type of delivery) :

- Boîte Postale (BP) - P.O. Box,
- Course Spéciale (CS) - Special Delivery,
- Service Public (SP) - Public Service,
- Tri Service Arrivée (TSA) - Sorting by department within the company receiving the mail

1.3. Coding of the address and mail sequencing

In most of the delivery offices of La Poste, mail sequencing is done by hand by the postmen from 6 o'clock to 9 o'clock am. More than 40 % of the work of French postal delivery workers consists in sorting the mail. Up to 1996, the automation of mail sequencing was impossible because the delivery line could only be encoded automatically by the sorting machine reader if it was printed or typed. In 1997, La Poste decided to equip its mail sorting centres with videocoding equipment: three tele-videocoding centres are being set up in France. At the end of this period of deployment, the complete address on all the mail processed by La Poste will be encoded by Optical Character Recognition of printed lines and videocoding of the delivery line for handwritten addresses. All the envelopes will be marked with a specific fluorescent bar code. At present, three types of marks can be affixed to the envelopes :
-
the routing code
- the delivery code
- the ID Tag ("chronomarque" in French), which assigns an identifying code to the address code of the letter, after videocoding.

The encoding of the delivery line for all letters will enable La Poste to plan automatic mail sequencing in both sorting centres and delivery offices. For the latter, the SRTP has designed a specific mail sequencing machine.

2. DESIGN OF TABOU

2.1. The requirement

The "TABOU" Research Project was implemented to answer the need for a sequencing machine for delivey offices with five objectives:
- low cost because this machine would be bought on a large scale by La Poste
- compact design because delivery offices have not a lot of floor space available
- adaptability in order to keep the possibility of manual mail sequencing
- ease of operation because the staff of delivery offices are not technicians.
- use of SRTP technology

2.2. The patent

Started in 1993, the TABOU project resulted, in 1995, in a 20-bin type model which enabled the SRTP to validate the technical feasiblity of the machine. In view of the encouraging performances of the model, in 1995, the SRTP decided to build a 39-bin prototype and to take out an international patent in order to protect its technology.
The patent referenced by the number 95.08022 consists of 28 claims which describe all the components of a very compact mail sequencing machine. Its compactness is due mainly to the fact that letters are conveyed widthwise and a suitable monitoring and control system is used :

- the TABOU feeder receives vertical stacks of envelopes to be sorted,
it is a hopper located at standard table height in the front section of the machine.
It is long enough to hold at least the number of mailpieces in one postal round.
This hopper comprises a bearing plane on which the longer edge of the envelope rests and a base plane, arranged as a shoulder to the upper section of the caisson and on which the shorter edge of the envelope rests. These two planes form a dihedron.

- the TABOU extraction head picks up the envelopes by suction, one by one until the feeder is empty.
It consists of a motorised, endless perforated belt, which emerges in a suction window located at the wall which constitutes the end of the hopper.

The TABOU conveyor, set between the extraction head and the pigeonholes, transports the envelopes with the same orientation as in the feeder (i.e. in portrait mode).

- the equipment for reading the indexing code, comprising a CCD camera designed to read the fluorescent or black and white bar-codes of the envelopes in parallel mode, given that the indexing bars are oriented in the direction of the motion of the envelopes.

The TABOU monitoring and control system, which comprises :
- computerised memorising of the sorting plan,
- computerised processing of information from the bar-code in order to allocate a pigeonhole to the envelope as soon as the

- computerised conversion of the allocation number of the pigeonhole ("sanction" in French) into the distance to be travelled in order to reach the right pigeonhole,
- an encoder pulse generator, working in conjunction with the envelope conveying system which translates the distances travelled by the conveyor into pulses,
- a photocell, located slightly downstream of the reading head, for triggering the pulse countdown.

2.3. The TABOU prototype

From April to June 1997, the SRTP tested the feasibility of the TABOU prototype and put it through field tests.

In July, the Mail Directorate of La Poste asked the SRTP to build two additional prototypes.

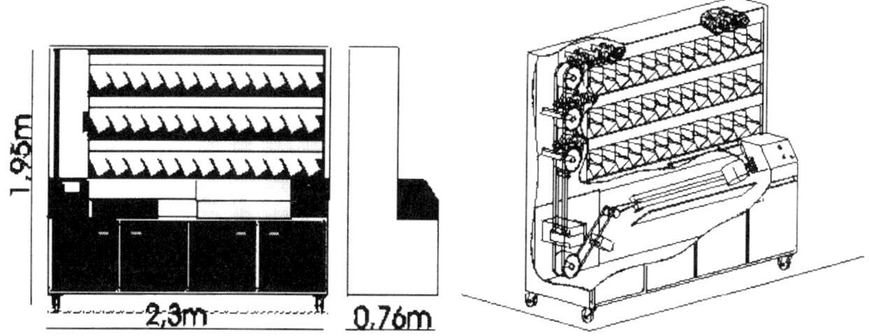

Fig.1 : Dimensions and views of the TABOU prototype

The features of the TABOU prototype are :

Dimensions :
Height : 1.95 m
Longth : 2.30 m
Width : 0.76 m
Weight : 650 Kg

Power: 220 v mono 50 Hz - 16 A
No special requirement for environment and power

Specification :
mobile machine on four wheels
Input feeder capacity: 1.6 m or 1500 letters
Singulator : 10 letters/s
Conveyor speed : 2 m/s
letters conveyed in portrait mode
Fluorescent bar code reader (route, delivery index, Identification tag).
Possibility of adding a black and white bar code reader
3 lines of 13 bins including 3 overflow bins
Bin capacity : 120 mm or 80 letters.
Operational control by PC (Pentium).
Operating system of the control PC : Windows-NT 4.
Compliant with European standards and safety regulation.

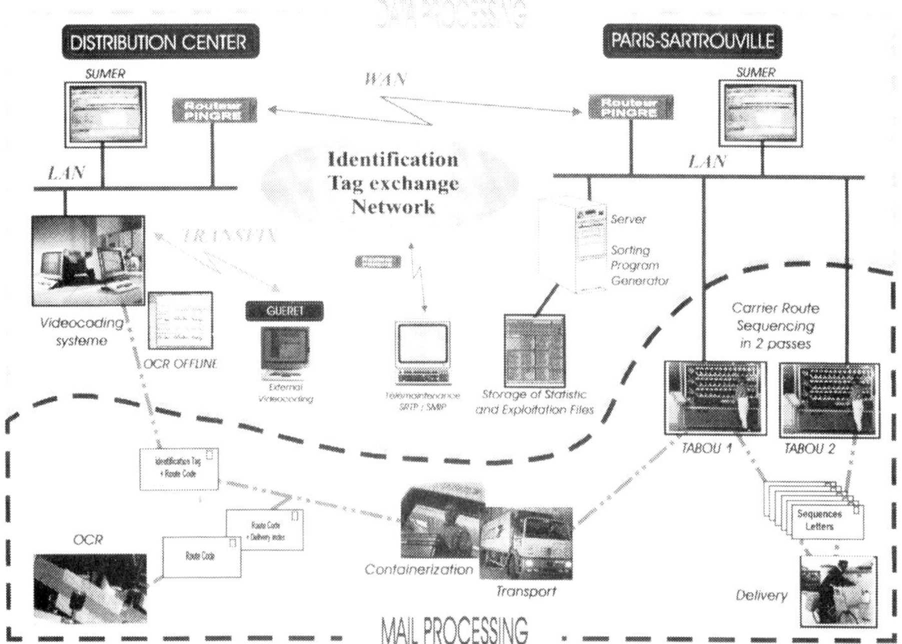

Fig. 2 :   Experiment in Sartrouville : mail and data processing

At the end of 1996, the mail directorate of La Poste asked the SRTP to build two TABOU prototypes in order to trial the machine in a delivery office.

In 1997, the delivery office of Sartrouville was selected as an experimental site. Situated in the suburbs of Paris, the town of Sartrouville comprises 33 postal rounds.

Since April 1998, the TABOU prototype has been sorting, with a single operator, 40000 letters per day, on average, for 31 postal rounds. The mail sorting in Sartrouville is carried out according to the following procedure:

Every night, the ID Tag file of the mail indexed by the sorting centre is transferred to the PC of the TABOU prototype. Every morning, except Sunday, two lorries come from the sorting centre of Saint Quentin and deliver mail to the delivery office at 6.00 and 6.30 am. All the addresses of the envelopes are marked with a fluorescent bar-code (the post code mark and possibly the ID Tag mark). Sorting plans are recorded on the PC of the TABOU prototype.
The sorting plans are updated every two months by loading into the PC server in the delivery office an extract of the national directory of addresses.

Between 6.00 and 6.30 am, all the standard mail (up to C5) is loaded into the feeder of the TABOU machine by the operator.

The mail is sorted in two sequential operations: :
Firstly, a TG3-type sort separates mail into 31 batches in one pass (one mail batch per postman)

Secondly, a TG4-type sort sequences every mail batch in two passes.
By using an optimal method of allocating the TABOU bins and overlapping, throughputs were up by 30% at least, last September.
Now, the average throughput for TG3 is 18000 letters per hour.
The average throughput for TG4 is 14000 letters per hour.

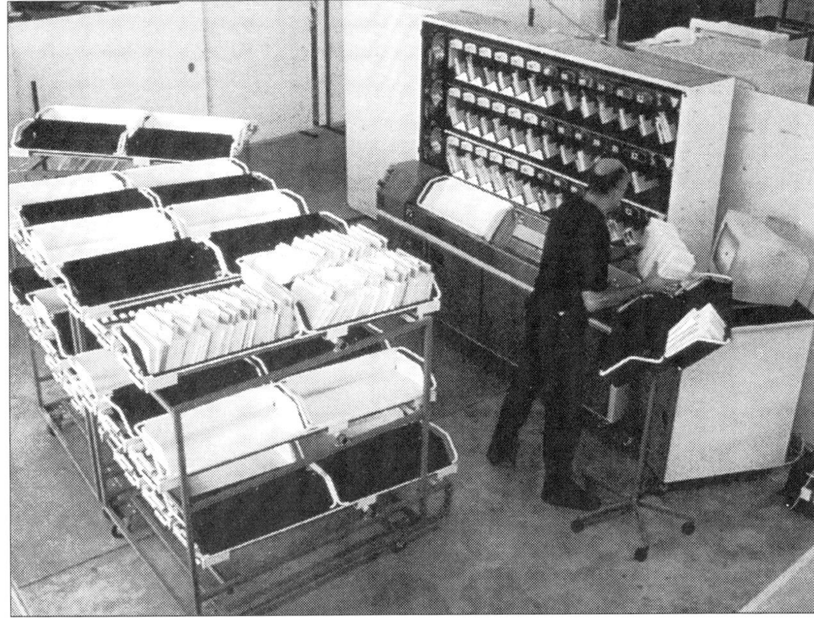

Fig. 3 : Sorting mail in the delivery office of Sartouville

4. FUTURE OF THE TABOU PROJECT

At present, ten TABOU prototypes are scheduled for delivery to La Poste, in 1999, in order to trial the TABOU machine in different sized delivery offices. The Mail Directorate of La Poste is carrying out an economic study to analyse the advantages of the TABOU machine in comparison with stacker type sequencing machines for delivery offices.
Moreover, the SRTP is studing the different solutions for reading a customer black and white bar-code in order to sort mail by department within the company receiving the mail.
We think several delivery offices will be equipped with the TABOU machine, in the future.
Who will be the manufacturer and the supplier of La Poste for the TABOU machine ?
Since November 1998, La Poste has been consulting all the main manufacturers of mail sorting machines in the world to select a supplier for TABOU machines and sign a patent licence agreement for the sale of TABOU machines to foreign Post Offices and mail operators.

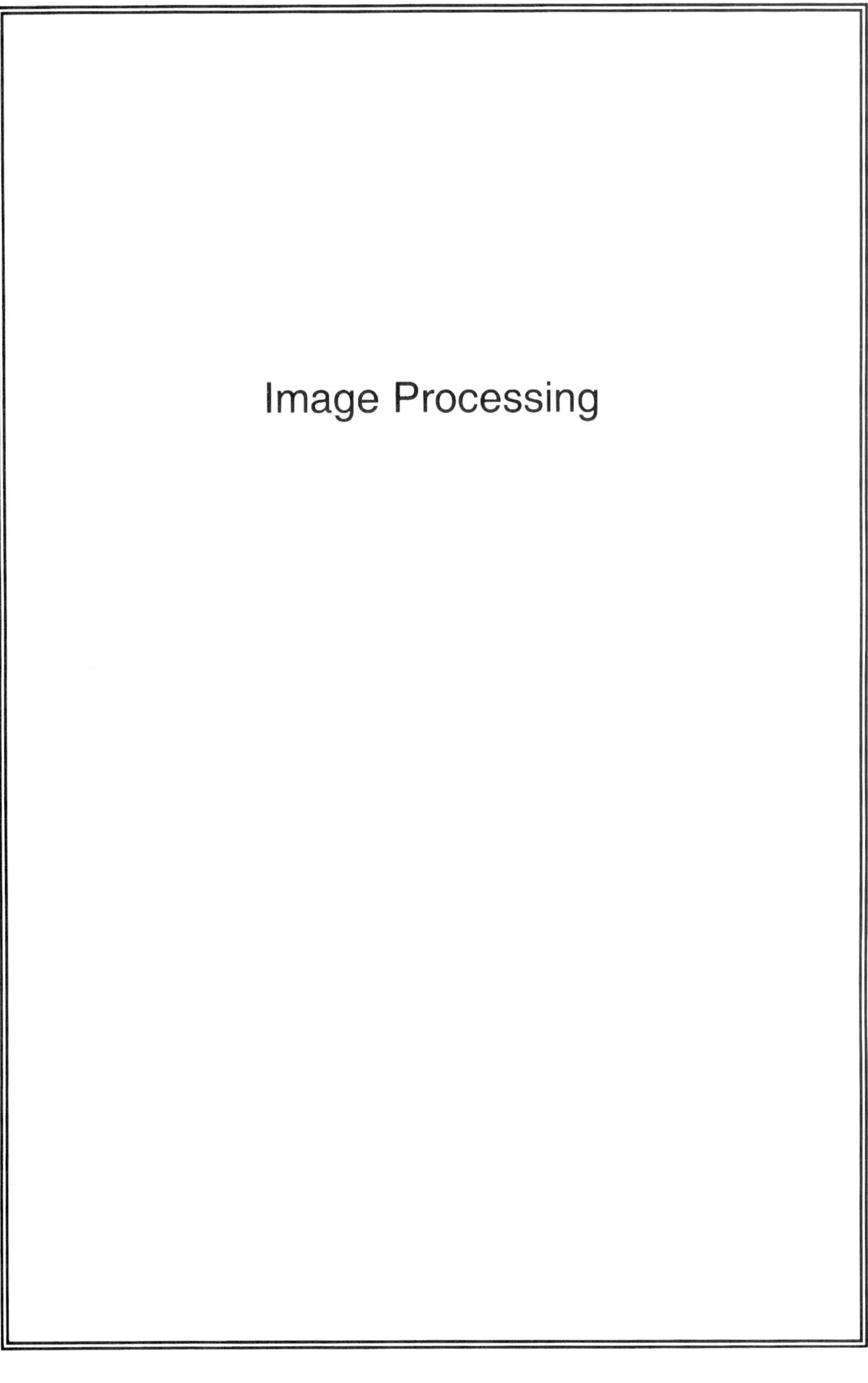

Image Processing

C568/062/99

Address interpretation

S N SRIHARI, W-J YANG, and **V GOVINDARAJU**
CEDAR SUNY at Buffalo, New York, USA

ABSTRACT

The task of determining the postal destination of a piece of mail involves not only the recognition of the shapes of the symbols written or printed on the piece but also relating the task to standard addresses contained in a database. This task can be said to be one of Address Interpretation (AI).

Handwriting is a skill that is personal to individuals. The task of reading handwriting is one involving specialized human skills. Knowledge of the subject domain is essential, as for example in the case of a pharmacist reading a doctor's prescription.

The Handwritten Address Interpretation, or HWAI, project that has been conducted at CEDAR since the early eighties explores the use of knowledge of the postal domain in the recognition of handwritten addresses. The task is considered to be one of interpretation rather than recognition since the goal is to assign the address to its correct destination irrespective of incomplete or contradictory information present in the writing. The work at CEDAR has led to a system that recognizes handwritten addresses that is currently in use by the United States Postal Service. We have since modified the system for use by other countries including Australia.

This paper describes the role of knowledge in handwritten AI. An information-theoretic study of address fields is suggested as an aid to AI.

1 INTRODUCTION

A tremendous number of mail pieces are handled every day by post offices in the world. In fiscal year 1997, the United States Postal Service (USPS) handled about 630 million pieces of mail a day, six days a week, for a total of 191 billion pieces. The national delivery network in the US now reaches nearly 130 million addresses [1]. To assist in processing this large volume of mail, automatic address interpretation is vital. Most of the previous work emphasizes overcoming word recognition problems [2, 3]. The street name lexicon from a postal address directory is useful for handwritten word recognition [4]. Statistical information from postal addresses is useful for address interpretation beyond the task of word recognition. However, statistical analysis of postal address fields and the study of applying results of the statistical analysis to address interpretation has not been done yet.

2 OVERVIEW OF AUTOMATIC ADDRESS INTERPRETATION

The address interpretation model is shown in Fig. 1. An address interpretation engine (AI) takes an image (x) as input, interprets the image content of destination address block and address fields, and refers to some information sources (S) for useful information in order to assign the most cost effective delivery point code ($I(x)$). The information source could consist of mail stream history and a directory of legal addresses.

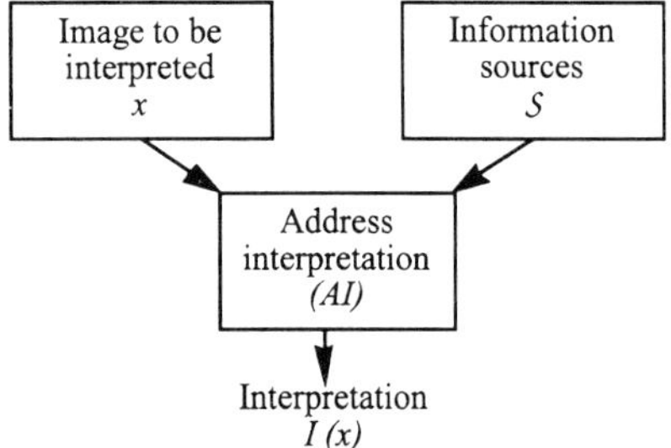

Figure 1: Address interpretation model

The address fields are different in different countries. For the address fields in a US postal address, the possible address fields are city name (f_1), state abbreviation (f_2), 5-digit ZIP Code (f_3), 4-digit ZIP+4 add-on (f_4), primary number (f_5), street name (f_6), secondary designator abbreviation (f_7), secondary number (f_8), and building/firm name (f_9). An example is shown in Fig. 2. It is possible that some address fields are not required for some addresses, and, in most cases, f_4 is not on a mail piece. For the example in Fig. 3, f_7, f_8, and f_9 are not required (or those fields are empty) for that destination address and f_4 is not provided.

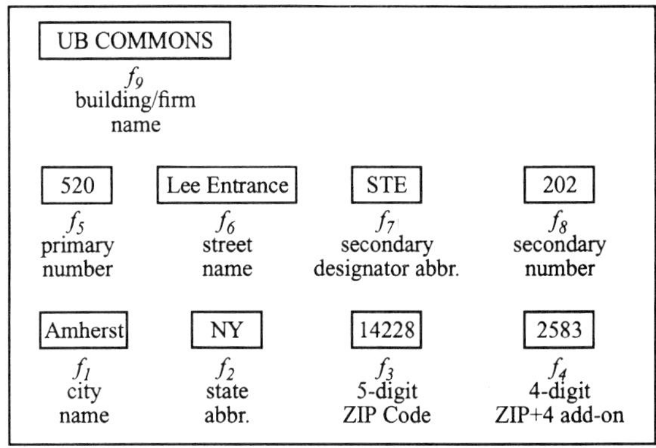

Figure 2: Address field example

Figure 3: Sample mail piece

The goal of automatic address interpretation is to construct the most cost effective delivery point code for a destination address. One possible simple strategy is to independently recognize each field and then find the best interpretation. This is what is done in reading machine-printed addresses. Such a strategy is not feasible for handwriting where recognition confidences for individual fields will be very low. An assignment strategy for determining a delivery point code is shown in Fig. 4 for the destination address shown in Fig. 3. After the destination address block is located, the ZIP Code (14221) is interpreted, then the primary number (276) is interpreted, and then the street name lexicon is generated for the (ZIP Code, primary number) pair by querying an address directory. The street name lexicon is expanded to include variant names (e.g., LN vs. LANE). The word recognizer is called to get the best match street name in the lexicon (i.e., MEADOWVIEW LN). Finally, the corresponding ZIP+4 add-on (i.e., 3557) is retrieved to construct the delivery point code (142213557).

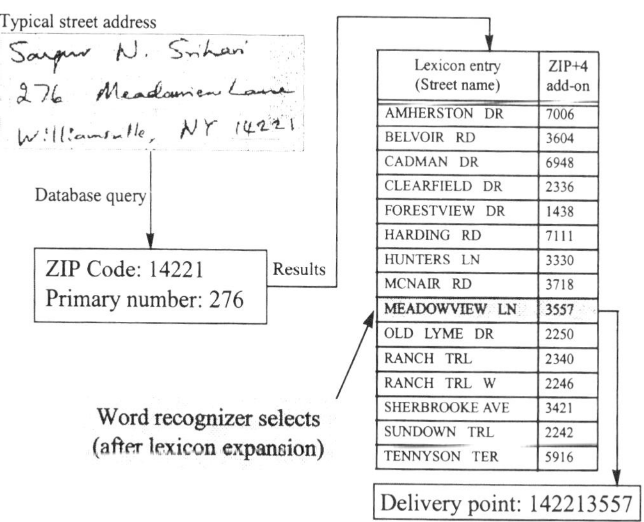

Lexicon entry (Street name)	ZIP+4 add-on
AMHERSTON DR	7006
BELVOIR RD	3604
CADMAN DR	6948
CLEARFIELD DR	2336
FORESTVIEW DR	1438
HARDING RD	7111
HUNTERS LN	3330
MCNAIR RD	3718
MEADOWVIEW LN	3557
OLD LYME DR	2250
RANCH TRL	2340
RANCH TRL W	2246
SHERBROOKE AVE	3421
SUNDOWN TRL	2242
TENNYSON TER	5916

Figure 4: An assignment strategy used in CEDAR's HWAI system

If, in a certain stage, a component cannot be interpreted correctly, the mail piece would be sent to less cost effective delivery point (e.g., 5-digit ZIP) or rejected to avoid expensive error cost. An information measure can help select the most effective component for recovering a damaged component and can suggest effective processing flow for determining a component value.

3 INFORMATION-THEORETIC ANALYSIS

Here we discuss information content in postal address fields and the use of that information for defining a control structure for automatic address interpretation. This is done by measuring information provided by combination of components and information interaction among components. Where a component can be a ZIP-Code digit or a five-digit ZIP Code. Postal address directories and real mail stream history are two examples of information sources. The quantity of redundancy between components can be computed from the information provided by these components. This information helps select a useful component for recovering a damaged or uncertain component. The uncertainty of a component based on knowing another component can be measured by conditional entropy. The analysis result helps understand information content of postal addresses and can be used to rank processing priority for determining a component value and to help recover a damaged component in address interpretation.

3.1 Measure of information

The concept of Shannon's entropy [5] for a discrete source of information is useful as a measure of information. The concept of entropy is closely tied to the concept of uncertainty embedded in a probability distribution. This concept is used to study the information content and interaction in postal addresses. Three kinds of information are defined and measured.

Definition: A *component* c is an address field f_i, a portion of f_i (e.g., a digit), or a combination of components.

- The information provided by a component x , whose i th value has probability p_i is given (in bits) by

$$H(x) = -\sum_i p_i log_2 p_i. \tag{1}$$

When all the values of x are equally likely (Entropy assuming uniform distribution):

$$\mathcal{H}(x) = log_2 |x| \tag{2}$$

- The uncertainty of a component y when a component x is known $= H_x(y)$ (i.e., conditional entropy). The quantity of $H_x(y)$ measures how uncertain we are of y on the average of known x. The conditional entropy $H_x(y)$ is defined as the average entropy of y for a given value of x, weighted according to the probability of getting that particular x.

$$H_x(y) = -\sum_{i,j} p_{ij} \left(log_2 \frac{p_{ij}}{\sum_j p_{ij}} \right) \tag{3}$$

Where x_i is a value of the component x ; y_j is a value of the component y ; p_{ij} is the joint probability of $p(x_i, y_j)$.

- Redundancy of components x to y is defined as follows.

$$R_x(y) = \mathcal{I}(x, y)/\mathcal{H}(y), \tag{4}$$

where

$$\mathcal{I}(x, y) = \mathcal{H}(x) + \mathcal{H}(y) - \mathcal{H}(x, y). \tag{5}$$

The redundancy quantity lies between 0 and 1. The larger quantity indicates that more information in the y component is shared by the x.

3.2 Example of information measure

As an illustration, suppose there are 3 fields (A, B, C) of interest in address interpretation and field B contains subfields B_1 and B_2. The possible values for each field are shown in Fig. 5. From the information measure defined in the section 3.1, the result is shown in Fig. 5. For example, there are 4 different values in the field A, therefore $\mathcal{H}(A) = 2$. It means that knowing the field A provides 2 bits of information. Since $p_{a10} = p_{a11} = p_{b11} = p_{c00} = p_{d91} = 1/5$, and $p_a = 2/5$ and $p_b = p_c = p_d = 1/5$, the uncertainty of field B by knowing the field A is computed by $H_A(B) = -(\frac{1}{5}log_2\frac{1}{2} + \frac{1}{5}log_2\frac{1}{2} + \frac{1}{5}log_21 + \frac{1}{5}log_21 + \frac{1}{5}log_21) = 0.4$ bit. Since $H_A(B)$ possesses the least amount in the column for $H_x(B)$, it indicates that the field A is the most valuable component to interpret in order to reduce the uncertainty of value in field B. Since $\mathcal{H}(A) = 2, \mathcal{H}(B_1) = log_23, \mathcal{H}(A, B_1) = 2$, the redundancy of field A to field B is $R_A(B_1) = 1$. The field B_1 is a completely redundant component to the field A. It means that if the field B_1 cannot be interpreted, the interpretation of the field A can recover the value in the field B_1.

	field A	field B		field C
		B_1	B_2	
Value sets:	(a,b,c,d)	(0,1,9)	(0,1)	(e,f)

Value of field A	Value of field B		Value of field C
	B_1	B_2	
a	1	0	e
a	1	1	e
b	1	1	e
c	0	0	f
d	9	1	f

Address records

$p_{a10} = 1/5,$

$p_{ae} = 2/5$, etc.

x	$\mathcal{H}(x)$	$H_x(B)$	$R_x(B_1)$
A	2	0.4	1
B_1	log_23	0.55	1
B_2	1	0.95	0.37
C	1	0.95	0.63

Information measure

Figure 5: Example of information measure

4 MEASURE OF INFORMATION FROM A US POSTAL ADDRESS DIRECTORY

4.1 Fields of interest

Two kinds of US postal address directories, National City State File and Delivery Point Files, are analyzed. The city name (f_1), state abbreviation (f_2), and ZIP Code (f_3) of every record in a National City State File are extracted and the duplicated data are eliminated. The ZIP Code (f_3), ZIP+4 add-on (f_4), primary number (f_5), street name (f_6), secondary designator abbreviation (f_7), secondary number (f_8), and building/firm name (f_9) of every record in Delivery Point Files are also extracted, and the duplicated data are eliminated. The resulting data are used for the measure of information. Any combination of f_1, f_2, and any combination of digits (f_{3i}) in f_3, is used to measure the information provided and conditional entropy for the f_3 field. Any combination of fields from f_3 to f_9 is also used to measure the information provided and conditional entropy for the f_4 field. In the measure of the conditional entropy, the probability value is from the occurrence of records of the fields of interest.

4.2 Information content in US postal address fields

- The information provided by each field is shown in Table 1. How much information is provided on the average for a field of an address can be found. For example, knowing f_1 provides 15.28 bits of information.

Table 1: Information provided by each field

Component	x	$\mathcal{H}(x)$
City name	f_1	15.28
State abbr.	f_2	5.95
ZIP Code	f_3	15.39
ZIP+4 Add-on	f_4	13.29
Primary number	f_5	20.14
Street name	f_6	20.22
Sec. designator abbr.	f_7	4.58
Sec. number	f_8	16.92
Building/firm name	f_9	19.85

- The assignment strategy for determining the delivery point code by our handwritten address interpretation(HWAI) system is shown in Fig. 4 [2]. The propagation of uncertainty of applying this strategy is shown in Fig. 6. In the condition of only knowing valid values of ZIP+4 add-on (f_4), 13.29 bits of information are provided. After a ZIP Code is known, the uncertainty for f_4 is reduced to 10.50 bits. When more information is known, less uncertainty for $f4$ is obtained.

  The average number of ZIP+4 add-ons when a ZIP Code is known is 1065.43. This number is reduced to 2.30 when a (ZIP Code, primary number) pair is known, and further reduced to 1.04 when a (ZIP Code, primary number, street name) is known. It is possible that a ZIP+4 add-on can be determined by knowing the value of one or two fields.

  The Fig. 7 shows the propagation of uncertainty of another assignment strategy which is (ZIP Code -> secondary number -> street name). The rate of uncertainty reduction is less than that shown in Fig. 6. It indicates that the assignment strategy is less efficient and more fields of information may need to be identified in order to construct a ZIP+4 add-on value. An alternate assignment strategy may be required for interpreting a mail piece in the case that there are some components (e.g., primary number) cannot be identified correctly which can be caused by the confidence not high enough in locating and/or recognizing these components.

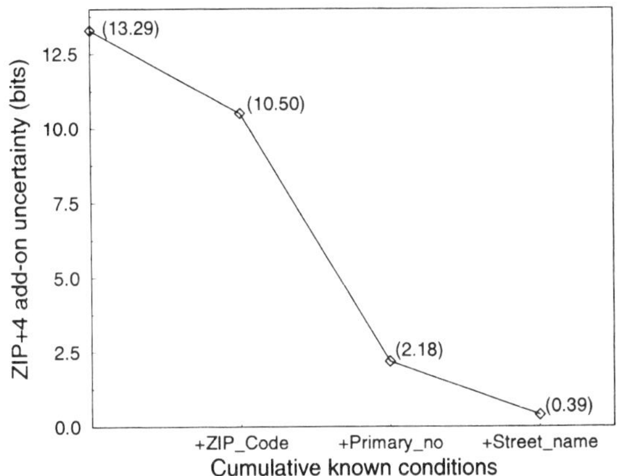

Figure 6: Propagation of uncertainty for an assignment strategy

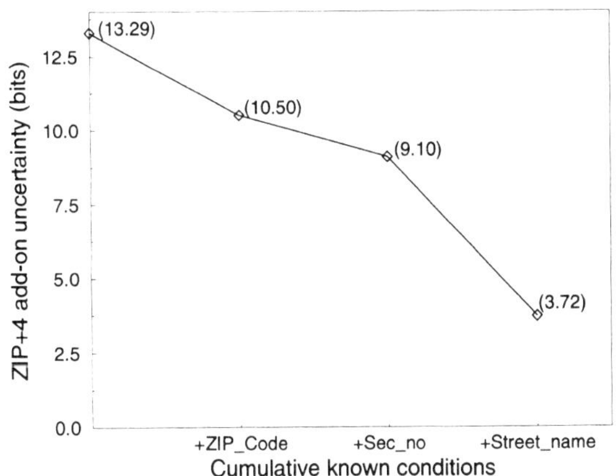

Figure 7: Propagation of uncertainty for another assignment strategy

4.3 Ranking processing priority for confirming ZIP Code

Assume that the components to be recognized for determining f_3 are f_{3i}, f_2, and f_1. Based on the available information at that time, the component to be recognized next, which is the most effective, can be determined from the measure of information. For the example of determining f_3 in the condition of known f_1, the most effective component (on the average condition) to be recognized next is f_{35} (Fig. 8). The value of f_3 (ZIP-Code) uncertainty is 0.63 bit which possesses the smallest value in the column of knowing 2 components. Since the uncertainty is low, it indicates that the f_3 value may be determined at that time (average number of f_3 candidates is 1.29, when f_1 and f_{35} are known), and if it is

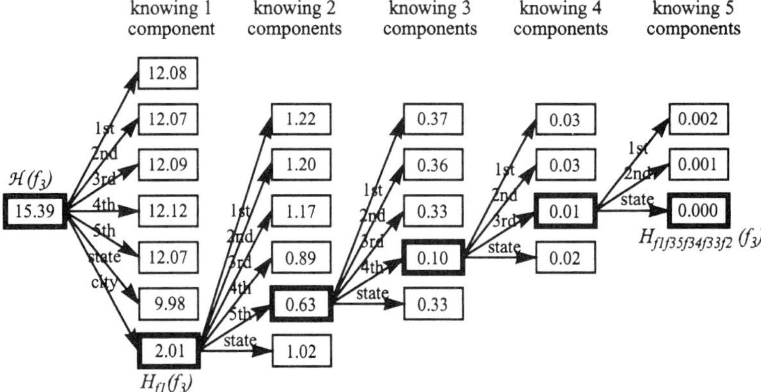

Figure 8: Most effective component to be recognized next for a known condition

determined, no further processing is required. If f_3 is still not determined, following the same rule, the next most effective component is f_{34}, then f_{33}, and finally f_2. One can notice that, while applying this processing flow, f_{31} and f_{32} need not be recognized since the value of $H_{f_1 f_{35} f_{34} f_{33} f_2}(f_3)$ is 0 after f_2 is recognized. The scheme of ranking processing priority can be applied for confirming other fields of interest.

The most effective processing path for determining a target component can be formed by ranking uncertainty values. By following this path and in any known condition, fewer candidates for the target component are left for choice compared to the other paths. If the target component needs to be recognized, the accuracy of recognition can be increased since the word recognizer or digit recognizer works better on a smaller size of lexicon [4]. It also means that a mail piece has higher possibility of being correctly interpreted in this case. Another advantage is that fewer components will need to be interpreted when the most effective path is used. The target component may be uniquely identified in the middle of the path. It indicates the advantage of improving processing speed and reducing error since fewer components need to be located and recognized. If a certain component cannot be located or recognized by applying the most effective path, an alternate component can be suggested by the measured information which is the second effective component to be recognized for a known condition. Hence, an alternate processing path is formed.

4.4 Recovering the first ZIP-Code digit from other components

If the confidence value of a candidate for a component from a recognizer is low or close to that of other candidates, it is possible to recover this component by knowing other components (e.g., knowing state abbreviation to help the determination of the 1st ZIP-Code digit as shown in Fig. 9). The most valuable component for recovering a component can be discovered by measuring the redundancy of other components to this component. For f_2 and f_{3i} to f_{31}, the information measure result with the ranking order is shown in Table 2. Since f_2 possesses the largest redundancy quantity, it is the most valuable component to be recognized in order to eliminate invalid digit candidates or used to verify the digit recognition result. From the database information, there are 62 state abbreviations. For 60 of them, the first ZIP-Code digit is unique. For the other 2 (i.e., NY and TX), there are 2 valid 1st ZIP-Code digits for each state abbreviation. The measure of redundancy indicates what components possess the most and least redundant information for a candidate component. The same scheme of redundancy measure can be applied to other fields of interest.

$$\underline{NY} \quad \textcircled{1} \quad \underline{4} \quad \underline{2} \quad \underline{2} \quad \underline{8}$$
$$f_2 \qquad f_{31} \; f_{32} \; f_{33} \; f_{34} \; f_{35}$$

Figure 9: Illustration of recovering the 1st ZIP-Code digit from other components

Table 2: Redundancy measure

x	$\mathcal{H}(x)$	$\mathcal{H}(x, f_{31})$	$R_x(f_{31})$	Ranking
f_2	5.95	6.00	0.99	1
f_{32}	3.32	6.64	0	2
f_{33}	3.32	6.64	0	2
f_{34}	3.32	6.64	0	2
f_{35}	3.32	6.64	0	2
f_{31}	3.32	-	-	-

5 MEASURE OF INFORMATION FROM THE MAIL STREAM AND COMPARISON OF RESULTS

The same measuring schemes are applied to measuring information from the mail stream. Eighteen sets, each from a mail processing site in the US, of mail pieces are collected (101,320 valid samples in total) to measure the information from the mail stream. However, in the measure of the conditional entropy, the probability value is from the occurrence of the fields of interest from the mail stream. Each set is measured separately. The results are shown on the average of these sets.

The statistical results show that the ZIP-Code uncertainty of a known condition from the mail stream is less than that from the address directory in most of the conditions. It indicates that the information from the mail stream is more effective to determine a ZIP Code than that from the address directory. Taking the most effective processing flow of using the ZIP-Code digits and a state abbreviation to determine a ZIP-Code value as an example. The decreasing rate of the ZIP-Code uncertainty is shown in Fig. 10.

Although the information from the mail stream seems more effective, the following statistical information should be noticed. The number of valid ZIP Codes from the address directory is 42,880. However, it is 1,129 on the average from the mail stream. The number of state abbreviations from the address directory is 62. It is 42.7 from the mail stream. It indicates that the mail stream does not cover all values of valid ZIP Codes and state abbreviations. One reason could be that some of the ZIP Codes and state abbreviations are less likely to appear on mail pieces for a specific site. However, another possible reason could be the shortage of samples (101,320 valid samples from 18 sites). More samples should be taken into account as a result to represent the statistical distribution.

6 SUMMARY AND CONCLUSION

The measure of information provided by knowing a component, the uncertainty of a component y when a component x is known, and the redundancy of components x to y have been defined. These information measures have been applied to US postal addresses. From the measured information, the propagation of uncertainty for an assignment strategy can be realized, and the efficiency of different assignment strategies can be compared. Since

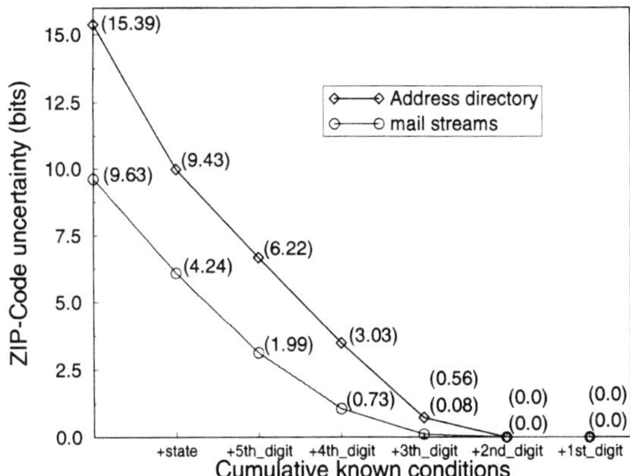

Figure 10: The decreasing rate of ZIP-Code uncertainty from address directory and mail stream

the most valuable component to interpret for reducing the uncertainty of a target component can be obtained by iterating this process of knowing more and more components, the most effective processing flow for determining the target component can be formed. This scheme provides the flexibility in the control structure for address interpretation and possesses the advantage of improving the processing speed and the interpretation accuracy. Furthermore, from the measure of redundancy, a component which possesses a higher possibility of recovering a damaged component can be discovered.

For a practical application, the processing cost of a component should be considered in forming the most effective processing path. The processing cost can include the error rate, accuracy, and speed, of location and recognition of the component.

The presented schemes are not restricted to either the use of US postal addresses or the sources of information. The information source can be from real mail stream or address directory. However, the information reliability and complexity of collecting information should be considered in order to enhance the effectiveness of the use of information.

REFERENCES

[1] United States Postal Service, "1997 annual report,"
   http://www.usps.gov/history/anrpt97/intro_everyday.htm, 1997.

[2] S. N. Srihari and E. J. Kuebert, "Integration of hand-written address interpretation technology into the United States postal service remote computer reader system," *International Conf. on Document Analysis and Recognition (ICDAR'97)*, pp. 892–896, Aug. 1997.

[3] S. N. Srihari, "Recognition of handwritten and machine-printed text for postal address interpretation," *Pattern Recognition Letters*, vol. 14, no. 4, pp. 291–302, 1993.

[4] G. Kim and V. Govindaraju, "A lexicon driven approach to handwritten word recognition for real-time applications," *IEEE Trans. on Pattern Analysis and Machine Intelligence*, vol. 19, pp. 366–379, Apr. 1997.

[5] C. E. Shannon, "A mathematical theory of communication," *Bell System Technical Journal*, vol. 27, pp. 379–423, July and Oct. 1948.

Advanced postal address readers with built-in learning capabilities

U MILETZKI
Siemens ElectroCom GmbH, Konstanz, Germany

SYNOPSIS

The scope of this paper is to give an outline of future Postal Address Readers* whose prominent characteristic is their learning capability, which can be expressed as *learning from letters*. In contrast to current reader systems with their predefined and frozen parameters and knowledge bases, they will be able to adapt themselves automatically to the changing characteristics of mail mix and writing conventions. The structure of such a system and the process of online learning is explained. Three cases of adaptiveness are treated: instantaneous, short and long term adaptiveness, i.e. learning from the same mail piece, from the immediately previous one and from all previous mail pieces. Such an adaptive system is beneficial for both the customer and the supplier, since it offers for the first time an optimized solution for each site at any time and it helps save a lot of engineering effort for tedious local system improvements.

1. INTRODUCTION

Postal Address Readers of today are, without doubt, intelligent machines. They can find addresses, which may be written in machine print or handprint or even in cursive script, read and interpret them without human aid. They can recognize all postally relevant information on a mail piece, see *figure 1*.

* The research and development work presented here was funded by German Federal Ministry of Education and Research (Project READ, continuation project ADAPTIVE READ is planned).

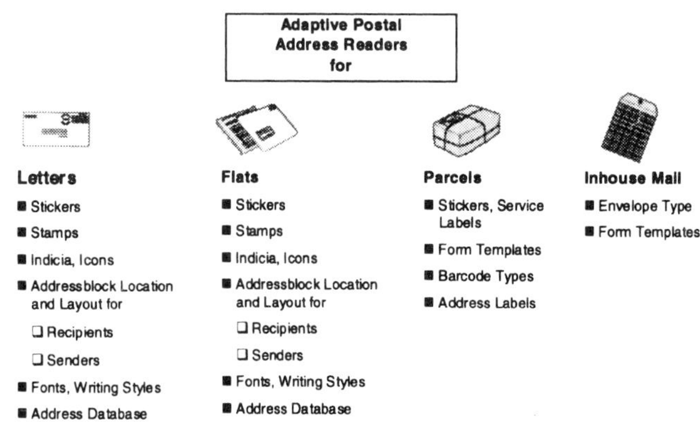

Figure 1: Overview of objects to be learned on different mail types

However they suffer from a basic shortcoming: their intelligence is predefined and they are not able to adapt themselves to changing reading conditions. Their adaptive phase has been finished prior to their delivery. They are condemned to repeat endlessly the same faults, since they cannot learn from them. In a fast changing world the mail mix and characteristics continuously change, the innovation of graphic design turns a mail envelope into an advertisement pillar. This is an ever growing challenge for a predefined layout analysis, which has to find the regions-of-interest carrying postally relevant information and to differentiate them from irrelevant adverts.

The ubiquitous desk top publishing industry creates new writing and printing conventions; the number of fonts is boundless. And last, but not least, in a modern dynamic society rapid changes in the address database are usual. It is the intention of this paper to show how this inherent deficiency of a predefined system can be overcome by a new approach of reading systems with built-in learning capability. An outline of such a system is given, its properties are described as well as the system in action. It is the intention of this paper to show that the prominent feature of the proposed system is to keep its optimized quality over time and at each site due to its online adaptiveness, see *figure 2*. While in the course of time a current system with a single learning process (SLP) degrades in quality due to the aging of the learning set and the resulting growing mis-match between interior and exterior world, a future system with a continuous learning process (CLP) will at least keep its initial read rate, see *figure 2a*.

Similarly an SLP-System is optimized to a fictive postal mail center, let's call it the *virtual site*, due to the chosen mix of learning samples which was adapted during the development phase. This system clearly will have deficiencies at individual *real sites* with deviating requirements. In contrast to this a CLP-System will be able to adapt itself optimally to site specific conditions, see *figure 2b*. Current nationwide solutions with a series of identical readers will be replaced by a new generation of readers which individually adapt themselves to site specific requirements.

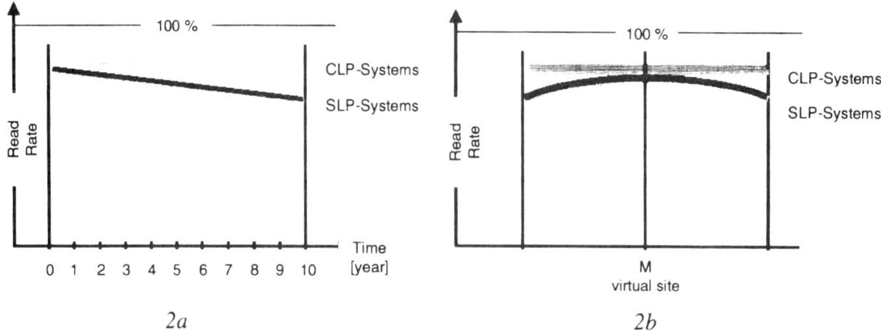

Figure 2: Degradation of read rate (a) time dependant, (b) site dependant

2. STATE-OF-THE-ART OF LEARNING SYSTEMS

Learning systems of today are typically back stage in the development labs of the suppliers of postal sorting equipment. The learning process takes place once-and-for-all during the development phase, typically lasts over days and weeks and is completely supervised. Large representative quantities of mail images are recorded by a scanning station already in place and in the development lab snippets of images from all kind of objects to be learned are extracted and each learning sample is provided with a label indicating its true meaning. This work is typically done by part-time students. It is time consuming and tedious, the quality requirements are very high, otherwise the system gets out of tune and produces more errors than is acceptable. Originally the objects to be learned were characters only, but state-of-the-art reading systems contain a large variety of classifiers for the different pattern recognition tasks on the way from image to understanding, see *figure 3*.

The Layout Analysis process of a mixed mail sorter runs a classifier for the classification of the mail piece type, e.g. a letter, flat, small parcel or an inhouse envelope: the module responsible for the detection of regions-of-interest contains a classifier for the different ROI-types such as recipient's and sender's address block but also many other postally relevant regions like indicia, stamps, and stickers. It further contains a hand/machine classifier for rejection of non-address text lines; the segmentation module contains several classifiers for decisions like word or character gap, proportional or fixed pitch, and cut-off positions of the individual character segments; the character recognition unit contains a bundle of classifiers for the different writing styles, and alphabets; the word recognition module contains a standard distribution classifier for intermediate symbols for the hidden Markov recognition approach; finally, the address interpretation holds grammatical rules for the recognition of the address structure and a nationwide dictionary with all valid addresses for the verification of the recognized address candidates.

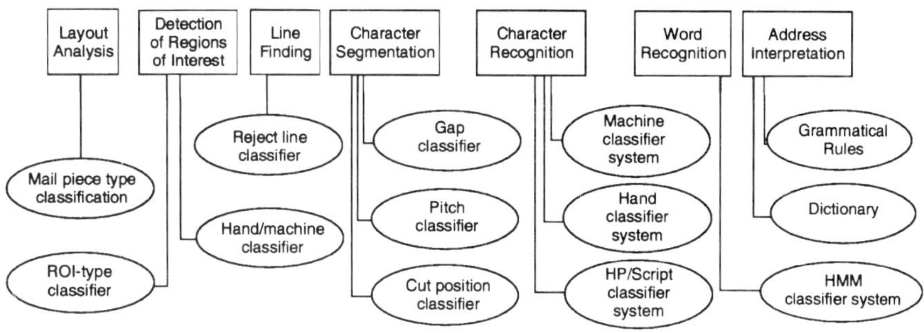

Figure 3: Predefined Classifiers and knowledge bases of a current reader system

3. ADVANCED READERS WITH ONLINE LEARNING CAPABILITY

Advanced reading systems coming up in the next years will incorporate the learning process as soon as the high standards of chip technology will allow it, to produce an adaptive system at reasonable market prices. Offline learning processes, which in the past lasted several days, can nowadays be performed on processors within a couple of hours. Highly efficient adaptation algorithms are meanwhile at hand, which can learn labeled samples of characters, words, or any other patterns in a sequential manner. What is still missing is the overall architecture of an adaptive reader in practice. The goal of this paper is to show what such an architecture can look like and how near it is, and also to show the pitfalls and risks to be avoided.

Before going into detail, it is important to emphasize that, with the appearance of online learning in the product, the offline learning will not disappear but rather play another role: it will, in the future, provide an initial classifier system for a nationwide application with default parameters which then will be the starting point of the online adaptive process - otherwise to start the learning process from zero would not be efficient enough and much too risky and not fit into the postal services' acceptance test procedures: the readers will have to show acceptable read performance from the very first day of delivery! The built-in learning processes in turn will gradually individualize the reader system by adapting it to site-specific conditions and over time it will keep track of the changing characteristics of the mail mix as well as font and script styles.

What does learning mean? Learning is an evolutionary achievement. Generally speaking it means an adjustment to environmental conditions which implies the ability to change one's behaviour and to exploit experience in order to avoid the repetition of errors. Transferred to technical systems, learning ability requires the generation of models for repetitive structures, and the measurement of deviation between model and real world. Learning includes changing of model parameters – which implies adaptiveness - and memorizing past situations in an adequate manner – which is the feature of experience from history. The acquired knowledge should then be activated in a given online situation – which is the changing of behaviour. Three different modes of learning performed by Postal Address Readers may be distinguished: inference of knowledge from other parts of the same mail piece, from the previous mail piece and from all previous mail pieces. This knowledge may likewise refer to layout, font type, or address structures. These three cases will be explained more in detail in the next paragraphs.

3.1 LEARNING FROM SAME MAIL PIECE

The simplest case is that of learning from the same mail piece. It contains already the aspect of adaptiveness but its experience is very rudimentary: the memory is ultra short and limited to the environment of the current mail piece image, as a rule less than a second according to the processing time of one mail piece. It knows no experience from "history". The technical solution is rather an inference mechanism than a complete learning method based on models gained from past experience. Adaptive processes like autofocussing belong to it as implemented in parcel readers. For accommodating the optics to to the varying distance between scanner and individual parcel the distance between object and image plane are measured and the sharpness of edges in the image is controlled. The dynamic binarisation process is another sample of adaptiveness that makes a decision between fore – and background. The foreground should carry all relevant information, the background should contain all the noise caused e.g. by "manila" colour of the envelope or by paper inclusions. An adaptive threshold does the decision dependent on the surrounding image and the characteristics of the mostly homgeneous mail piece background are inferred to the current image portion to be binarised. Ultimately, for the decision of one pixel to belong to fore- or background, the whole image information has to be considered. Also, since one mail piece comes from one source, font and writing style information in one line of text presumably can be applied to the next lines within one and the same homogeneous text block, e.g. an address block, or a preprinted text for automatic forwarding. But the possibilities of inference are very restricted and quickly exhausted.

To summarize, the background of an envelope is, as a rule, the same over the whole area with certain exceptions like windows. Typically, only one or two fonts are applied to sender's and recipient's address. Further fonts can occur in advertising text blocks. This homogeneity can be exploited by an adaptive process to infer knowledge gained from one part of a mail piece to another.

3.2 LEARNING FROM PREVIOUS MAIL PIECE

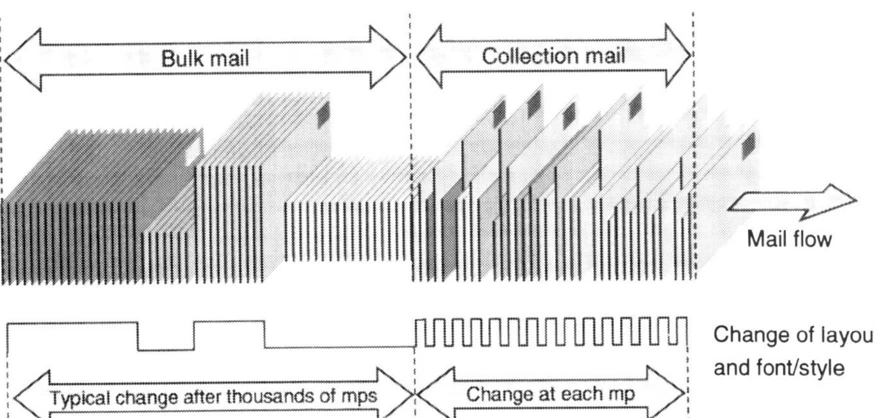

Figure 4: Typical mail flow with bulk and collection mail

The second case is much more interesting : to learn from the previous mail piece. When we look at the daily mail flow as a whole, we can clearly distinguish between bulk and collection mail sections see *figure 4*. During collection mail processing it normally makes no sense to learn from a previous mail piece since learning implies repetitive features while in this case each mail piece comes from a different source with completely different properties. But during bulk mail processing all mail pieces of a sequence have been delivered from one supplier, i.e. from one company or mail server. Mail pieces are printed with the same printer and the same addressing software. So their characteristics are highly homogeneous: same paper and envelope type, same layout, same font and address structure, and even same indicia and possibly same advertisement imprints. Imagine such a bulk mail stack is a large flick book then one can easily see, the only thing that differs between successive mail pieces is the current address. But even here one can expect a certain regularity, and assume that all addresses start at the same position, and are sorted according to a given sorting scheme, e.g. recipients in alphabetic order or postal codes in ascending numeral order. All these features can be observed on the previous mail sample and applied to the current one.

But how can the system know whether bulk mail or collection mail is processed? The conventional solution is that the operator pushes a button and runs the reader equipment in the so called batch mode, which traditionally allows him to define the address block location via terminal. Much more elegant however is an adaptive reader system which automatically recognizes the type of mail, in this case bulk mail, applies its inference mechanism, makes its conclusions for the current mail piece from the current and previous situation and finally makes the traditional manual batch mode superfluous. *Figure 5* outlines a system which can automatically detect bulk mail and infer knowledge from previous to current mail pieces:

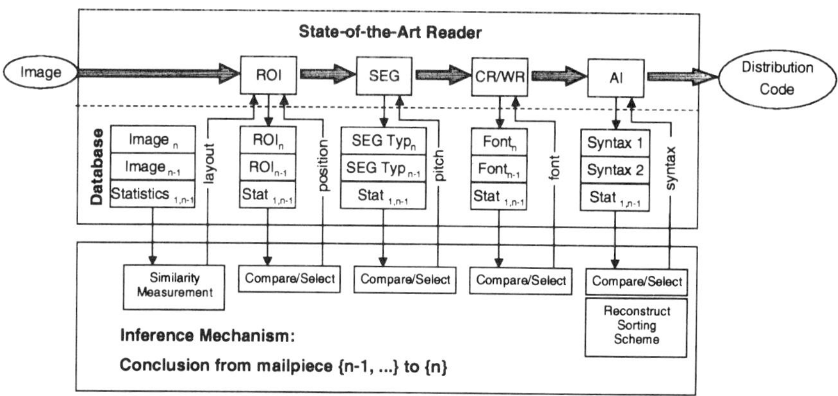

Figure 5: learning from previous mail pieces in bulk mail

Above the dashed line the state-of-the-art reader is depicted, which processes a mail piece image through the usual stations for detection of region of interest (ROI), Segmentation (SEG), Recognition of Characters (CR) and Words (WR), and finally interprets the intermediate results from the individual recognizers in the Address Interpretation module (AI).

This conventional structure is expanded by a database for the images from the current mail piece (n) and its predecessor (n-1), and for observed features assigned to them such as:

- ROI - parameters (coordinates and type)
- SEG - type (fixed or proportional pitch or script)
- Font - type (font family, hand print or script)
- Syntax (rules).

Additionally the database memorizes the tendency of the features of mail piece 1 to n-1, i.e of the stack from beginning to now, thus delivering the average position of all previous ROIs of the stack, the average SEG Type with a score telling how often it has been confirmed as target address. Font specification and syntax rules, are treated likewise.

A third part called inference mechanism is responsible for the recognition of a stack, its beginning and end. Note that a stack in this sense may consist of no less than two mail pieces of same origin at a minimum and many thousands at a maximum. For the duration of a stack it inherits the accumulated bulk mail knowledge to the current mail piece:

- the similarity between image n and n-1 is measured and if it is higher than a given threshold, we can be sure both mail pieces are of the same origin i.e. they are of the same stack, and the inference mechanism on bottom of figure 5 is activated.
- ROI_n and ROI_{n-1} are compared, and the average position and orientation of all ROIs of the already processed part of the homogeneous stack are sent to ROI Detection unit.
- similarly the segmentation type of corresponding address blocks of mail piece n and n-1 is compared and checked against the average result. Based on these results the estimated segmentation type is transmitted to the segmentation unit.
- The font types recognized by a font classifier of the last two subsequent mail pieces are compared and the results are verified by the average results and sent to the recognition unit.
- The syntax of the addresses of the stack is reconstructed while the mail pieces are compared two by two. The results are likewise sent to the address interpretation unit. Additionally a sorting scheme if present can be reconstructed and can be used as plausibility check during the interpretation process.

The state-of-the-art-reader produces statements based on the observations of the current mail piece while the inference mechanism determines the averaged characteristics from previous observations of the stack which often will be identical but may also lead to different statements for layout, address block position, font type, and address syntax. Both statements have to be harmonized in the individual processing stages in deliberate decision. Therefore it is necessary that not only the conventional intermediate recognition results are provided with confidence levels, but also the statements concluded from the past, and a soft-decision based on these confidence levels should be performed (in case the two statements are identical, they support each other and their total confidence will increase, but in case they contradict to each other, the more confident statement should be taken, or if they are equally confident, the mail piece should be rejected). In case the conventional reading process of its own was without success, the inference mechanism may help by inheriting its knowledge. This case will probably cause the highest increase of read rate. The remaining cases are trivial.

To summarize we can say that a batch detecting mechanism may remarkably increase the reading ability of this reader through the inheritage of "historical information." The traditional batch mode with manual entry of ROI-position can be completely replaced by this much more efficient method.

3.3 LEARNING FROM ALL PREVIOUS MAIL PIECES

This case is the most general, and hence the most interesting one. It recognizes from the continuously flowing mail stream the changing habits in layout and writing of mail envelopes, and automatically updates its knowledge bases and classifiers continuously or in short cycles. In contrast to the method described in the last paragraph, this generalized approach improves the reading of all kind of mail. It optimizes the system at all times and at any site, as already mentioned in the introduction. How does such a system work? Among all known recognition methods, statistical methods seem to be most apt for incorporation in a reader product. The modeling of objects to be recognized is done statistically, no source code or software configuration changes are necessary. All changes are kept in a set of coefficients i.e. the classifier, which are the result of a mathematical optimization process. Comprehensive experience from offline learning systems is available and many steps of the online learning process are similar. Statistical methods are already in use for a variety of pattern recognition tasks, see *figure 3*, and all of them can be made adaptive in the same manner. The principle is as follows, see *figure 6*.

During the standard reading process a screening and filtering process takes place, which selects learning samples over the training period. The latter may last from a day to a week. Within this period all critical patterns are collected and stored in a learning sample database. They are the basis for continuous improvements. But also a smaller portion of good quality patterns have to be selected and stored and must be retrained in order to avoid the risk of

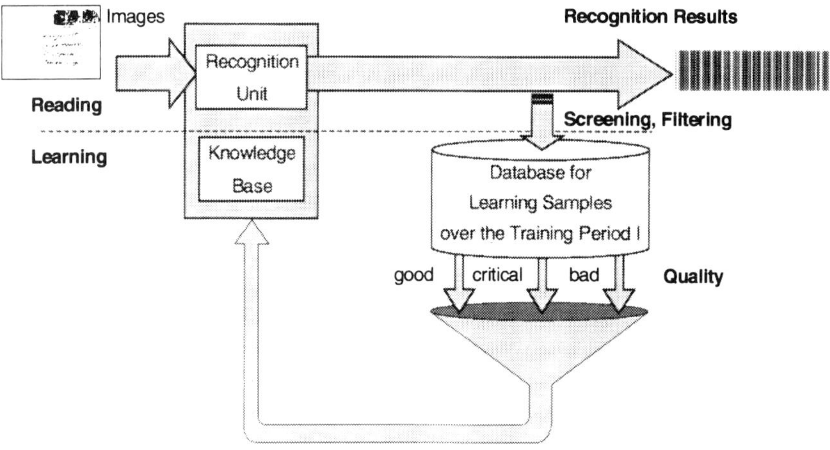

Figure 6: learning from samples

unlearning. In fact samples of all quality categories – good, critical, and bad – have to be stored and fed to the current classifiers according to a heuristically determined mixing formula.
In principle, all statistical methods may be used for adaptation, see (1), but since in an online adaptive reader processing time is very limited, it is advantageous to have a very efficient

method at hand with high convergence speed and high generalization ability. The stepwise improvement classifier method described in (2) and (3) is such a method, where the classifier is iteratively adapted thus guaranteeing an optimal recognition rate for the learning set. There are also many risks to by-pass in order not to reduce the system quality:

- poor images from a scanner in bad service condition must not be adapted, otherwise next day after repair the reader reads even worse!
- a truthing label must be attached to all learning samples. It is produced by a higher instance in the reading process. Typically the true meaning must be verified by the address interpretation, and the label must have a high confidence level, otherwise it is too dangerous to accept it for the adaptation process.
- the individual modules must automatically be harmonized with each other. Otherwise the adaptation of one isolated unit may lead to a deterioration of the overall reading performance.

The structure of an online learning component is shown in *figure 7*. It consists of a classical recognition unit except that the classifier must be a writeable file which will be periodically overwritten by a better matching classifier. It contains an adaptation database which stores the learning samples selected from the continuous dataflow by a monitoring unit. The heart of the learning system is the adaptation module, which trains the learning samples subsequently, after having generated an appropriate mix of good, critical, and poor quality learning samples according to a mixing formula. Finally it contains a classifier database for the first, the so called proto-classifier, the current optimized one, and the new classifier. Last but not least the learning component must provide a test set database holding the test data which correspond exactly to the classifiers, i.e. not-learned characters from the same pattern generation process. A comprehensive self-test procedure after each training period guarantees the selection of the best classifiers available at any time. A detailed description is given in (4).

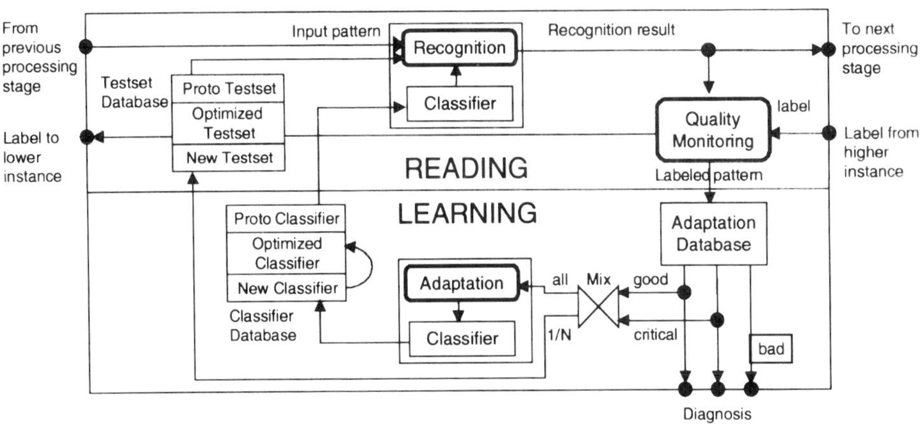

Figure 7: on line learning component

4. SUMMARY

An outline of a continuously learning system was given. The conditions of learning were discussed and practical solutions, which are compatible with the operational schedule of a postal address reader, were proposed. Particularly the cases of learning from the previous mail piece in batchmode and a generalized learning approach for all mail pieces were discussed. With the Stepwise Improvement Method (SIC) of character classifiers, an essential component of an adaptive reader is at hand. It can also be applied to continuous learning of other patterns recognition tasks arising during the process of image analysis, such as distinction between individual writing styles and segmentation.

One of the next steps is to gather experience as regards the adaptation of these patterns by means of the SIC-Method. For the learning of syntax rules and dynamic dictionary adaptations new methods will have to be developed. Given the current state-of-the-art, the time has come to tackle the problem of a continuously learning system. The computational effort necessary for online learning can be supplied already by current processor types or at least by those announced for the near future. Experience from today's offline learning systems can be transferred to a certain extent to online learning systems. Nevertheless, a lot of research and development effort will be necessary until the next generation of Postal Address Readers will be endowed with reliable and effective online learning capability. It is planned to support this work by ADAPTIVE READ, another publicly funded project, as a successor of the READ-Project mentioned above.

REFERENCES

(1)   Jürgen Schürmann,
        Pattern Classification
        "A unified view of statistical and neural approaches", John Wiley & Sons, inc, 1996

(2)   Rainer Lindwurm,
        "Lernfähige Algorithmen", Deliverable No. 213 of the project READ, which was
         funded by the German Ministry for Education and Research (BMBF) under
         grant 01IN503C, during 1995-1998.

(3)   Jürgen Franke,
        "Expanding the performance of polynomial classifiers by iterative learning",
        proceedings of the sixth international workshop on frontiers in handwriting
        recognition, pages 315-323, Taejon Korea, 1998

(4)   Udo Miletzki, Thomas Bayer, Hartmut Schäfer
        Continuously Learning System
        Postal Address Readers with built-in Learning Capability
        Publication planned in the Proceedings of ICDAR'99, Bangalore

E-mail address of author: Udo.Miletzki@kst.siemens.de

C568/023/99

Towards the full hand-written address processing

G ARCAS-LUQUE
Mannesmann Dematic Postal Automation, Gentilly, France

2   ABSTRACT

Mail sorting must process two kinds of objects, handwritten and machine printed addresses, each having different levels of difficulty in terms of distribution and recognition. Indeed, handwritten mail represents a small part of the whole mail stream, and is characterised by a wide range of writing styles. For these two reasons the processing of machine printed mail has usually been favoured. However this is changing due to technical and commercial reasons and the processing of handwritten mail is evolving towards the recognition of all the fields within the address.

This paper presents the Mannesmann Dematic Postal Automation hand-written reader: this application is a state of the art industrial OCR enabling the recognition of the address delivery field. A description of the word recognition technique is provided, including the techniques used to match lexicons of some thousand elements.

3   INTRODUCTION

Mail sorting must process two kinds of objects, handwritten (HW) and machine printed (MP) addresses, each having very different levels of difficulty in terms of distribution and recognition. Indeed, handwritten mail represents a small part of the whole mail stream, and is characterised by a wide range of its writing styles. For these two reasons, the processing of machine printed mail has usually been favoured and is much more complete: all the different fields in the address are used to obtain the information required for outgoing and delivery sorting. On the other hand, handwritten processing has mainly been used for outgoing sorting, including foreign country detection and recognition.

Nowadays, this situation is changing for both technical and commercial reasons. From the latter point of view, handwritten mail represents an important part of the OCR rejects at the delivery level, which must then be sorted manually or videocoded. By chance, the technological improvements and the increase in computing power now makes it possible to implement the complex algorithms required for processing handwritten delivery fields.

The goal of this paper is to present the state of the art in terms of HW technology. This is done through the description of the HW address reader developed by MDPA (1). This application, abbreviated to MHR (Mannesmann Handwritten Reader), is implemented in an OCR product. The outgoing processing of this application is used at the present time by more than 10 postal administrations. The delivery processing is fully implemented for French mail, and is currently under development for other postal administrations.

This presentation begins with a comparison of the economic interests and the difficulties of HW mail and MP mail sorting. Then, the MHR functions are described, demonstrating the need for a powerful word recognition module, especially in the case of delivery processing. Finally a description, in greater detail, is presented of the function used by the MHR. This description shows that the technology applied to word recognition may lead to further functionalities, which, up till now, were considered unrealistic.

4   HANDWRITTEN VERSUS MACHINE PRINTED MAIL SORTING

Two factors explain why HW mail recognition and sorting have not progressed in the same manner as MP mail. The first one is, of course, the very high variability of HW styles and drawings; the second is the lower HW mail frequency, and consequently the lesser economic interest of HW sorting.

As an example of HW versus MP complexity, let us consider the problem of character segmentation. The following figure illustrates the conditional probabilities *p (word length / segmented length)*, for MP and HW word databases of similar size and composition.

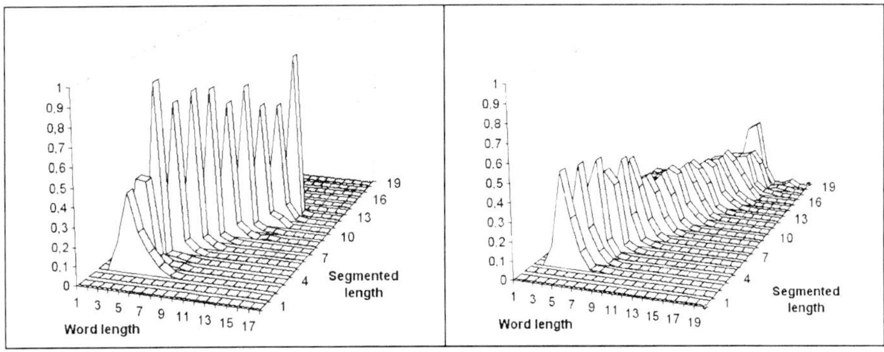

Figure 1 - MP and HW word length conditional probabilities

                                                                    C568/023/99

These distributions are computed with the word segmentation modules of the MDPA OCR. Although they do not have strictly identical functionalities - the HW segmentation being more of a grapheme segmentation - the comparison allows us to anticipate the potential difficulties of HW word recognition. For instance, 98% of MP city names are segmented with an error less or equal to one character. In the case of HW words, the same proportion is obtained for an error of 6 characters. Therefore, segmented length is a powerful lexicon filtering technique for MP words. This is not the case for HW words.

The solution to process HW addresses must therefore be found in more powerful algorithms, requiring more expensive implementation hardware. This leads us to the problem of economic interest, which is estimated through the relative proportion of HW and MP mails. For example, the proportion of HW mail is about 20% for French and UK mail. This has led postal administrations to favour MP sorting and select hardware for MP processing implementation. As a consequence, HW processing has long been limited to what could be done with the available computing power, i.e. code and city name recognition, neglecting delivery processing.

This is changing now for two reasons. Firstly, the cost of computing power is decreasing, allowing the implementation of more powerful algorithms. Secondly, postal operators are more and more concerned by remote sorting. From this point of view, HW processing is now favoured. It currently represents the main part of the main OCR rejects and furthermore, with time constraints being relaxed, the computing power needed to process HW delivery becomes available.

5   HANDWRITTEN READER FUNCTIONS

The MHR is used for a large category of mail and is adapted to processing mail from several countries. This has been made possible thanks to the definition and integration of some parameters rendering it possible to modify the MHR working. From an ideal point of view, the adaptation of the reader for a new country should only require:

- a description of the address syntax,
- the adaptation of the recognition models,
- the formatting of the postal files to take into account the structures expected by the reader.

This goal is achieved in the case of the outgoing processing, although some specific rules can still be needed in some cases.

5.1   Outgoing functions

The general syntax processed by the MHR is defined as follows in Backus Form, with SP and CR meaning respectively space and carriage return, "|" meaning "or", "*" meaning several times and "[]" meaning optional:

outgoing_lines	::=	*outgoing_line	outgoing_line CR country_line*
outgoing line	::=	*code [CR] cityname	cityname [CR] code*
code	::=	*[prefixe] SP|- postcode*
postcode	::=	*(alpha | numerical)**

prefixe   ::=   *ISO_2_characters | variation*
cityname   ::=   *canonical_name | variation*
countryname   ::=   *canonical_name | variation*

Although the MHR is able to handle the whole syntax, it is interesting to constrain some elements. This allows the reader to be fitted to the needs of the postal administrations and available computing power. For instance, the reader can be adapted to look for and recognise postcodes of varying length, containing numerical digits as well as alphabetical characters. Obviously, this is power consuming and superfluous if the mail stream is constituted by addresses presenting postcodes with a stated syntax.

5.2 Delivery functions

In the actual state of development, the recognition of street lines and P.O. Box lines on HW mail have been fully integrated for French addresses. Although it is effective, it has not been developed to the point where the adaptation of mail from another country could be achieved only by the modification of parameters, models and postal data. Nevertheless, the most important algorithmic tools are available, namely word recognition and large lexicon matching.

The syntax currently processed by the MHR is described as follows:

delivery_line   ::=   *locality_line | street_line | pobox_line*
locality line   ::=   *locality name*
street_line   ::=   *[street_number] street_descriptor street_name*
street_descriptor   ::=   *canonical_name | variation*
street_name   ::=   *canonical_name | variation*
pobox_line   ::=   *pobox_name pobox_number*
pobox_name   ::=   *canonical_name | 2_letter_abbreviation*

The concerned syntax presents a very large diversity. However, the delivery module does not have to cope with the whole syntax at a time. The process is currently driven by the results of the outgoing recognition, i.e. the postcode hypothesis and its type which may be:

- not affected by the delivery process - in this case, a dependent locality is searched -,
- geographic, i.e. associated to the street syntax,
- "separation", i.e. associated to a P.O. Box like syntax.

In these conditions, the performances obtained on collection mail are:

	Read Rate	**Error Rate**
Street name lines	> 40 %	< 2 %
P.O. Box lines	> 50 %	< 0.5 %

As one may guess, the delivery processing is strongly concerned by the recognition of words or expressions. The technique used in the MHR is fully described in the next section.

6 WORD RECOGNITION

6.1 Domain

The word recognition module intends to recognise words or expression compounds of several words by matching them to a lexicon. In the MHR, it is used for the recognition of several alphabetical fields, the only difference in the process being the origin of the lexicons. The following list describes these fields, with the constitution of their associated lexicons and the heuristics used to localise them within the address.

6.1.1 City names

The city name is searched for to the right or the left of the postcode, which is itself searched for on the two lowest lines of the address. The city name lexicon is dynamically inferred from the postal files. Generated from a low number of postcode hypotheses, the lexicon includes less than one hundred city names. The lexicon contains not only the names found in the lexicon, but also variations constituted by the omission of words, or the substitution of words by abbreviations. For instance, in French, the word "Saint" is also considered with its abbreviation "St".

6.1.2 Country names

The country name is supposed to be alone on the last line of the address. Its associated lexicon is static. It contains country names with variations, including the names in the languages of departure , arrival and English.

6.1.3 P.O. Box names

In France, the delivery lines can contain P.O. Box references written in full, for instance "Boîte Postale" or abbreviated "BP". They are considered to be on the line above the city name line. The lexicons are derived from the postal files and include a few dozen elements.

6.1.4 Street names

The street line is considered to be the one above the city name line. The street name lexicon is also inferred from the code hypothesis. All the street names corresponding to this postcode are included in the lexicon. As a result, the lexicon can have an important size. Furthermore, it is completed by some variations generated from the canonical expression. For example, these variations relate to:

- the street type, for instance "Boulevard" may be abbreviated as B, BV, BVD, BLD,
- titles, like "Président" or "Docteur" which are respectively abbreviated in "Pdt" and "Dr",
- first names, which may be omitted, or substituted by their first letter.

All this leads to lexicons with a mean size of about 3000 elements. Of course, such lexicons require some optimisation, which is described hereafter, to be processed in a real time environment.

6.2 Pre-processing

The input of the word recogniser is a rectangular bitmap, extracted from the envelope image, and supposed to contain the field of interest. The first step of the process consists in describing the image as a structural representation. This is possible in several ways, for example with contours or skeletonization, the latter being used for this presentation. Each connected component is transformed into a graph, the arcs of which encode the strokes of the pattern with Freeman codes (2). This representation is not absolutely necessary for word recognition but it gives some advantages, particularly in a real time environment:

- The representation is very compact.
- This compactness allows all further processing to be carried out very quickly. For instance, pseudo contours can be computed instantly from the arc Freeman descriptions and the order of the arcs around the graph nodes.
- This representation gives fast access to some structural properties of the pattern. For instance, the cyclic number of a single component graph defined as $N_{arcs} - N_{nodes} + 1$, is equal to 0 if the graph has no loop.

To minimise as much as possible the variability of writings, two normalisation processings are then carried out, seen in Figure 2. The first one is intended to normalise the skew of the line, the second one to normalise the slant of characters. Both are shear transforms, operating on Freeman codes, and are defined as follows.

6.2.1 Line skew normalisation

$x' = x$ with the line skew θ estimated by linear regression on the lower

$y' = y + x \tan \theta$ extremal points of the contour.

6.2.2 Character slant normalisation

$x' = x + y \tan \theta$ with the character slant θ estimated from a polygonal

$y' = y$ approximation of the Freeman description, the slant being computed as the mean of the near vertical segments.

---

a) Original bitmap

b) Skeletonised image

c) Slant normalised image

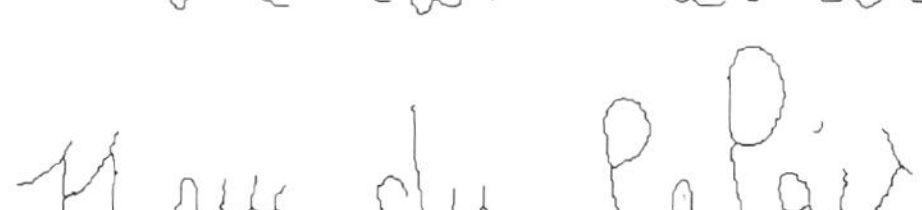

Figure 2 - Image pre-processing

6.3 Grapheme segmentation

After normalisation, the word to be recognised is segmented into smaller components, called graphemes, which could be defined as the writing basic component. Actually, with the process being approximate, the best definition should rely on the operating mode of extraction. The latter splits the full word at the minima of the upper contour situated near the lower base line, see Figure 3. This leads to graphemes which may be pieces of characters, like the parts of the letter U, or characters, like the letters e or i, or associations of characters when there is a drawing overlap. This is not very important as long as the training stage is able to learn how characters are detected as isolated, cut or associated with other ones.

a) Base lines detection and
   split point detection

b) Grapheme boxing

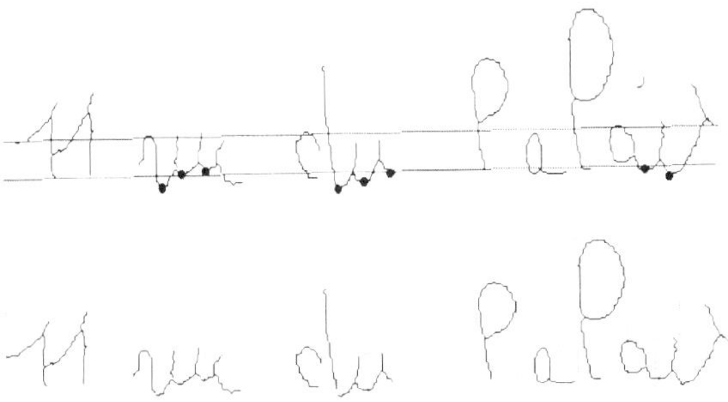

Figure 3 - Grapheme detection

6.4   Primitive extraction
The word recognition technique used by the MHR requires as input a symbolic description of the pattern to be recognised. The function of the primitive extraction stage is therefore to assign to each grapheme a symbol describing its structure. The choice has been made to combine several representations, like the two described hereafter.

6.4.1   Structural grapheme representation
This symbolic representation relies on an intuitive description of graphemes. They are described in terms of small or big loops, ascenders and descenders. Considering some joker symbols, this leads to an alphabet of 19 symbols describing the structure of the grapheme.

6.4.2   Rough character classification
This symbolic description uses the results of rough hand printed character recognition. In this case, the goal is not to recognise precisely the characters, but to proceed to some vector quantisation of the grapheme space. In fact, we use a Nearest Neighbour Technique whose reference patterns are learnt in a hand printed character database. Considering a joker symbol and without distinguishing lower and upper cases, this leads to an alphabet of 27 symbols describing the grapheme in some way.

6.5   Markov recogniser
The MHR uses the Markov technique to model the way a given word can be represented by a list of discrete primitives (3), (4), (5). In this formalism, words are identified to discrete sources emitting sequences of symbols. A Markov model is an abstract random process, with the same statistical properties as the source, allowing the probability to be computed for a sequence of symbols emitted by the source.
Many structures and variants of Markov models exist. The one used here has the following characteristics:

* The model of a word is the concatenation of the models of its characters.

- The model of a character is a Hidden Markov Model (HMM), taking into account the way a character may be represented by one primitive, a pair of primitive or not represented at all.

- The models of characters represent hand printed as well as cursive letters. This doesn't alter the performances of recognition and allows some simplifications :
  - no discrimination between hand printed and cursive writing is needed during the recognition process,
  - no discrimination is needed during the training stage.

- Uppercase and lowercase letters are modelled. The model handles the letter case by matching the primitive list to the following compound models :
  - all uppercase letters,
  - all lowercase letters with initials in uppercase or lowercase.

- Spaces are modelled by considering the probabilities that a space, according to its length, is a space between words or between characters.

- The Viterbi algorithm makes the match between a primitive list and a model. This algorithm uses dynamic programming to find the best path between primitives and characters, i.e. the one that gives the highest probability.

6.6 Large lexicon matching

The basic technique of matching must be optimised in order to handle efficiently large lexicons, e.g. lexicons with several thousand words. This has been done in three ways: dynamic programming optimisation, computation factorisation and lexicon filtering.

6.6.1 Dynamic programming optimisation

Some classical techniques to optimise dynamic programming are used:

- Considering the hypothesis that the matching path must not be too far from the shortest path, a window is used, avoiding the need to compute the elements $M_{i,j}$ with abs $(i - j) > K$.
- The structure of the character models implies some constraints on the matching path. For instance, a character is not represented by more than two primitives. As a consequence, all the elements $M_{i,j}$ with $i > 2j$ do not require to be computed or initialised.
- The naive approach to initialise the dynamic programming matrix consists of initialising all $M_{i,j}$ elements for $i <$ number of primitives and $j <$ number of characters. This is a surprisingly time consuming process, particularly if it is done several thousand times. The matrix initialisation has been reduced to a minimum by setting in each line only the first element needed by the computation.

6.6.2 Computation factorisation

A general technique for computation optimisation is to take advantage of the characteristics of the data to be processed. In the case of French street lines, a strong characteristic is the presence in the lexicon of several variations of each street name. All these variations have the particularity that the last word is not altered and remains constant over all the variations. As a consequence, if the computations during dynamic programming are made from the end of the street name to the beginning, all the computations made for the last word may be used for all variations.

This is of course extended to variations ended by more than one common word.

6.6.3 Lexicon filtering
Even if the word recogniser is optimised, the word recognition stage remains time consuming
if the lexicon is large. It could therefore be interesting to find some criterion allowing to
exclude a solution for a computing cost less than the Markov recognition one. In the case of
MP words, many solutions use the segmented length and the results of recognition at character
level (6). This is of course difficult or impossible in the case of handwritten, and it is
necessary to extend the data used for filtering to word level. After experimentation, it has been
noted that the recognition of the last word allows a filtering ratio of 3 to 1 or more without
loss of system performances. An additional advantage is that the recognition of the last word,
not only allows filtering, but can also be used in the subsequent computation to match the
street name and its variations.

7   FURTHER FUNCTIONS AND DEVELOPMENTS

So far, this presentation has illustrated a new algorithmic tool, namely the real time HMM
matching of large lexicons, and its application to HW delivery. Of course, it is now possible to
go one step beyond, and apply this technique in other contexts or for additional functionalities.
We present in the following the list of developments and applications which could be
available in the short term.

7.1   HW outgoing processing enhancements
As already stated, the HW outgoing processing is postcode driven, i.e. the city name lexicon is
generated from a few code hypotheses. This leads to lexicons with a size less than one
hundred city names. This allows of course economical implementation but leads to some
limitations. For instance, if the code is not detected, the system gives a reject decision even if
the city name could be well recognised. The solution is to recognise independently postcodes
and city names. This implies to match the whole country city name lexicon and can be
achieved with additional filtering concepts.

7.2   HW delivery processing enhancements
The main characteristic of the MHR delivery module is the recognition of the delivery line as
a whole, without looking for individual words. This choice was made for simplicity reasons
but does not allow to process explicitly some rather uncommon syntax. Word detection, or
key word detection, would allow to increase the possibility of the delivery module, but also to
consider some criteria of lexicon filtering relying on key word recognition.

7.3   New functions
The HW address recognition can now be enlarged to other fields, e.g. company and person
names. Both cases may add redundancy to other address fields, allowing for a more complex
but also more reliable decision stage. From another point of view, the recognition of these
fields can add some new functionalities to the MHR. In particular, the recognition of person
names would allow the re-addressing of mail by using adequate databases.

7.4   MP processing enhancements
As a consequence of its relatively easy recognition, MP mail processing has been mainly

character driven. This gives satisfying results as long as the character segmentation has sufficient performance. The limitations of this approach tend to identify MP and HW processing. For this reason, the techniques presented here should improve the performances of the MDPA MP reader.

8   CONCLUSION

This article has presented a state of the art HW mail reader, including delivery processing. Evolving from postcode to street line reading, this application demonstrates that HW reading functions are closer to those of MP processing. This is due to three factors which have already been mentioned : evolution of hardware, evolution of algorithmic techniques and increase of the postal operator needs.

On the other hand, MP mail reading seems to have reached the asymptote set up by the intrinsic limitations of the recognition techniques used so far. To overcome these limitations, it is necessary to progress from character level recognition to word level recognition. This will of course lead to the use of algorithmic solutions very close to the HW ones.

Mail processing is therefore characterised by the convergence of needs and techniques for HW and MP mail. It is perhaps too early to consider the realisation of an OCR using the same modules for HW and MP processing, with perhaps some adaptation of the parameters and models. However the concept of reading HW and MP addresses by the same means seems to be a promising paradigm to progress and to fill in the gap between human reading and machine reading.

9   REFERENCES

(1)   B.K. Benyoub, "Une application industrielle de reconnaissance d'adresses", Colloque National sur l'Ecrit et le document (CNED 96), 1996, pp. 259-266.
(2)   A. Rosenfeld and A.C. Kak, "Digital Picture Processing", Academic Press, 1982.
(3)   L. Rabiner, Bong Hwang Juang, "Fondamentals of Speech Recognitoin", Prentice Hall Signal Processing Series, Alan V. Oppenheim Seris Editor.
(4)   F. Jelinek, "Continuous Speech Recognition by Statistical Methods", Proc. IEEE, vol. 64, no. 4, pp. 532-556, Avril 1976.
(5)   L.R. Bahl, F. Jelinek, R.L. Mercer, "A maximum Likelyhood Approach to Continuous Speech Recognition", IEEE Trans. On PAMI, vol. 5, no.2, 1983.
(6)   W.B. Cavnar and A.J. Vayda, "Using Superimposed Coding Of N-Gram Lists For Efficient Inexact Matching", Proceedings of the Fifth USPS Advanced Technology Conference, Wash. D.C., 1992, pp. 253-267.

C568/008/99

Korean flat sorting

E BRUZZONE, E FURFARO, B GHIOLO, and **S SPAGGIARI**
R & D Department, Elsag spa, Genova, Italy

Abstract

This paper addresses the problem of sorting Korean flats by means of an ICR (Intelligent Character Recognition) system. The flat is encoded by extracting and recognizing the zip code, composed of six digits. No recognition of Hangul character is performed. The basic problem in Korean flat encoding is the random position of the zip code. Differently from Western addresses, Korean ones can be located in a wide area on the flat surface and the zip code line is not in a standard position within the address block. The zip code might be constrained into six adjacent black ink boxes, although their position is not fixed. Its recognition is based on character segmentation, character classification and contextual analysis by means of grammar rules and lexicons. The system has been tested on the application field, being installed on flat sorting machines to be delivered to the Korean Ministry of Information and Communication for nine postal sites.

1. INTRODUCTION

Korean address interpretation is a challenging goal due both to the random position of the zip code with respect to the remaining part of the address and to the Hangul writing style. The Korean flat sorting process can be performed by means of the zip code identification. Korean zip codes are composed of six digits, often divided by a dash (e.g. 500-150). Sometimes, a special symbol, namely the *Korean Postcode Symbol*, can appear on the left side of the zip code. The first three digits (left component) are used for outward sorting. They describe a province or large city (digit number one), a geographic area (digit number two), and a district, country, or small city (digit number three). The last three digits (right component) are used for sorting delivery routes and represent a city block, a town or P.O. Box or a building receiving large amount of mail pieces per day. The number of available zip codes is about 10.000.

The position of the zip code is not fixed in the address and the zip code can be constrained or unconstrained, i.e. contained into six boxes of fixed dimensions or free on the flat surface in a right-bottom area. A Korean flat is classified as *constrained*, if it has a sequence of six-boxes filled with characters; it is classified as *unconstrained*, otherwise. The boxes might be printed with dropout ink, although very often they are printed in black ink. The zip code might be written outside of the preprinted boxes. The maximum size of the area containing the zip code is about 360x256 mm. The entire area is scanned in order to produce the input image.

Under this assumption, the zip can be floating on the surface and mixed with Hangul characters as well as indicia, stamps or other data printed on the flat. As a consequence, the Korean zip code cannot be extracted as first or last line of an address located with a standard address block locator for Western languages, where the key address lines are the last two/three lines of the address block. Techniques generally used in form processing application, such as geometric structures location and removal, have to be applied to deal with constrained flats.

The appearance of the items include envelopes, open magazines, newspapers, letters, polywraps, folded, tyvek, plastic wraps, postcards, digests, and flimsies. Hence the quality of input images can range from fairly good to very poor. The images in Figure 1 show examples of hand-printed and typewritten Korean addresses. The flat is acquired as a gray level image and transformed into a bitmap. It is scanned in the front position with the wider side on the bottom. The address can be oriented either horizontally or vertically (see right-top image in Figure 1).

The final goal of the sorting process on Korean flats is to locate and decode the zip string, i.e., a six-digits number referring to the mailpiece delivery address by means of a well-defined lexicon. The lexicon is composed by the collection of the existing zip codes. The flat surface may contain a customer (i.e., preprinted by the sender) or a destination (i.e., printed by a Manual Infeeder line) barcode. The flat encoding process does not require the zip code reading when one of the two barcodes is correctly decoded.

In Section 2 we describe the ICR system used for extracting and recognizing the Korean zip code, based on the Elsag core technology for postal and form processing applications. In Section 3 some experimental results on life mail are presented, while in Section 4 the integration of the ICR system into a mail sorting machine is addressed. Finally, in Section 5 conclusions and further developments are described.

2. ZIP CODE RECOGNITION

The ICR system for Korean addresses interpretation is based on the Elsag core technology, called *Reader Engine* (1), available both as a software-only product and equipped with dedicated hardware for high speed performances, in Windows NT, Windows 95, AIX and SCO operating systems. Reader Engine is the result of the work carried on at our labs in the last two decades and has been used for different applications, such as zip codes, check amounts and form field recognition. Reader Engine handles images and, for each image, produces the recognition results on the required fields. The recognition results can be provided in wide format. In this case a string is not described by a sequence of ASCII values, but by a set of characters from which alternative sequences of ASCII characters (i.e., strings) can be created. The flexibility of Elsag Reader Engine has been basic in the development of the Korean application, in fact, black-ink boxes, which are the peculiarity of the Korean application, cannot be removed during the acquisition phase. Hence,. image processing techniques, typical of form processing ICR systems, have to be used to locate and remove them.
The ICR system for Korean flat is composed by three modules, which operate as follows. A preprocessing phase is in charge of barcode decoding, address block location and noise filtering. The preprocessing phase provides the bitmap feeding Reader Engine to extract the text lines as ASCII strings. Finally, the Reader Engine output is processed by a post-processing module for contextual analysis (2).

Figure 1: Examples of b/w images of Korean flats (left-top: constrained horizontal hand-printed zip code, right-top: constrained vertical hand-printed zip-code, left-bottom: unconstrained horizontal typewritten zip code, right-bottom: unconstrained horizontal typewritten zip code).

2.1.   Preprocessing

The preprocessing on Korean flats is based on a standard procedure, although some ad hoc solutions have been developed. The barcode reading procedure is composed by location and decoding. If the customer or the destination barcode is found and identified, the ICR process is not performed. On the contrary, the address block location (ABL) is executed. For Korean flats, the address block location is not a basic step, as it is in the case of Western postal objects, due to the random position of the zip code. It is useful for identifying typewritten addresses printed on horizontal or vertical labels. Otherwise it simply reduces the input image to a smaller area making it easier to locate the zip code. A noise filtering procedure is used to eliminate small fragments from the image background. Also, deskew operations are performed in order to correct the skew angle, when a significant skew value is detected. The ABL process can produce multiple hypotheses. In general, it produces a number of typewritten hypotheses ranging from one to three, and does the same in the handwritten case. The Korean application is based on two typewritten hypotheses and one handwritten hypothesis. The handwritten area corresponds roughly to the right-bottom third of the flat, as well as one of the typewritten areas.

2.2.   Reader Engine

The Reader Engine component handles each image provided by the preprocessing module in order to extract test lines in the form of alternative sequences of characters associated with ASCII codes. The first step concerns the classification of the flat as constrained or unconstrained. The flat classification is based on the location of one box satisfying well-defined tolerances, such as maximum and minimum width and height or maximum and minimum thickness. When one box is found, a horizontal (vertical) strip is isolated and the remaining boxes are searched for on that area (see Figure 5). The geometric values resulting from the first box search are used to start a procedure that locates a sequence of six adjacent boxes, positioned into two groups of three equally-distance placed boxes. A dash often divides the two groups. For horizontal constrained zip codes, the right most and the left most vertical lines of the strip and the first right-bottom horizontal line are located. Then, on the refined area the box set is found. If six adjacent boxes satisfying the geometric constraints are found and at least one box is filled with a digit, the flat is classified as constrained and the recognition process is carried on in *constrained* mode. On the contrary, the flat is classified as unconstrained and the recognition process is performed in *unconstrained* mode. A similar search is activated for vertical constrained zip codes.

In the constrained mode, the zip location is strictly related to the box-printed area identification. The found boxes are then removed by means of a box removal procedure that does not affects the characters, i.e., boxes are removed without any character restoration. The resulting bitmap represents one text line, on which a standard character segmentation process is activated, in order to isolate the six digits. The segmentation process provides multiple hypotheses, i.e., alternative sequences of characters. Each character is classified with respect to the set of ten digits and a ranked list of ASCII values is associated with it. The result is then used as input to the post-processing.

In the unconstrained mode, no boxes are available. As a consequence, the zip code text line cannot be straightforward located. A line finder procedure is applied to the input image and all the text lines are extracted. Each text line is segmented into alternative sequences of characters and each character is classified with respect to the ten digits. The selection of the text line containing the zip code is then carried on by the post-processing phase.

2.3.   Post-processing

The post-processing phase is based on the analysis both of the results of the preprocessing module and of Reader Engine. The strings isolated and recognized from the typewritten and the handwritten ABL hypotheses are analyzed to identify the zip code. The general output of Reader Engine on each single image is composed by a set of recognized text lines for unconstrained flats. In constrained flats the text line is one per image. Each recognized line is formed by character segmentation hypotheses. Each character is associated with its recognition values, i.e., a ordered list of ASCII values.

Given the images provided by the preprocessing phase (ABL hypotheses) together with the recognized text lines, a lexicon containing all the available Korean zip codes is used to extract/create the correct hypothesis. The general idea is to extract the best line from each ABL hypothesis and then to compare the best guesses associated with all the images. Given an ABL image, the best line is found using syntax rules and comparison with the lexicon elements. The left and right components of the zip are analyzed separately. With respect to the real case of three digits, each located component is assumed to contain one extra digit or to miss one, due to segmentation failures. The match with the lexicon codes is executed allowing the removal or adding of one character for each component when necessary. Hence each component is formed either by two or three or four adjacent characters. The component is generated by means of character segmentation hypotheses. Also substitutions of ASCII values is allowed among the list of values associated with each character. Only one mistake per component can be corrected during the lexicon match.

The best match among the text lines of one ABL hypothesis is computed by means of cost functions based on the presence of the found code into the lexicon, the geometric position of the text line within the address block, the level of syntax match, and the rank of the ASCII values for each character. The final best guess is extracted among the best matches of the three ABL hypotheses by considering also the score provided by the preprocessing phase on the images.

Note that constrained flats are processed easier and faster than unconstrained ones and their encoding process is correct if the user has correctly filled up the boxed field. For flats detected as constrained, the zip extraction is executed by the Reader Engine module, while for flats detected as unconstrained, the zip is processed together with other text lines, hence it is isolated by the post-processing phase.

3. EXPERIMENTAL RESULTS

The Korean ICR system has been tested on a test deck of about 1000 Korean real flats, all of them containing the six-digit zip code (3). The set is composed of about 62% of hand-printed and 38% of typewritten addresses. The general quality of the flats is not good, being the flat surface ruined by wrinkles and scratches or colored with high contrast. The constrained flats are usually associated with customers providing large amount of mails. There are constrained samples where the zip code is written out of the six preprinted boxes. Also in same samples the boxes do not satisfy the predefined dimensions. The hand-printed digits may be very different in shape from the Western ones (e.g., digit 2 or 5 or 9). As a consequence, the character classification process has been properly tuned.

Experimental results show that the zip code is often located in the portion produced by the preprocessing phase that corresponds to the right-bottom image corner, rather than in the standard address block. The box elimination performs well, although it fails when the boxes are very corrupted. In very noisy images, the character segmentation step produces erroneous character hypotheses (either a character is missing or fragmented into parts). On the right-bottom example depicted in Figure 1 the system extracts the zip code from the straightforward hypothesis of the preprocessing phase (i.e., the right-bottom part of the input image) analyzing several text lines (see Figure 2). In the example depicted in Figure 3 the second hypothesis of the preprocessing phase (i.e., an address block) provides the zip code (see Figure 4).

In the left-top example of Figure 1 the zip code is constrained and extracted by means of the boxes location (see Figure 5). Note that the preprocessing phase eliminates some noise on the image. The results of the boxes removal as well as of the segmentation process are shown in the top right frame of Figure 5. Another example of constrained zip code is depicted in Figure 6. In spite of the noisy background, the ICR system can locate the six boxes and recognize the zip code.

The recognition rate of the Korean ICR system on the selected test deck is about 84% on typewritten mail pieces and 50% on hand-printed ones. The error rate is about 2% and 3% respectively (3).

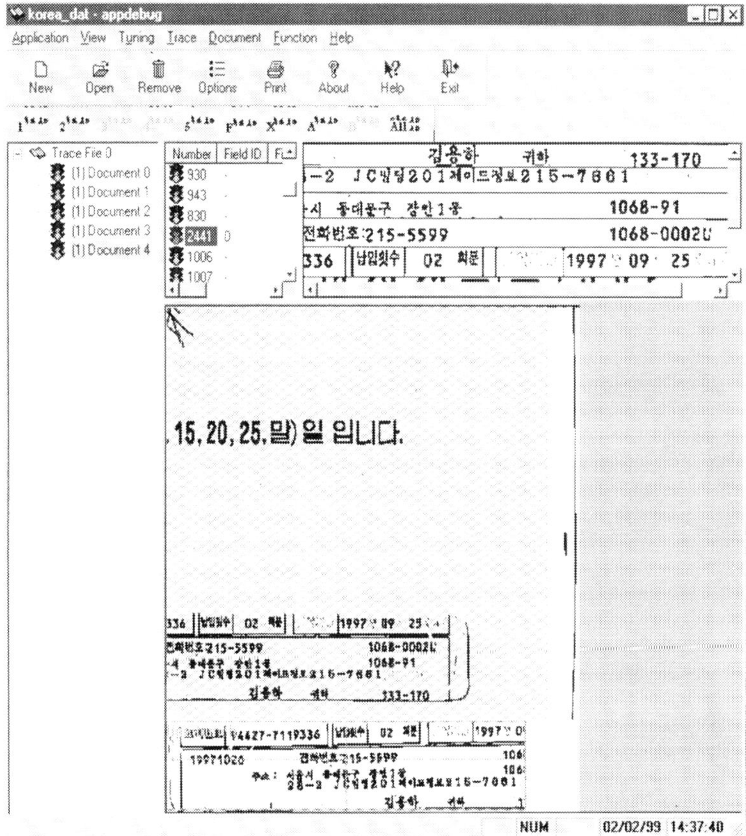

Figure 2: Partial results of the zip code extraction on the right-bottom example depicted in Figure 1. The gray-patterned area represents the surface of the flat due to the first preprocessing hypothesis. The upper frame represents the extracted text lines.

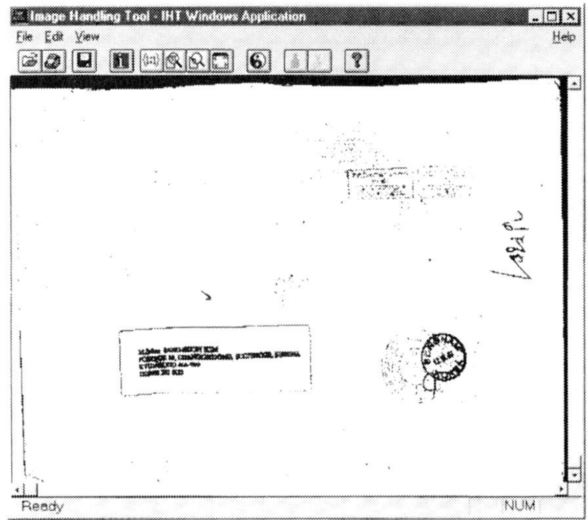

Figure 3: An example of unconstrained typewritten Korean flat, with address printed on label.

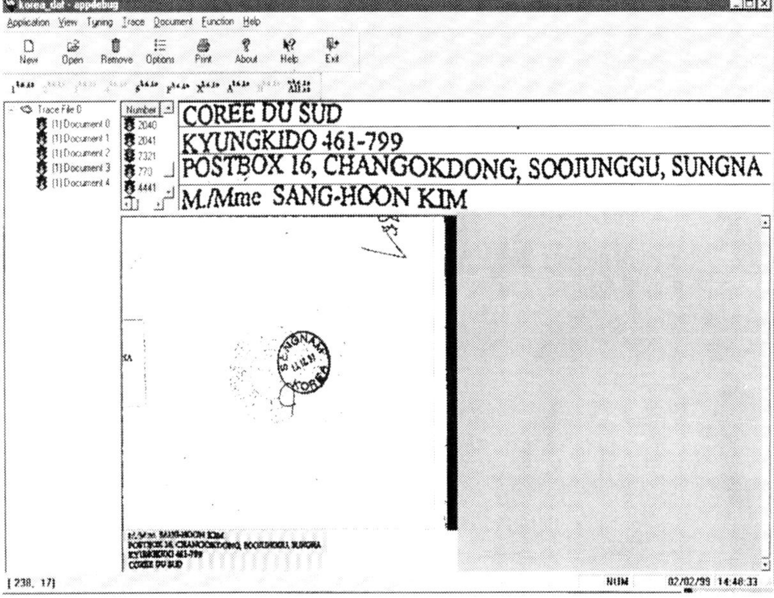

Figure 4: Partial results of the zip code extraction on the example depicted in Figure 3. The gray-patterned area represents the surface of the flat due to the second preprocessing hypothesis. The upper frame represents the extracted text lines.

C568/008/99

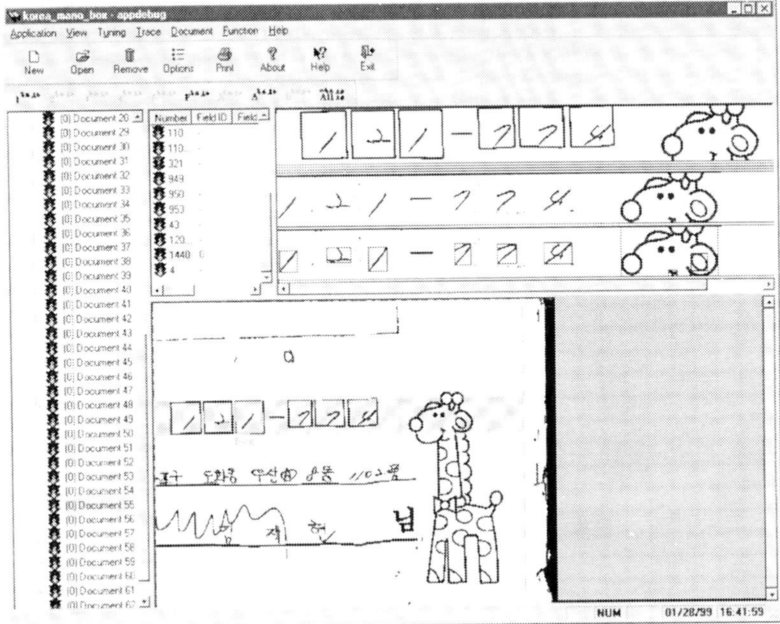

Figure 5: Partial results of the zip code recognition of the left-top example depicted in Figure 1. The gray-patterned area represents the surface of the flat where the boxes have been located. Boxes removal and character segmentation results are shown in the upper frame.

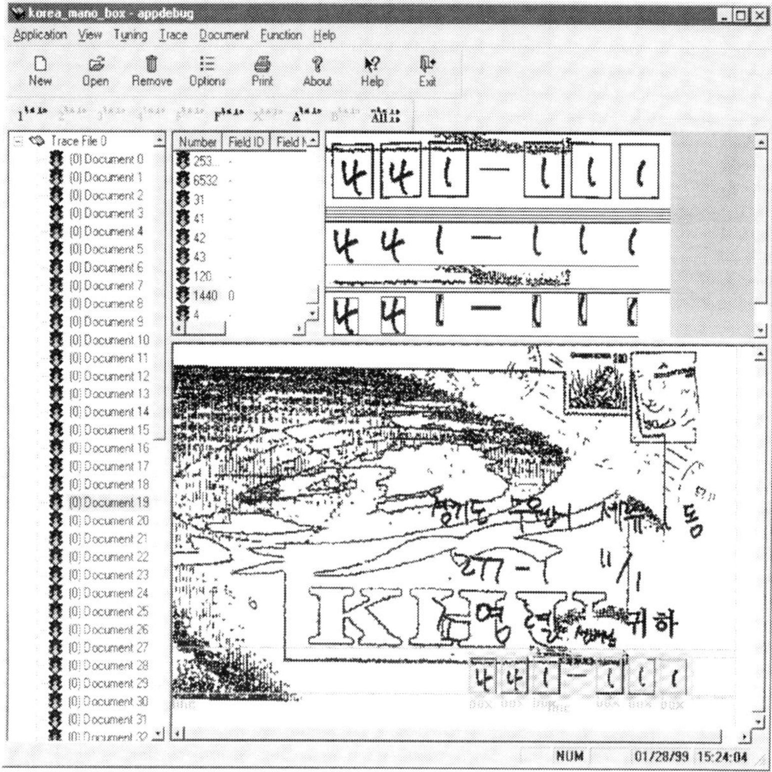

Figure 6: Partial results of the zip code recognition of a constrained hand-printed flat. The gray-patterned area represents the surface of the flat where the boxes have been located. Boxes removal and character segmentation results are shown in the upper frame.

4. INTEGRATION IN A FLAT SORTING MACHINE

The described recognition system for Korean addresses has been integrated into the general-purpose Elsag coding engine architecture, namely POE^2 (Postal Object Encoding Engine). POE^2 is a software system able to code any postal object, such as letters, flats and parcels. POE^2 is flexible with respect to the way a postal object is encoded and it can host several types of coding devices. At present the coding devices are ICRSs (Intelligent Character Recognition Systems) and VCSs (Video Coding Systems). POE^2 is used in the Elsag sorting machines to handle a postal object from the moment the object enters the feeder until it reaches an exit tray. Both an ICR and a VCS use an object image to decode it. POE^2 has been designed so that a single VCS handles images rejected from different ICRSs.

Each sorting machine is equipped with several components according to the client's specifications. The Korean sorting machines are CFSMs (Compact Flat Sorting Machines) composed by one or two CFSs (Compact Flat Sorters), two AIs (Automatic Infeeders), two MIs (Manual Infeeders), one POE^2 with four videocoding stations, together with supervision systems (4). The Korean flat ICR system operates on the image acquired after the flat has been received by the acquisition system. The preprocessing component can decode the barcode already printed by another sorting site or the flat is encoded via zip recognition. If the ICR system fails, an id tag is associated both with the physical object and the image. The image is stored in a database and submitted to a videocoding station. The VCS can work both on-line, i.e., it encodes the flat before it exits the sorting machine, or off-line, where the flat is stored in a reject tray and in a following and independent step it is submitted to the AI again.
In 1999, Elsag spa will delivery nine mail sorting machines to the Korean Ministry of Information and Communication (MIC) for flat sorting, equipped with the described ICR system. Five machines are composed by one CFSs and four machines with two CFSs. The required speed rate is about 18000 flats/hour, while the required recognition rate is about 80% for typewritten addresses and about 70% for hand-printed addresses.

5. CONCLUSIONS

We presented the ICR system developed for Korean addresses interpretation for flat sorting applications. The ICR system is based on a general-purpose core technology and on some *ad hoc* processes. No recognition of Hangul characters is performed. The address interpretation is based on the extraction and understanding of a six-digit zip code. The main peculiarity of the problem is that the address can be randomly located on the flat surface and the zip code text line does not have any fixed position within the address block. The developed ICR system is able to extract the zip code both on constrained flats, where the zip code is contained into six adjacent boxes, and unconstrained flats, where the zip code is extracted analyzing all the text lines of the ABL hypotheses. For constrained zip codes, the zip extraction is straightforward once the boxes have been located. Grammar rules and a lexicon-guided approach are used to understand the zip code. For unconstrained flats, cost functions are used to select the best text line representing the zip code, among the text lines of the ABL hypotheses.
The Korean ICR system is a component of a general-purpose coding engine, namely POE^2, which is the software plugged into the Elsag mail sorting machines used for postal object encoding by means of independent and different encoding tools (e.g., ICR or videocoding systems).
At present, tests on live mail pieces are carried on in order to increase the ICR performance to fulfill the accuracy requirements. The zip code extraction might be increased looking for the special symbol that, according to the specifications, may appear on the left of the zip code. Character segmentation is investigated to improve the zip recognition. Finally, the post-processing phase has to be tuned on large amount of live mail pieces in order to achieve the best average result.

Acknowledgements

The authors are grateful to the anonimous referees and to the R&D Elsag spa group for the helpful suggestions and comments.

REFERENCES

(1) Bruzzone, E., Carossino, A., *EB-Reader Description,* Tech. Report N. RK4.3028.00367, Elsag Bailey, Genova (1997).
(2) Elliman, D.C., *A Review of Segmentation and Contextual Analysis Techniques for Text Recognition* , **Pattern Recognition**, Vol 23, pp. 337-346 (1990).
(3) Furfaro, E., Ghiolo, B., Scaiola P., *The Korean Flat Application*, Tech. Report N. RK4.3748.016.00, Elsag spa, Genova (1998).
(4) Dallorso, P., *Korean Sorting Machine Specifications*, Tech. Report N. RQ4.9717.042.00, Elsag Bailey, Genova (1998).

The role of speech recognition of address coding

T BAYER
Siemens ElectroCom GmbH & Co, Konstanz, Germany

Address reading is one of the central tasks of mail processing. The address data is either automatically extracted via Optical Character Recognition (OCR) and address interpretation or, in case of rejects, manually keyed in by video coding operators. Input by speech is an alternate way of entering data because sufficient hardware resources are now available and speech recognition algorithms have became more robust. Moreover, speech operation does not generally require specific skills. Cumbersome or time-consuming coding tasks can be accelerated by using speech recognition. Inward coding in general and particularly coding for east Asian countries having a non-alphabetic script have been identified as first promising applications. When using speech recognition in addition to a keyboard, estimations show the potentials of this technology: Processing time can be reduced up to 16% under good conditions. A worst case analysis – essentially low recognition accuracy -- however leads to an extension of the overall processing time. However, first applications emerge, such as in-house mail processing where speech recognition will supplement current coding processes.

INTRODUCTION

One of the central tasks of mail-piece processing is address reading. The address information is either automatically extracted by a reader employing image analysis, OCR and address interpretation, or is entered manually by coding personnel if the reader cannot determine the desired information. Thus, the reading work is shared between man and machine (fig. 1).

The requirements for address coding are twofold: the address information must be extracted reliably – the error rate must be low, in general below 2% -- and fast. Automatic and video coding must fulfill both requirements. An automatic address reader has to cope with both requirements; the essential demand for video coding is speed (high throughput) in order to minimize labor costs. Reliability is not negligible, but operators are typically experienced enough to achieve the desired productivity.

In order to meet the speed requirements, coding is organized as video coding in most applications. The image of the mail-piece is transferred to a terminal and the address elements necessary for the coding task are keyed in. Instead of exclusively using keyboards, the presented approach proposes speech input: the corresponding address elements are spoken and a speech recognition system transforms the utterance into a string of address elements from which the sorting information can be inferred.

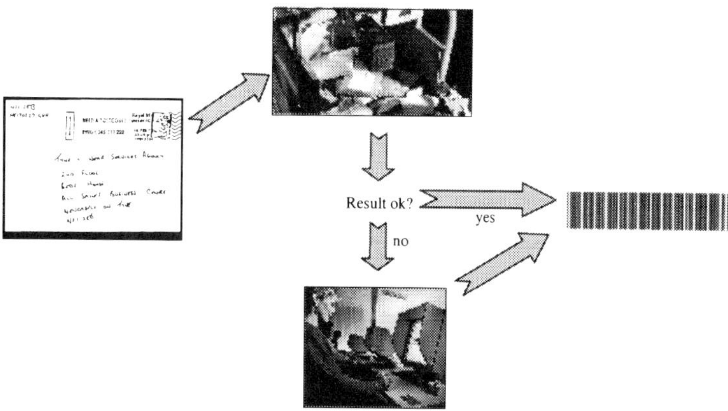

Fig.1: Work share of man and machine for the address coding task

The paper first presents the basic properties of video coding and takes a look at technological issues of speech recognition. Coding scenarios are discussed and the feasibility of speech recognition is shown along with the obstacles must still be circumvented. Finally, results of a preliminary case study are presented.

1. VIDEO CODING

Video coding incorporates the human operator into the process of reading. The common task is coding the destination address in order to sort the mailpiece; other tasks, such as defining the region of interest, particularly in case of flats processing, are not considered here. Two main tasks of sorting and coding can be distinguished, outward and inward sorting.

1.1 Outward Sorting

The code for outward sorting is generally represented by a postal code; digits are standard, sometimes supplemented by letters (like Canada, UK). Based on a five-digit code, the speed of a coding operator is 3000–5000 items per hour using an ergonomically designed keyboard. If the postal code incorporates alphanumeric characters, a standard keyboard is used which, however, reduces data entry speed significantly. Error rates in video coding are less than 1%.

1.2 Inward Sorting

The code for inward sorting includes the postal code and other data, such as the street, P.O. box, apartment number, floor, name, etc. Inward coding may require sophisticated rules depending on specific requirements which must be acquired by operators with intensive training. Thus, inward coding is not as straightforward as outward coding.

Extraction coding was introduced to avoid unnecessary keying and to speed up coding. An extraction code describes a lexical entry, e.g. a street name, by a small and defined subset of characters leading to a significant reduction in the number of the keystrokes needed. Since each extraction code must uniquely represent one lexical item, the extraction code itself is inferred from the operational dictionary. For example, streets might be coded using the first four and the last two characters of the name. Another example for speeding up keying is completion mode; in this case, the string is completed automatically when the character prefix uniquely defines an element of the dictionary. This rule, however, does not prevent the operator from keying the minimum of characters since he or she does not know in advance when the prefix typed becomes unique. Since the size of the dictionary may be large and the contents may change over time, collisions – a code refers to more than one lexical item – cannot be excluded. Therefore, and since contradictions in resolving the inward code may happen, the operator may be forced to further interact with the system, e.g. distinguish between a few alternatives and select the correct one in a pop-up menu.

Time needed for data entry strongly depends on the actual application and ranges between 3 and 10 seconds per item, and is thus significantly greater than is needed for outward coding. The next major difference compared to outward coding is the complexity of the coding rules. Where the rules are simple in the first case, both the coding and extraction rules may be complex for inward coding and require well trained operators.

The rising economic importance of East Asian countries in near future means that concepts for inward coding must be developed for non-alphabetic scripts, such as Chinese or Korean. Outward coding is usually as simple as in western countries because numerals are used. Inward coding keying would be more labor intensive because one grapheme requires a sequence of keystrokes. Additionally, the rewriting rules of a grapheme into a sequence of characters are not common knowledge, but only known by skilled personnel.

2. SPEECH RECOGNITION

Recognizing and understanding human speech by a machine has been a goal in cognitive science and in Artificial Intelligence for several decades. Although a large amount of research effort had been spent in the past, today's speech recognition systems are still far from reaching humans' abilities.

There are still too many obstacles which prevent speech recognition systems from working reliably well. The most important obstacle is the nearly unlimited domain speech has to deal with: The common vocabulary of a language is huge, particularly when considering inflection of verbs and nouns, or specific domains like medicine, engineering, philosophy, etc. Moreover, many lexically different words share the same meaning (synonyms) and many words comprise different meanings (homonyms, like bank). In addition to these language-dependent properties, speech carries specific phonetic and syntactic properties which significantly complicate recognition, such as dialect, pronunciations, speaker-dependent pitch (men vs. women), ungrammatical sentences, noise caused by the speaker (e.g. slight coughing), background noise, etc.

Although the general task of recognizing and understanding speech is far from being solved because of its tremendous complexity, most of these difficulties can be overcome in controlled domains consisting of a limited dictionary and of limited syntactic rules. Moreover, as recognition algorithms have matured and hardware becomes more powerful, the overall performance has been significantly improved in both recognition speed and accuracy. Examples of applications that successfully employ speech recognition include:

> *Control*: several command words or a fixed set of phrases controls specific functions of technical systems.
>
> *Information access (via telephone)*: Speech recognition within telephone service applications are widespread, for instance for telephone banking, or information systems; such systems guide the user by a simple dialogue where, typically, numerals are spoken and recognized.
>
> *Dictation systems:* Text passages are spoken and converted into electronic text.

Current speech recognition systems distinguish between input of a single word (e.g. for control or information access) and continuous speech (e.g. dictation system), and between speaker-dependent or speaker-independent systems. The application scenario determines which kind of recognition is appropriate. Obviously, speaker-dependent, single-word recognizers work more reliably than speaker-independent systems for continuous speech.

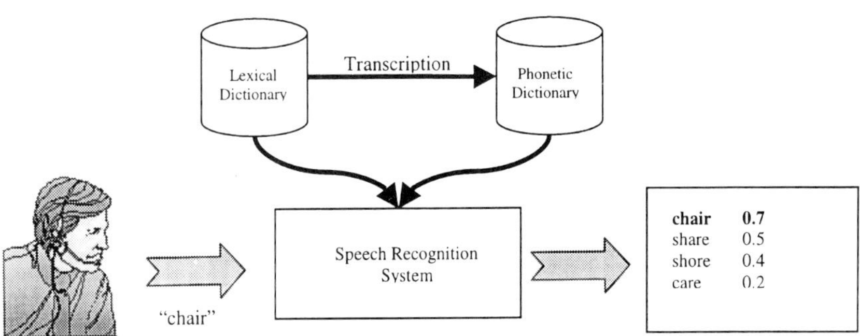

Fig. 2: Speech recognition: Transforming an utterance into a list of hypotheses; the top choice is the most probable result.

Advanced systems rely on a phoneme-based representation of speech and use Hidden Markov Models (HMM) for representation of phonemes and of words. The domain is defined in a sort of dictionary, composed of pairs of a word in lexical form and its corresponding phonetic representation. When recognizing a word utterance, the recognizer matches the phonetic representation of the utterance with all existing models and returns a list of word hypotheses ranked due to decreasing certainty (see fig. 2). In case of continuous speech, language models are integrated, defining transition probabilities between subsequent words. These models enable the system to reliably select the proper candidate from the hypothesis list for each word.

When using speech recognition in a specific application, there are two major aspects to consider:

- The reliability of this technology: How many errors does the system make?
- The language the system has to deal with since a phonetic description of each dictionary entry is required.

Assuming a cooperative operator, reliability essentially depends on the size and composition of the dictionary. If the dictionary contains words that are phonetically similar, they can hardly be disambiguated. Speaker's behavior (pitch, pronunciation, and so on) and background noise cannot be neglected, but these factors can be compensated to a high degree. Also, speaker-independence plays a minor role, because today's systems do not deteriorate too much when accommodating varying speakers. Thus, it depends on the domain in whiche the speech recognizer is employed and in the possibility of using suitable context information for reducing the lexicon (e.g. language models).

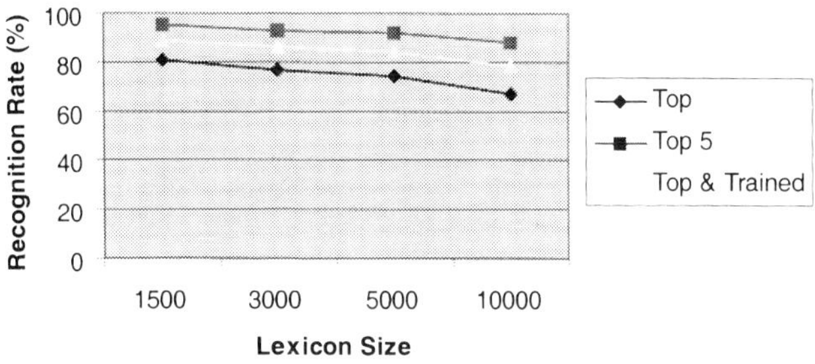

Fig.3: Dependency of dictionary size and recognition rate.

Fig. 3 shows the dependency of recognition rate (mean value of different speakers) from dictionary size of a dedicated speech recognition system. The domain of this experiment was German street names. 1500 examples had been selected for test, and successively 1500, 3000, 5000, and 10.000 entries were used as dictionaries sizes, where each dictionary contained the 1500 test entries. The meaning of *Top* is that only the first choice is considered; *Top 5* considers the first 5 choices; *Top & Trained* is a system which was specifically adapted to the 1500 test words.

The second issue is the language the recognizer has to deal with. Before recognition, each word of the dictionary must be transformed into its corresponding phonetic description. Automatic transcription algorithms (with sufficient precision) exist for certain languages, e.g. German. However, for many languages, including English, algorithms exist, but the result needs careful manual proofreading. Independent from the language, the transcription task may be further complicated in the case of proper names because they may require specialized hand-crafted rules. Consequently, the transcription result of each word must be checked for correctness and occasionally be transcribed, both manually.

3. SCENARIOS OF SPEECH RECOGNITION FOR ADDRESS CODING

After the presentation of current coding properties and the discussion of performance of speech recognition, the question now is how speech recognition can support existing coding processes. We must identify scenarios in which this approach can be used to speed up existing keying procedures.

The essential requirement for video coding is speed. As long as people are employed for video coding, reliability is not relevant because they make very few errors. However, when considering speech input, reliability - basically the error rate - becomes an issue. Speech recognition accelerates standard video coding when the following requirements are met:

(1) Speaking is less time consuming than keying, and
(2) Reliability of speech recognition is high (or speech recognition reliability is low but can significantly be raised by exploiting contextual information) and
(3) Costs of speech recognition (both SW license and engineering effort, e.g. for language adaptation) do not exceed present costs.

Considering speed and the discussion in section 2 and 3, it is obvious that keying the outward postal code is always most efficient. For inward coding, on the other hand, speech recognition indeed represents an alternative since processing time for video coding currently requires several seconds per item. Speaking required address elements like street, P.O. Box, etc., takes approximately one second each. However, looking solely at time for speaking is insufficient because not all words are recognized correctly. Hence, depending on the recognition rate, a certain portion of the mail must be further processed, increasing the total processing time. Therefore, the complete coding process including speech recognition must be examined for checking requirements (1) and (2).

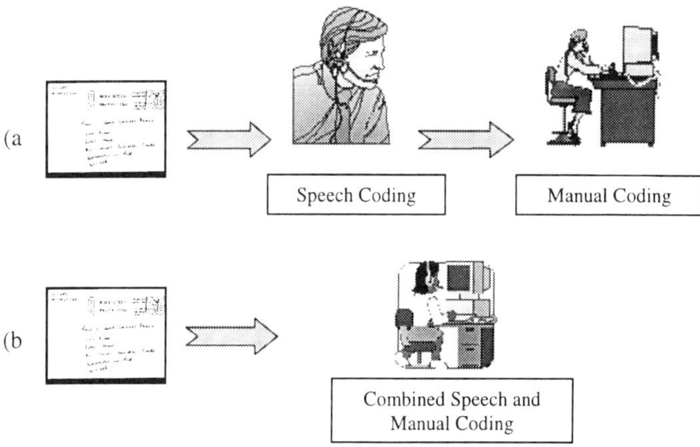

Fig. 4: Two ways of integrating speech recognition into existing coding processes

There are two ways to embed speech recognition into existing coding systems:

- Separate speech and video coding In this scenario (see Fig. 4a), the speech recognition system supplements existing manual coding. After having spoken the text necessary for coding (e.g. street name and house-number), the system considers only the first choice of the alternative list (mode *Top* in Fig. 3). If the certainty is above a predefined threshold, the result is accepted. If it is below that value, the mail-piece is passed to the standard video coding terminal. In this mode, the recognition system must be adapted to produce fewer than 2% errors (comparable to automatic readers). Consequently, the recognition rates of fig. 3, mode *Top*, cannot be reached and drop by approximately 20%.

- Integrated coding (fig. 4b): Here, speech recognition and video coding are combined in one workstation. If the speech recognition system was able to recognize the utterance successfully, the task is done. If it cannot recognize the words, the alternative list comprising about 3 to 6 elements is displayed. If it includes the correct element, the operator speaks or keys the number of this element and the code is determined. In case of conflicts, or if the list does not contain the correct element, the code is keyed as one would in standard video coding.

The basic difference between the two modes is that, in the first scenario, the coding operator only speaks the phrases that are necessary for coding. He or she is not bothered with possible rejects and potential change of input mode. Experience gathered from existing working places for coding proposes that the simple repetition of the same actions as most efficient mode.

Both scenarios help to get a better feeling for potential timesaving in coding. Timesaving is determined by the ratio of mail items that can be completely processed by speech input. In the following, best and worst case assumptions were made in order to infer reasonable values for the first scenario of consecutive speech and manual coding.

For example, the number of all German street names, for instance, exceeds 100,000. From the discussion above, it is obvious that one out of 100,000 entries cannot be recognized with sufficient reliability (far below 50% in *Top* mode). However, additional information is generally available which reduces the number of potential candidates significantly, to within the range of a few hundred. For example, the postal code on the mail piece, either from the automatic reader or from outward coding, is used.

Let's assume that, 95% of the time, the correct postal code is present and the dictionary can be reduced to several thousands or hundreds. The recognition rate of the speech recognition system is estimated to lie between 40% and 60% with 2% maximum error. Then, in best case, it can be expected that up to 57% (0.95 x 0.6) of the mail can be coded in this mode; at worst, around 38% (0.95 x 0.4). The second source to be estimated is the time the operator needs to enter the data. The average processing time for speech coding takes presumably 2 to 3 seconds, and 4 to 5 seconds for video coding on average. In a progressive estimation (2 for speech, 5 for manual keying, 60% recognition rate), the total time of combined speech and video coding for N items requires (1.0 x 2)N + (0.43 x 5)N = 4,2N seconds which normally would take 5N seconds when using video coding alone, achieving a 16% reduction. In a worst case scenario (3 seconds for speech, 4 for manual, coding, with a 40% recognition rate), time is not reduced, but extended, to 5.48 = (1.0 x 3) + (0.62 x 4)) in comparison to 4 seconds when using only video coding. These exemplary calculations show how sensitive the scenarios react when varying recognition reliability and average processing time. Further studies have to constrain the assumptions made to obtain more reliable statements.

For the second scenario, recognition performance is not as predominate as in the first, since the operator is presented a list of hypotheses. From fig. 3 (mode *Top 5*), we know that for even 10.000 entries, the rate lies around 90%, Thus, we anticipate processing more mail using speech than in our first scenario. However, speed improvement may be compensated because of the change in input medium; ergonomics of possible user interactions must therefore be evaluated.

The third and last issue to consider is costs, which consist of license cost for the speech recognition module and engineering costs for adaptation to each application. License costs are assumed to be low in a widespread installation of speech systems. Engineering costs for this task mainly comprise the transformation of words into their corresponding phonetic transcriptions. In general, a standard vocabulary of high quality transcriptions is already available. Since the coverage of the vocabulary of postal applications is probably not very high because of proper names with possibly uncommon pronunciations, missing ones have to be transcribed manually. In the worst cases, all lexicon entries may have to be taught by speaking them several times.

On the other hand, training of operators is negligible when using speech recognition. In the case of isolated words, there are no rules. For phrases, e.g. combinations of street and house number, the only rule is the sequence of items to be spoken.

4. EXPERIMENTAL STUDY

Speech recognition was assessed in the context of in-house mail processing using a proposed sorter for mixed mail (see fig. 5). The task is divided into the automated sorting of incoming mail and of in-house mail, and in the preparation of outgoing mail. The preparation process is not discussed further in this paper.

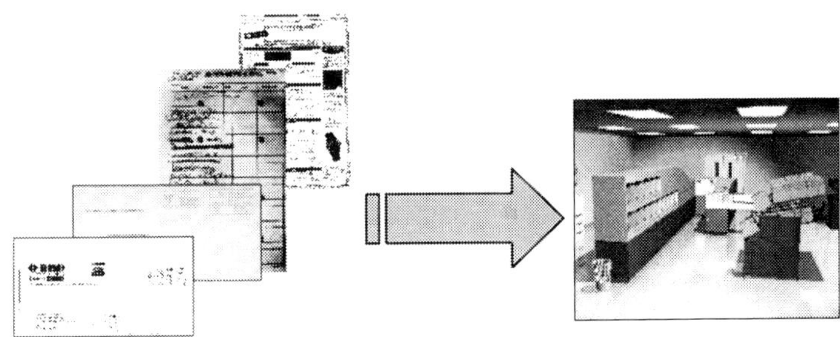

Fig. 5: A proposed mixed mail sorter and the types of mail to be sorted.

Sorting of incoming and in-house mail aims at distributing the mail to the most detailed level possible, preferably to the employee, or more roughly to the employee's department or to the building in which the office is located. Consequently, the data to be extracted for this sorting process consists of three elements, the name, composed of last name and first name, the department name or the mail stop. The process model does not currently incorporate elaborate video coding desks, instead, rejected mail will be hand sorted.

The study assumed a future scenario in which speech recognition is available for coding operators in order to facilitate the manual sorting process. The coding task was not approached in full detail but was reduced to extraction of last and first name of the employee, thus omitting department names and mail stops. We implemented a prototype of a multi-media coding system based on Windows NT. For speech recognition, IBM's ViaVoice was selected (speaker-independent, English language) because it is one of the advanced commercial systems and could readily be integrated into our system environment. We were aware that ViaVoice is really intended for dictation applications; therefore, we did not expect to get best recognition performance for the task of recognizing names.).

Its recognition rate shall be briefly commented upon, particularly with respect to the rates attained by the one of fig. 3: It is reported to reach a recognition rate up to 97% using a lexicon of 40.000 entries. However, this figure must be carefully interpreted. The recognition rate (in *Top* mode) for isolated spoken words is actually much lower than the experimental study shows (see section 5 below). The improvement to rates above 90% can only be gained because an elaborate language model is able to select the proper word candidate in the context of a sentence. Thus, 97% is most likely to be the upper limit that can be reached after several trials. On the average, the recognition rate will be much lower.

The size of the lexicon comprised some 3,400 last names and 1,500 first names; they represented approximately 4,200 distinct combinations of last and first names in the context of a U.S. company. We conducted several experiments to test the recognition performance in relation to the dictionary size, quality of phonetic description, and reduction of the dictionary by assessing preliminary OCR results. The experiments are summarized in the following table:

	Lexicon size 1000	Lexicon size 100
Phonetic description automatically generated	44% / 28%	44% / 8%
Standard phonetic description		69% / 10%
Speaker adaptation		74% / 4%
Context evaluation (reduction to 20 entries in mean)	80% / 0%	

For each experiment, the recognition and error rate is reported. Besides the expected result that the smaller the lexicon, the better the rates, the first interesting observation is that good quality of phonetic description (automatic, manually generated, adapted to speaker) improves the rates in case of the 100-word dictionary. Thus, automatic transcription for the English language has not yet reach a satisfactory quality. Engineering effort, and thus additional costs, is indeed an issue.

The second observation is that exploiting context resources for reducing large dictionaries, like OCR results, further increases recognition performance. Although these values (4th line) are not as well founded as the others because of fewer experiments, they are probably representative. Results show that the dictionary size could be reduced to a very small size without loosing the correct item in most cases.

Hence, furnished with good quality phonetic transcription and the possibility of reducing large dictionaries by assessing OCR results, even the dictation system ViaVoice gained a recognition performance which would be a satisfactory starting point for further process optimisation. Since speech recognition systems are available which clearly outperform ViaVoice in such dedicated applications (system represented in table in fig. 3), we are confident of successfully applying speech recognition in forthcoming coding systems.

5. CONCLUSION

In mail processing, address data is either automatically extracted via OCR and address interpretation or, in case of rejects, entered manually by video coding operators. Currently, the necessary data is keyed in via keyboard. Speech recognition is an alternate way of entering data because adequate hardware resources are available and recognition algorithms have become more robust. The advantages of speech are its natural mode and its easy operation.

The essential requirement for speech recognition is recognition accuracy because the higher the recognition rates, the more mail can be processed with speech input and, hence, the more time can be saved. Dedicated systems currently attain a recognition accuracy of some 80% on a lexicon size of 1000 entries, a result which deteriorates to 70% in case of 10000 entries (German, speaker-independent). Off-the-shelf products, like ViaVoice or Dragon Dictate, are not suitable for the task of address coding as their recognition rate is too low for this highly specialized application.

When matching the power of current speech recognition technology with the requirements of coding, it is evident that outward coding is best handled by manual keying. Inward coding, however, has been identified as an application in which speech recognition can accelerate coding. Currently, several seconds are spent for this task. Using Speech recognition, optimistic estimations show that processing time can be reduced up to about 16% under certain assumptions; a worst case analysis however, leads to an extension of the overall processing time. Further experiments have to examine the scenarios in more detail in order to enable a better assessment of the value of speech recognition. In particular, recognition accuracy as an essential influencing factor must be analyzed.

In addition to recognition accuracy, language engineering tasks have to be taken into account, because it may be necessary to manually improve automatic transcription results for certain languages. East Asian languages with non-alphabetic scripts, in particular, have to be examined. These are promising candidates for inward coding using speech.

Speech technology has reached a level of maturity that allows it to be economically utilised in many applications requiring easy interaction with humans. Postal coding will certainly belong to these applications.

Digital postage mark verification

T BIASI, R A CORDERY, S JOSHI, and **L PINTSOV**
Pitney Bowes Inc, Connecticut, USA

SYNOPSIS

Many Postal Authorities are now faced with the need to implement systems to verify Digital Postage Marks on mail pieces that pass through the postal process. The cost of a verification system must be balanced by the value of the fraud detected or prevented. This paper is concerned with the economics of verification systems designed to detect and produce evidence of fraud. Three types of verification systems are analyzed: manual, distributed automated and centralized automated.

1 INTRODUCTION

Digital Postage Marks (DPM) are part of pre-payment systems for postal services. Pre-payment systems enable very effective revenue collection, particularly for collection mail deposited into street letter boxes [1]. In order to maintain control over revenue collection a pre-payment system necessitates a postage evidence verification system.

With the introduction of the DPM as a payment evidencing method, many Postal Authorities are now faced with the need to implement systems that verify such marks on mail pieces that pass through the postal process. These verification systems must be designed such that their implementation cost is balanced by the value of the fraud detected or prevented.

This paper is concerned with the economics of verification systems designed to detect and produce evidence of fraud. The mail stream under consideration is anonymous deposit, collection mail (i.e. not controlled acceptance mail) imprinted with a DPM by the mailer.

Three types of verification systems are analyzed: manual, distributed automated and centralized automated. The cost to operate each of these systems is modeled as a function of the anticipated mail volume and the optimal sampling rate taken for verification [1].

Numerical values for basic parameters in our model are established based on the postal communication system as it exists in the US. We believe, however, that the overall methodology and our essential conclusions are equally applicable to postal systems in the industrial world, and, with some moderate changes, to postal systems in the developing world.

The cost analysis in this model is based on assumed average values for the parameters used. Our sensitivity analysis indicated that variation within a meaningful range of values for critical parameters does not change any of our conclusions.

It should be emphasized that the cost of verification systems must be considered in the context of the entire payment system, not just as a counterbalance against fraud. A properly implemented verification system not only enables effective detection of fraud but also provides deterrence effects and strengthens confidence within the mailer community. Verification systems should be viewed as necessary mechanisms to enable effective payment systems for all mailers.

2 APPROACH

Before a system can be designed, a determination must be made regarding the purpose of any verification effort. There are two modes of operation for any verification system, 1) detect and reject or 2) detect and evidence. Each mode has different operating assumptions and implementation requirements.

The "detect and reject" approach implies that the verification system finds any mail pieces with invalid DPMs. Once found the mail piece will be outsorted and returned or simply destroyed. In a market segment with low per piece revenue value and a relatively low fraud level this is probably not a viable approach due to the value of most customers to the Postal Authority and the importance of deterrence based on apprehension and prosecution of fraud. In addition, the sampling rate would necessarily be high (and therefore costly) due to the need to find all occurrences.

"Detect and evidence" implies that the system must detect fraud at some acceptable level as determined by cost analysis and produce evidence of fraud to support prosecution. The deterrence effect created through the use of prosecution and a measured, appropriate penalty will prevent most fraud in this mail stream. This is the preferred mode of operation for a verification system since the deterrence effect is proven and works with an adequate detection rate and appropriate penalties imposed. An economic justification for this approach is shown where "the equilibrium value of a game with suboptimal equilibria can *increase* (even dramatically) if the game's payoff function is *decreased* in some appropriate fashion. Our calculation also shows that for minor offenses of this type to be suppressed, the penalty for the offense, multiplied by the probability of apprehension, must exceed the gain obtained by committing the offense." [2]

2.1 Other Assumptions

The fraud rate within the collection mail segment is initially assumed to be 1%. We assume that the DPM is a (highly readable) 2D barcode and that the read rate of any scanning equipment is 100% and therefore include any unreadable DPMs in the fraudulent category.

The total volume of the collection mail segment in the US is about 20 billion pieces per year.

Certain legal aspects of evidence collection are essentially forensic and manual in nature and will not be considered. For example, discrimination between legitimate and duplicate indicia or between natural and deliberate causes of unreadability. The activities involved in this determination are not included in the cost calculations. It is entirely within our methodology to account for such costs, but due to the secondary effect of these costs and space limitations we do not address these in the present paper.

3　THE VERIFICATION PROCESS

There are two parts to the verification process: detection and evidencing. In the detection phase the physical mail piece is scanned, the image is parsed (to find the DPM) and the DPM is analyzed. This analysis interprets the DPM data and for a failed or suspect piece generates an output file with information about the piece.

In the evidencing phase, a set of data is collected that can be used to convince a 3rd party (i.e. a judge) that certain fraudulent activities took place during the process of creating mail. The information in the output file created during the detection phase is used for this purpose. This information may include the name, address and ID of the mailer, the monetary value of the piece, formatting, font and address block location in addition to some forensic information. It is possible at this point that the mail piece image and output file may be used as a primary tool for detection rather than the physical mail piece, especially for suspect pieces that have already been delivered to the recipient.

Depending on the type of fraud suspected different amounts of information might be necessary. There are three types of fraud anticipated for DPM mail.

The DPM may be <u>counterfeit</u>. This type of fraud is evident from an inconsistency between information within the DPM and other mail piece information (i.e. destination address, date, meter serial number, etc.)

The DPM may be a <u>duplicate</u> of one already submitted. If two or more identical DPMs are found there is evidence of this type of fraud. There is then a need to determine which is the legitimate DPM and which is the duplicate. A determination of whether there is more than one mailer involved is also necessary.

The DPM may be <u>unreadable</u>. This type of failure may not be fraudulent but some investigation is necessary to determine if there is a natural cause for the unreadability.

4　SYSTEM ARCHITECTURE

4.1　Postal Mail Processing Environment

For the purpose of this analysis a simple postal mail processing environment is assumed as described below. (See Figure 1.)

At the induction or origination postal processing site, mail is processed through the facer/canceller (F/C) and then moved to the Optical Character Recognition system (OCR) for address recognition and barcoding. (Note that this step may include Remote Video Encoding.) Once barcoded the mail is sorted (Origination Barcode Sorter) and transported to a destination postal facility. At the destination site, the mail is automatically sorted (Destination Barcode Sorter) and moved to the delivery office where it is manually sorted and prepared for delivery.

4.2 Data Center Environment

In all architecture alternatives, a remote data center is connected to various vendor Key
Management System (KMS) servers and a postal Certificate Authority (CA) server. This
center is responsible for performing certain off-line functions. These functions may include
DPM post-processing, key updates and certification. DPM post-processing involves analysis
of duplicate DPM information and compilation and maintenance of suspect lists. Key
management involves storage of keys, updates and maintenance of revocation lists. The
certification is mainly to provide authenticated public keys for DPM verification and
revocation.

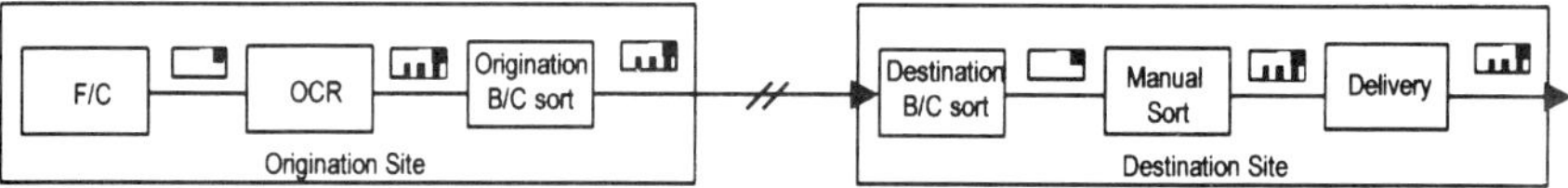

Postal Mail Processing Environment

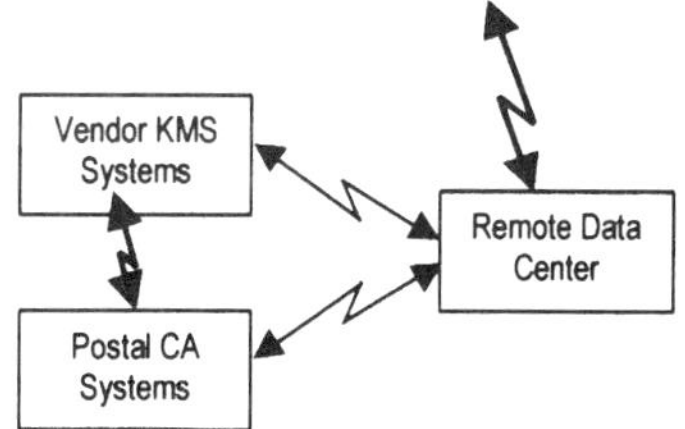

Data Center Environment

Figure 1 - Mail Processing and Data Center Environments

4.3 Where to Detect

The location of verification process relative to the source of the mail is an important economic
consideration since all manual steps associated with the prosecution and legal evidence
collection are more costly if removed from the origination point. There is also a loss of time
in the beginning of the investigation process.

Also, if detection is done at the delivery point, the cost of mail processing has already been
incurred and there will be additional costs associated with returning the mail piece to the
origination point.

4.4 Process Alternatives

4.4.1 Manual

In the manual verification process, the mail stream is sampled manually after facing and
cancelling. Pieces are scanned with a hand scanner that is connected to a laptop computer
with DPM processing software and a database of verification keys. The computer can be
connected to a remote data center and can receive keys, a list of suspect device identities,
revocation list and other information applicable to that origination site updated at regular
intervals.

In this system, there may be several scanning workstations at each origination site (dependent upon the volume of mail to be scanned). Each station will receive a daily download of key management information and suspect information. As each piece is scanned the DPM is processed and either accepted or rejected. Accepted pieces are returned to the mail stream and rejected pieces are manually processed.

Compared to an automated process there are several factors inherent in a manual process that make it less desirable:

- higher per mail piece costs (than an automated process) due to the labor involved

- limited ability to process the required number of sampled pieces in an appropriate timeframe

- 2 orders of magnitude lower productivity

Therefore for the same level of lost revenue control (as a function of sampling rate) there is a higher verification cost. (See Section 5.)

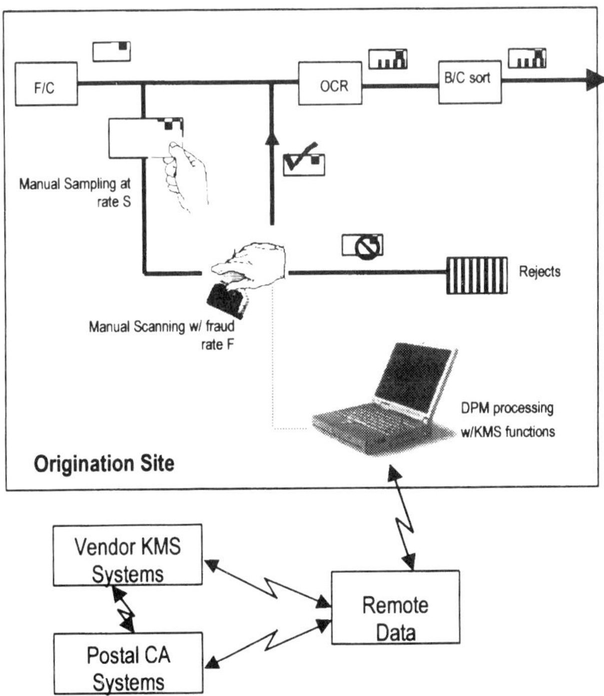

Figure 2 - Manual Verification System

4.4.2   Distributed Automated

In the distributed automated process, the envelope face is scanned as part of the mail sortation process. This image data is also sent to the DPM processing system. Depending on the sample rate selected, a certain number of these images will be checked.

Some DPM processing can be done quickly (non-encrypted information, revocation list, readable/unreadable, etc.) while some take more time (encrypted information, comparison to suspect information). At this point, the time for this analysis is unknown but is assumed to be approximately the same as the OCR address processing time. Therefore, since the time for a mail piece to arrive at the barcode sorter is on the order of hours, we assume that a mail piece DPM can be processed while the piece is still within the origination site. This allows a rejected piece to be processed before it is passed through the rest of the mail process and incurs the costs associated with that processing.

There is one DPM server with KMS hardware per site. Each station will receive a daily download of key management and suspect information from a remote data center as in the manual system. This data is stored in protected fashion locally.

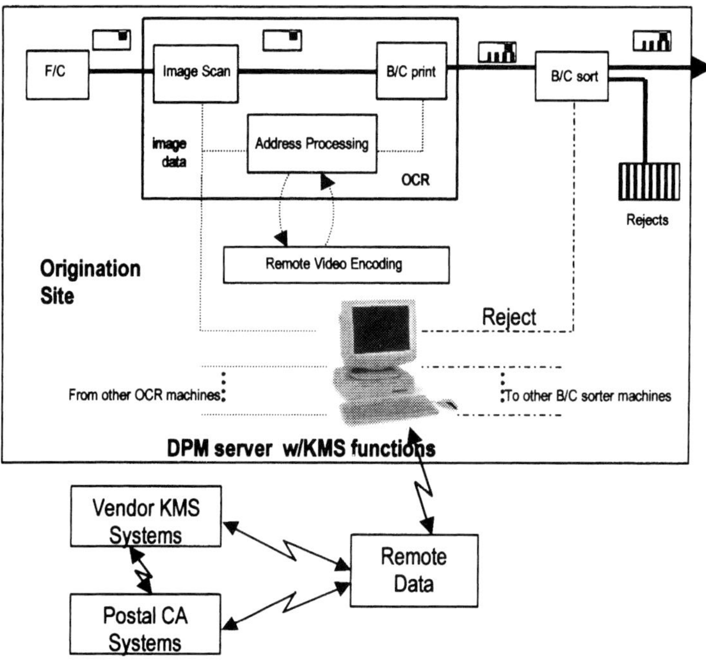

Figure 3 - Distributed Verification System

4.4.3   Centralized Automated

In the centralized automated process, the envelope face is scanned as part of the OCR process. This image data is also sent to a centralized remote DPM processing system. Depending on the sample rate selected, a certain number of these images will be checked.

The off-line DPM processing uses only the mail piece images to identify fraudulent DPMs. Suspect mail identification characteristics are accumulated and downloaded to the OCR systems at the origination sites on a daily basis. The OCRs are capable of rejecting and outsorting mail with the suspect characteristics.

The centralized system will allow some pieces to pass through the system that are later identified as fraudulent. The detection of fraudulent pieces is used to identify suspect mailers so that their mail can be scrutinized further, either by OCR outsorting or increased sampling.

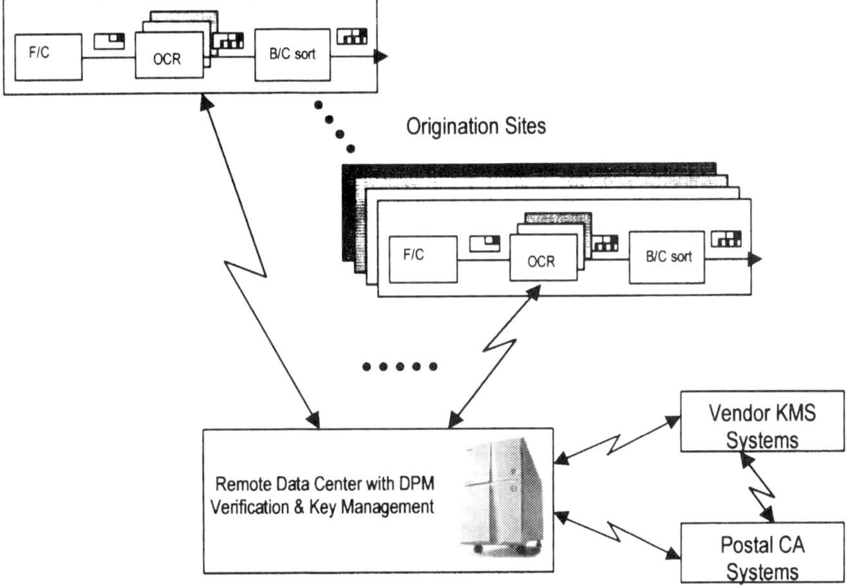

Figure 4 - Centralized Verification System

5 MODEL

Variable	Value or Range	Assumptions
Total annual collection mail volume system wide (V)	20 billion/year	Approximately 20% of total US mail volume is collection mail
Adoption rate for DPM (A)	5% min.-60% max.	Variable parameter
DPM fraud rate (F)	variable	Actual fraud + unreadable DPMs
Mail postage value (P)	$0.33	
Sample rate (S)	Calculated*	As a function of fraud rate [1]
DPM read rate (R)	100%	High quality print of 2D barcode
Depreciation period	10 years	Equipment and initial development costs are depreciated over 10 years.
Equipment maintenance costs	15% of initial equipment cost	All equipment has maintenance charges
Number of processing sites	600	
Number of OCR machines	5000	

In this paper we studied two different methods to achieve the optimal sampling rate. One was based on the assumption that the loss due to fraud follows an exponentially decaying curve. That is, as the sampling rate increases, the loss due to fraud decreases exponentially based on some decay parameter, D. [1] provides a good explanation for this analysis. The other approach was to consider the number of fraudulent mail pieces generated by a mailer until the fraud is detected, using a geometrically distributed random variable. In this model, the probability that the mailer gets caught in a fraudulent activity is a function of the sampling rate. The loss due to fraud is determined based on the fraud rate, the sampling rate, postage value for each mail piece and the number of fraudulent mail pieces that a mailer can get through without being caught. The optimal sampling rate is determined from this curve. The reader is referred to [3] for a detailed analysis. The second method computes the loss due to fraud as inversely proportional to the sampling.

For the first method, let L be the loss due to fraud at no (0%) sampling, and D be the decay parameter of the loss curve. (D is assumed to be 0.015 for this study.) The optimal sampling rate, minimizes the combined cost of verification and loss. The cost function is

$$V \cdot A \cdot S \cdot C_{DO1} + L \, e^{-S/D}$$

Note that the loss due to fraud is assumed to be an exponentially decaying as the sample rate increases. Standard derivative method yields the optimal sampling rate S^*, computed as

$$S^* = -D \ln(V \cdot A \cdot C_{DO1} \cdot D/L)$$

For the second method, let N_M be the number of mailers (all other variables and parameters are the same as in the previous method). The Postal Authority would need to minimize the total cost consisting of operating cost and the cost due to fraud.

$C_{DO1} \cdot V \cdot A \cdot S + (1/S - 1) \cdot F \cdot P \cdot N_M$ is the minimization function. The first term is the verification cost and the second term is the loss due to fraud.

Standard derivative method yields the optimal sampling rate S^*, computed as

$$S^* = \sqrt{(F \cdot P \cdot N_M) / (C_{DO1} \cdot V \cdot A)}$$

Under both of these methods, for models and assumptions we used, the computed optimal sampling rates were in close vicinity of each other (approximately 1%).

5.1 Manual

Variable	Value or Range	Assumptions
Remote data center (RDC) development cost C_{RD1}	$2M	No DPM processing at RDC
RDC equipment cost C_{RE1}	$20K	
RDC operating cost C_{RO1}	$200K/year	
DPM processing equipment development cost C_{DD1}	$300K	
DPM processing equipment cost C_{DE1}	$5K	At least one system per site depending upon volume
DPM operating cost C_{DO1}	=Labor rate / # of	Amount of labor is a function of the

| | pieces sampled
$.246 per piece | volume of mail sampled |
| | $50K/year | Labor rate for DPM operator
Productivity for an operator
sampling mail pieces and scanning
DPMs is 2 pieces /min for 70% of an
8 hr day, working 302 da/yr (365
less 52 Sundays and 11 holidays) |

The combined annual cost of the manual system is:

$$C_M = [(C_{RD1}/10) + (C_{RE1}/10) + C_{RO1}] + [(C_{DD1}/10) + (C_{DE1} \cdot 600/10) + (C_{DO1} \cdot V \cdot A \cdot S)]$$

5.2 Distributed Automated

Variable	Value or Range	Assumptions
Remote data center (RDC) development cost C_{RD2}	$2M	No DPM processing at RDC
RDC equipment cost C_{RE2}	$20K	
RDC operating cost C_{RO2}	$200K/year	
OCR upgrade kit cost C_O	$20K/kit	
DPM server development cost C_{DD2}	$300K	
DPM server equipment cost C_{DE2}	$15K	One system per site
DPM operating cost C_{DO2}	$9K/year	25% of a man year @ $36K

The combined annual cost of the distributed automated system is:

$$C_D = [(C_{RD2}/10)+(C_{RE2}/10)+C_{RO2}] + (C_O \cdot 5000/10) + [(C_{DD2}/10)+(C_{DE2} \cdot 600/10)+(C_{DO2})]$$

Note that C_D is not a function of mail volume.

5.3 Centralized Automated

Variable	Value or Range	Assumptions
Remote data center (RDC) development cost C_{RD3}	$2M	RDC w/ KMS and DPM processing
RDC equipment cost C_{RE3}	$220K	
RDC operating cost C_{RO3}	$500K/year	
OCR upgrade kit cost C_O	$20K	

The combined annual cost of the centralized automated system is:

$$C_C = [(C_{RD3}/10)+(C_{RE3}/10)+C_{RO3}] + (C_O \cdot 5000/10)$$

Note that C_C is not a function of mail volume.

6   COST ANALYSIS

It is worthwhile to remember that absolute values for comparable costs in these system alternatives are valid only in as much as our cost assumptions are valid (Section 5).

Since the cost of the manual verification system is a function of volume and DPM adoption rate, the selection of one of the alternatives on an economic basis will depend upon the assumptions made about these parameters in this market segment. In order to compare alternatives we have calculated the annual costs for each alternative as a function of <u>DPM volume</u> (rather than time or overall volume). These costs are shown graphically in Figure 5.

Based on the cost equations Figure 5 shows that the annual cost for centralized ($35.8M) and distributed ($43.1M) systems remain constant as the sampled volumes increase, but the cost for the manual system increases with volume. Since sample volume is a function of total collection mail volume, DPM adoption rate and sample rate, an increase in any of these factors will increase the manual system cost. Also, since the optimal sampling rate is a function of the fraud rate, higher levels of fraud will result in higher manual system cost. (See Figure 6.)

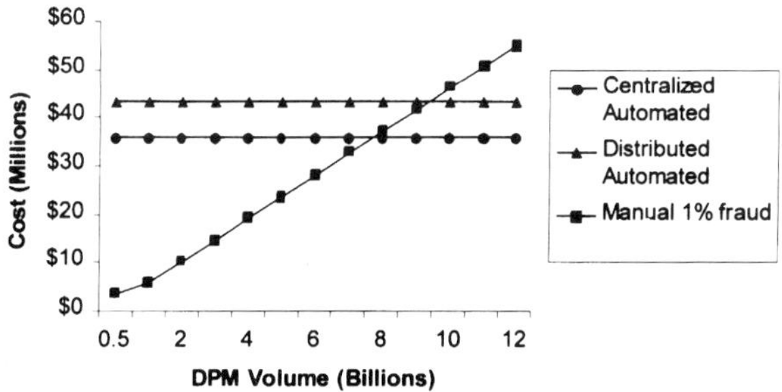

Figure 5 - Verification System Costs (w/ 1% Fraud)

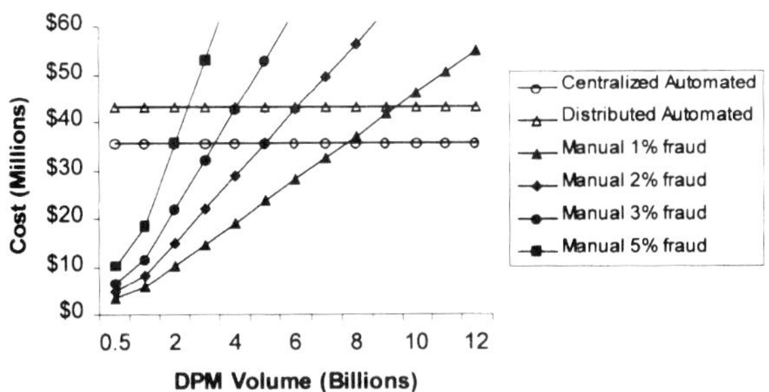

Figure 6 - Verification Costs w/ Variable Fraud

Based on this analysis, the choice of system depends greatly on the DPM volume estimates and the anticipated (or measured) fraud rate. Although the centralized system has a lower annual cost than the distributed system, the qualitative differences described in Section 4.3.3 may argue for the selection of the more expensive system.

This picture changes slightly if only ongoing maintenance and operational costs are considered and not initial development and set-up costs. As shown in Figure 7, the crossover point between manual and distributed systems now moves approximately 20% to the left (7.7 billion pieces/year rather than at 9.7 billion as shown in Figure 5).

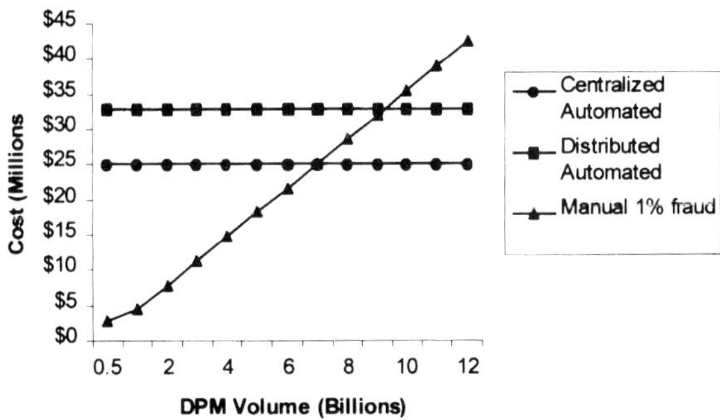

Figure 7 - Verification Costs w/out Setup

Another important consideration is the breakeven point between revenue loss due to fraud and verification system costs. Figure 8 shows that for 1% fraud, the manual system never reaches the breakeven point. The automated systems reach breakeven between 10 and 14 billion pieces per year.

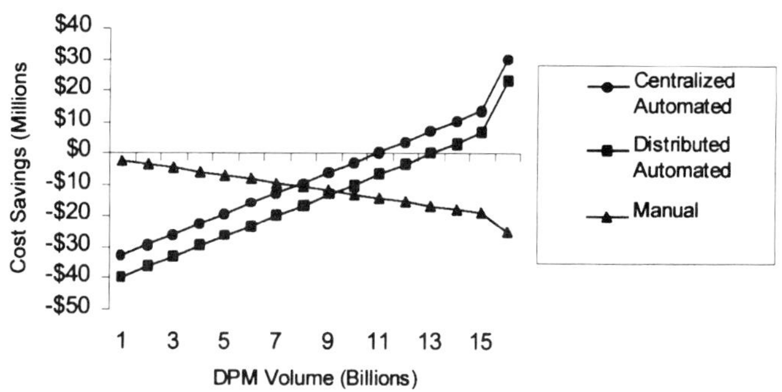

Figure 8 - Cost Savings

7 CONCLUSION

We draw several conclusions from this analysis.

1. A manual verification system is more cost effective at lower DPM volumes. The transition to a centralized automated system should be expected as DPM volume reaches higher levels (approximately 8 billion pieces/year).

2. A sample rate of 1% for this mail stream (i.e. DPM collection mail) is supported by two different analysis methods.

3. The cost savings analysis (figure 8) shows that even at low volumes, the manual system does not recover enough fraudulent revenue to cover the cost of the system. However, an implicit assumption in the cost model is that verification results in the recovery of (only) the revenue associated with the detected fraud. If, however, verification activity (and subsequent prosecution) imposes fines exceeding the value of the loss [2], the cost conclusions will be different.

4. While this analysis shows that the automated systems are independent of volume, it should be noted that we have not considered the labor associated with the investigation of fraudulent pieces (Section 2.1). If the cost for this processing is taken into account there will be some effect on the conclusions.

Although the absolute values computed in this analysis are only as valid as the original cost assumptions (Section 5), we feel that the model and methodology are applicable across a wide range of postal environments.

Our analysis should not be interpreted as a suggestion that verification systems cannot be cost justified based on expected fraud revenue recovery but, on the contrary, should be viewed as an attempt to understand the efficiency of various verification system architectures enabling effective system-wide payment solutions.

8　REFERENCES

[1] L. Pintsov, S. Joshi, T. Biasi, *Economics of Postage Payment and Mailer-Post Interface*, Emerging Competition in Postal and Delivery Services, ed. Michael Crew, Paul R. Kleindorfer, 1998

[2] Mark Kac, Gian-Carlo Rota, Jacob T. Schwartz, *Discrete Thoughts: Essays on Mathematics, Science, and Philosophy*, 1992

[3] R. Cordery and L. Pintsov, *Statistical Analysis of Cryptographic DPM Verification*, Pitney Bowes Inc. Internal Report, 1998

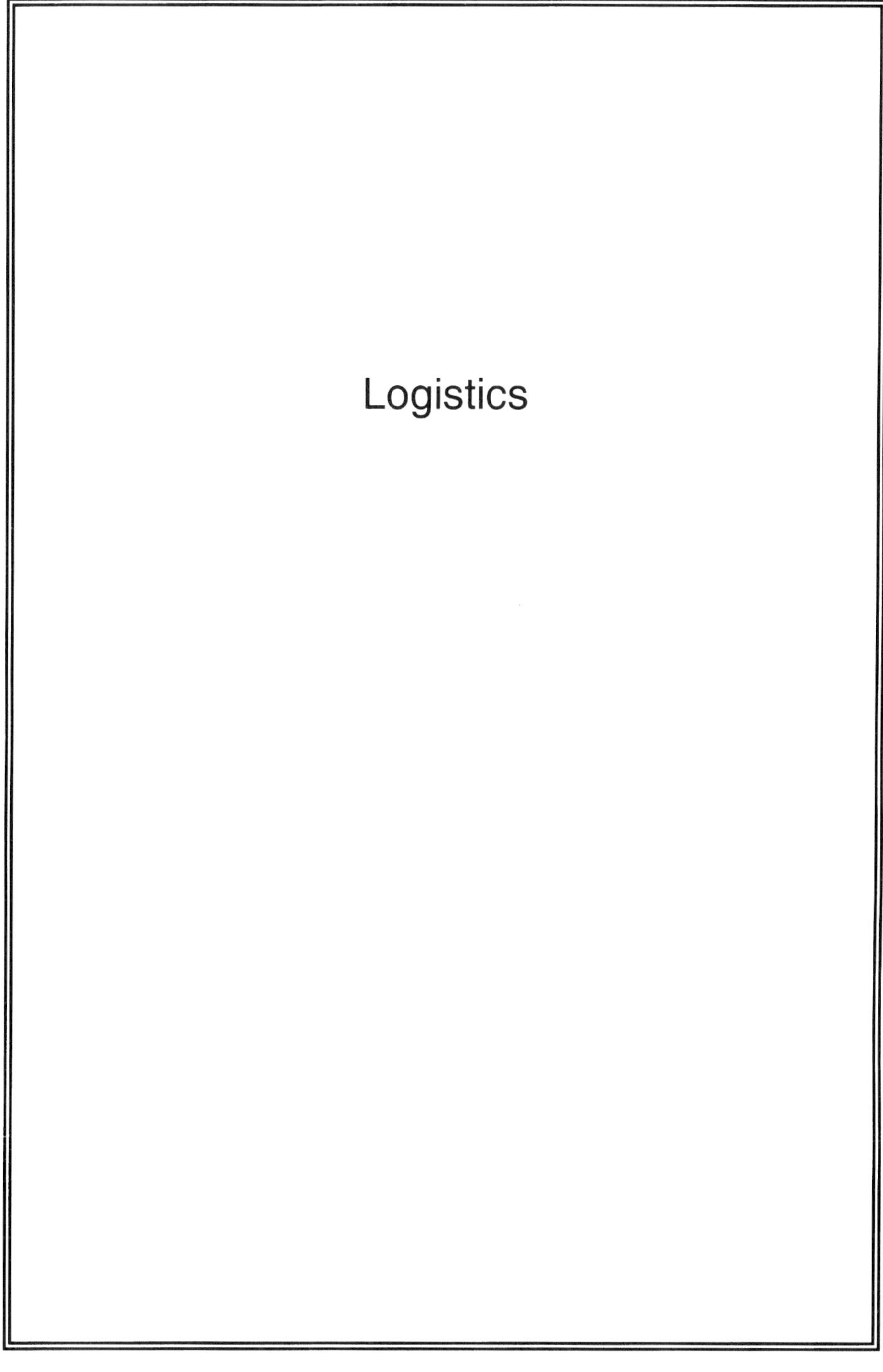

Logistics

Integrated supply chain management in Y2000+ postal companies

M CLASSEN and **D PALDER**
Gemini Consulting, Bad Homburg, Germany

1. REASONS TO REWORK THE POSTAL SUPPLY CHAIN

The competition among postal and logistic companies in Europe is becoming more and more obvious. At least three big players are heading for market leadership: Deutsche Post (including Danzas, DHL and others), Dutch Post (including TNT and others) and Schenker (including BTL and others) while additional actors are claiming their stakes on home markets or focusing on market niches. European CEP-markets are in a significant move.

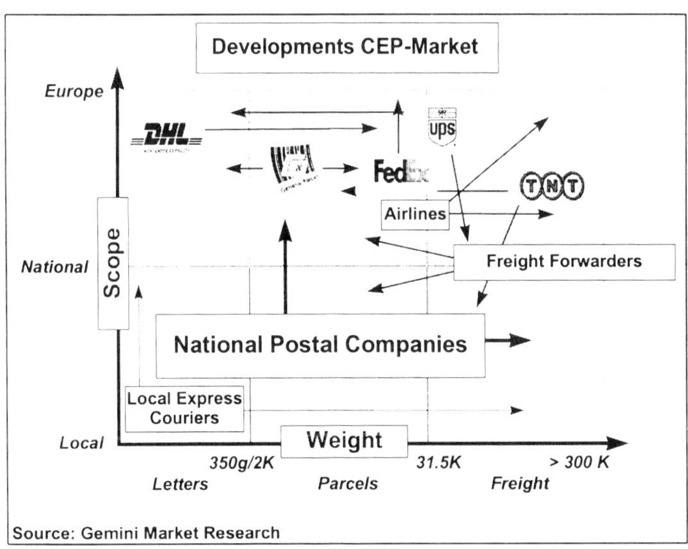

Surviving this complex and dynamic developments will require further process optimisation for offering superior customer and shareholder benefits. By looking at their core process - the supply chain from collection to delivery - postal companies face three major challenges:

1. Increasing separation of activities along the supply chain due to organisational changes which lead to partial optimisation.
2. Increasing need to integrate different networks into an efficient and customer-oriented system due to acquisitions.
3. Increasing activities in telco and logistic markets with an enlarging cross-fertilisation of activities and interrelated IT-networks.

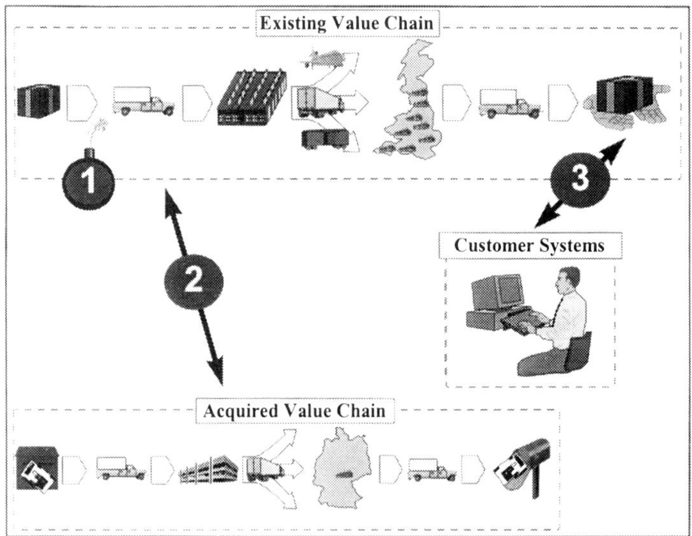

Marketing and production trends such as price sensitivity, logistic speed-up, technical innovation, outsourcing necessities, capacity overshoots require higher effectiveness, efficiency, and efficacy. In addition, parallelisation of product and information flows needs integrated not isolated business processes. Today's competition does not take place between products but between business models.

Whereas yesterday's task was to look more or less separately at organisational subunits tomorrow's challenge will be to optimise in a cross-sectional and cross-functional mode. The approach of solving these and related issues can be summarised as integrated supply chain management (ISCM). Optimisation of isolated business processes is not longer sufficient. Intelligent integration is key. However, the final objective is the profitability and sustainability of the business. Postal companies which apply ISCM have significant higher returns on investment and equity as well as higher sales. Consequently they are better prepared for competition.

The guiding principles of ISCM are threefold: First, direct linkage between business strategy and supply chain. Second, integrated process-oriented planning and management of flows of products, informations, and finances over the whole supply chain. Third, integration of all network partners by breaking down information barriers along the supply chain.

　　　　　　　　　　　　　　　　　　　　　　C568/004 © IMechE 1999

2. INTEGRATED SUPPLY CHAIN MANAGEMENT

2.1 Leveraging Synergies Along the Supply Chain
As a consequence an integrated supply chain management is one of the key challenges of Y2000+ postal companies. This challenge will move along two dimensions: People and systems. On its way it will touch a number of management issues. We will focus on four of them which seem to make a real difference and where learnings from other industries can be demonstrated :

1. Service level agreements between organisational units
2. Established knowledge-management (KM) in the company
3. Interface-focused management information systems (MIS)
4. Integration of IT-Systems between organisational units

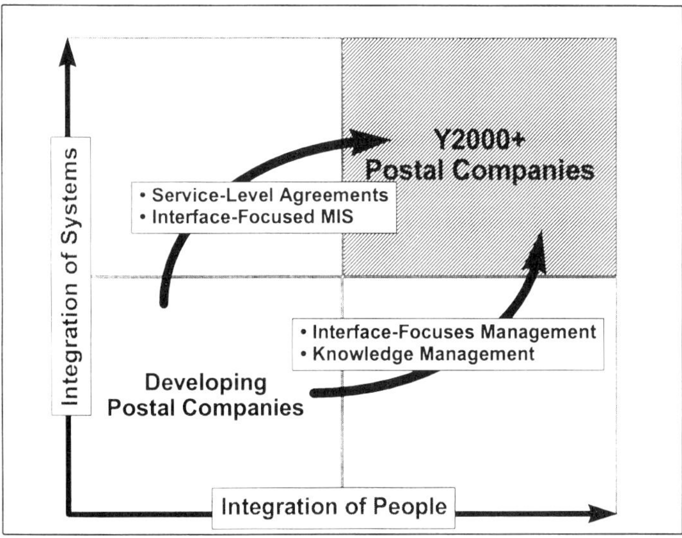

ISCM is ...
- Integrated since it applies a holistic perspective
- Supply-oriented since it addresses customer's demand
- Chain-related since it covers the whole process
- Management since it focuses on long-term profitability

2.2 Service Level Agreements Between Organisational Units
Increasing concentration of a company and its profit centres on their core businesses in combination with decreasing depth of value-adding activities is dramatically changing business culture. Former suppliers and organisational "neighbours" have become system partners. The formal frontiers outside and inside companies are more and more vanished by establishing a dynamic and time-limited network with other entities. Managing these virtual enterprises means not to stare at silos but to optimise processes between independent units. It also means to balance the give-and-take interactions by defining the optimal level between high quality and low costs.

However, classical contracts between the partners with legally formulated details are too inflexible. While formal aspects should be fixed in a skeleton contract a dynamic and flexible kind of agreement is necessary for collaboration, i.e. the service level agreement (SLA).

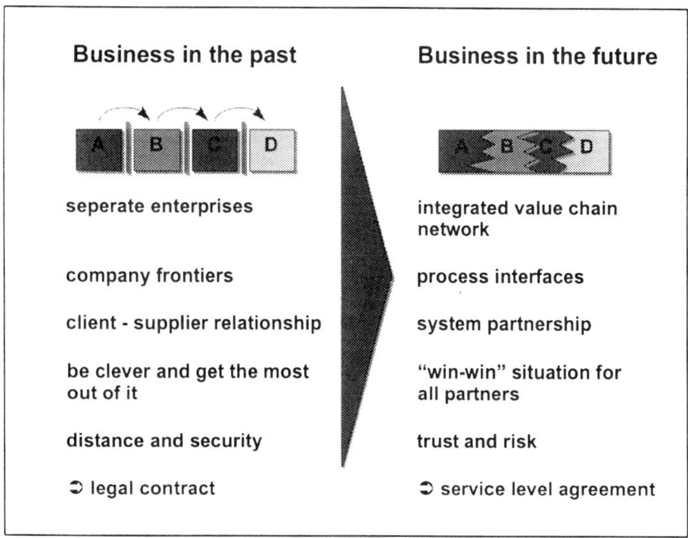

A SLA is an arrangement which defines the parameters of the agreed service components (i.e. quantities, schedule, delivery), the agreed quality level, and the related money transfers. In addition, a SLA fixes the measurable criteria and the feedback processes. They can be arranged between companies or between organisational units.

Even if a SLA often looks similar to a legal contract it differs with respect to its underlying partnering philosophy. Three points have to be taken into consideration in order to create a good SLA:

1. All factors of the agreement should be ambitious (stretched objectives) but realistic (fulfillable objectives). The performance should be controllable and therefore quantifiable and measurable. As a result it comes close to the management-by-objectives idea.

2. In order to support flexibility the SLA-structure should not be too rigid. This means no precise figures but agreed intervals (both in terms of quality and costs). At best this service levels have to be created in a common team with employees of all business partners.

3. Trust, fairness and transparency are key success factors. In contrast to the classical salesman attitude a SLA will create a real "win-win" partnership. This comprises inter-organisational access to all data which is related to the supply chain.

SLA can be seen in many logistic enterprises like Deutsche Post with Neckermann and General Parcel amongst its partners. They cover more than a simple ramp-to-ramp transport and a related data exchange. However, for the development of a real long-term partnership the entities have to be muutually adapted (i.e. staff transition). Roles and responsibilities have to be clearly defined. The management of the interfaces is the senior executives business. Otherwise the big picture often gets lost.

2.3 Interface-focused Management Information Systems
A fundamental need for effective ISCM is an IT infrastructure which guarantees two things: First, continuous flow of information to and from all network units (inside and between organisational units). Second, a management information system (MIS) which delivers valued data for managers and not extensive figures for "number crunchers".

The first issue is a tough one in most enterprises. As a result of historical island solutions heterogeneous and individualised IT-systems hinder cross-unit and cross-company information exchange. This leads to inconsistency and redundancy as well as inflexibility and inefficiency. Huge efforts must be put into the development of middleware solutions, internet connections and standardised interfaces (i.e. STEP, EDI, ALE) allowing information flow over the whole network.

The second issue is not an easier one. It touches the question of managing the links along the supply chain. Often the chain elements are optimised while links in between are ignored. The above mentioned SLA is only a part of the solution.

Over the last years ERP-Systems (Enterprise Resource Planning) like SAP and others have become the backbone of ISCM solutions. However, the next generation can be seen at the horizon: APS (Advanced Planning System). It will allow to optimise the supply chain by creating an integrated model for strategy, sales, production and distribution planning as well as order disposition. All processes are real time and in case that unexpected things are happening alternative scenarios can easily be checked.

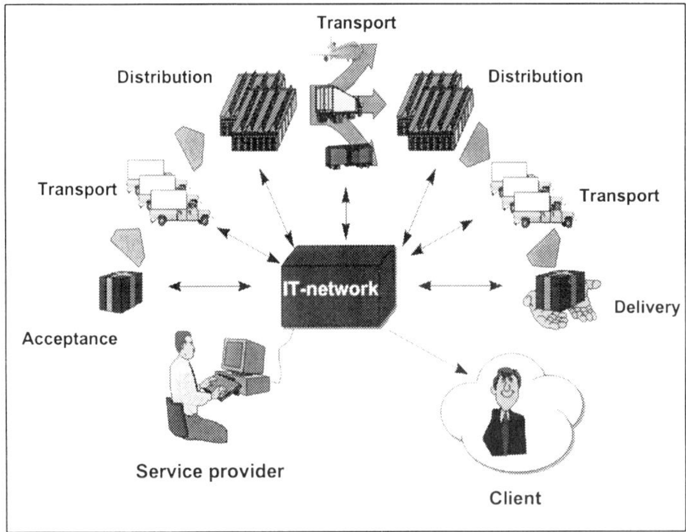

Examples for an ISCM with a full leverage of IT-systems cannot be found inside the postal industry. Interested managers should study for instance the processes at the retailer Wal Mart which is known as a virtual enterprise. Although most value adding processes have been outsourced, Wal Mart is still controlling them. The company is running one of the largest data warehouse networks in the world. Any single cash desk in every store is directly connected to the system. All suppliers have direct access to the system, getting continuous information which of there products has been sold and where, allowing them to refill the shelves during the same night. Wal Mart does not maintain any intermediate storage sides and pays suppliers not until the goods have been paid by the end consumer. The benefit of Wal Mart is a higher cash flow due to the substitution of stocks by information and just-in-time delivery. Similar revolutions to the established logistics will take place in the postal industry in the next years.

2.4 Interface-focused Management
An interface-focused MIS is not sufficient. Senior executives have to apply a corresponding "across units" and "cross border" management style. Strict directives and one-eyed solutions are no longer possible. The traditional client-supplier relationship is past. ISCM means to manage all relevant interfaces from a helicopter perspective. However, interests and conflicts

still exist. Solving them at court or at an external arbitration board may lead to clarification but at high costs: The end of the relationship. Today a clearing committee (established by all partners) with decisive competencies in case of disputes is necessary. This leads to a fair balancing of interests and moderation of conflicts in order to establish a real partnership. In addition, clear roles and responsibilities have to be dedicated to individuals. Activities should be at the lowest hierarchical level (principle of subsidiarity). The standard decision making process will be short and simple.

Recently, Michigan State University has investigated success factors for ISCM. They found twelve dimensions which are listed below (in order of importance).

Key success factors to process partnership

1. Employees of the supplier are members of, or participants in, the buyer´s project team

2. Direct cross-functional, inter-company communication

3. Shared education and training

4. Setting up common and linked information systems (e.g. elec-tronic data interchange, e-mail etc.)

5. Co-location of some of the buyer´s and seller´s personnel

6. Share of technology

7. Setting up formal trust development processes

8. Information exchange about customer requirements

9. Information sharing about technology

10. Physical sharing of assets (plants and equipment)

11. Formalised agreements on the sharing of risk and rewards

12. Agreement on the measurement of performance

The basis for all cooperations has to be mutual trust and partnership

Examples for trust and partnership can be found in many industries. Sometimes (as in private life) a co-operation fails. However, leveraging the above-mentioned ground rules and applying them to the concrete situation significantly increases the probability of a true success story with ISCM.

2.5 Establishing Knowledge Management (KM)

Knowledge of employees is seen as *the* differentiating production factor of the future. As a managerial consequence the company has to establish a KM-system where learning becomes normal and efficient. Otherwise the interconnections along and across the supply chain will be overlooked. If isolated in their unit and with no access to information from "abroad" people tend to ignore the whole system. This can be avoided by horizon-widening KM-systems which are easily accessible and open the scene to individual learning needs. An intelligent approach to approach this challenge contains three dimensions. This double S-curve has been proven successful in leading knowledge-firms in the telco and computer industries.

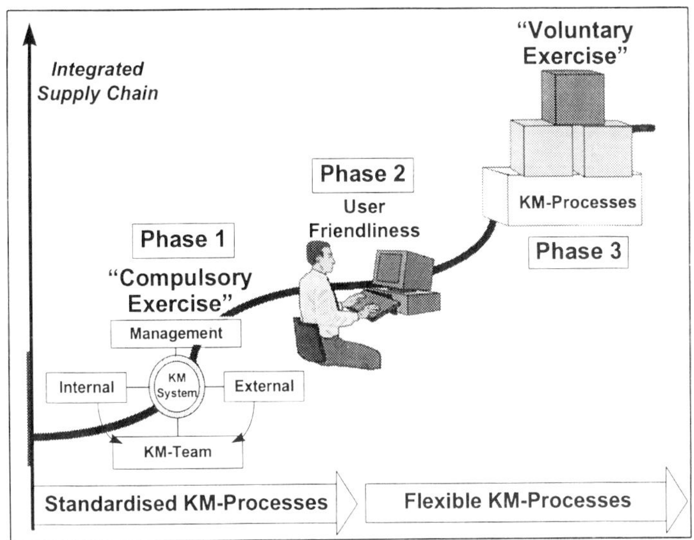

During Phase 1 knowledge must get into the KM-system (input-dimension). This means to focus the content on strategically selected issues. It also means to create real value by capturing experiences from the past, by integrating research work, by filtering superb from useless input, and by updating with new insights. It finally means to integrate the information by data warehousing and intranet groupware. It is a compulsory exercise and will result in a standardised KM-process. Many managers think that this is enough - it is not.

Unused knowledge is useless. In Phase 2 the user-friendliness of KM-systems has to be improved significantly. It mainly means to comfort access and to ease applications. This is heavily needed in most existing KM-systems. Here we touch very practical aspects like keyword lists, file linking and browser systems. Computer nitty-gritty is a time-consuming detail, but user-friendliness is key and any significant improvement pays off twofold.

However, knowledge has to be applied in an intelligent manner (output-dimension). The second S-curve adds implicit knowledge and "high-touch" among individuals to the KM-system (Phase 3). This gives the answer to learning needs and learning routines of individuals. They differ significantly and must be addressed to really add value through learning out of a system. Here most of the leading companies in other industries are still in an experimental stage. It is a voluntary exercise and will result in flexible processes. After all this opens the door largely to establish sustainable competitive advantages.

3. MANAGEMENT IMPLICATIONS FOR POSTAL COMPANIES

ISCM is not an objective in itself - it is an enabler. With ISCM a Y2000+ postal company can establish superior processes and market superior products. The final objective is the profitability and sustainability of the business.

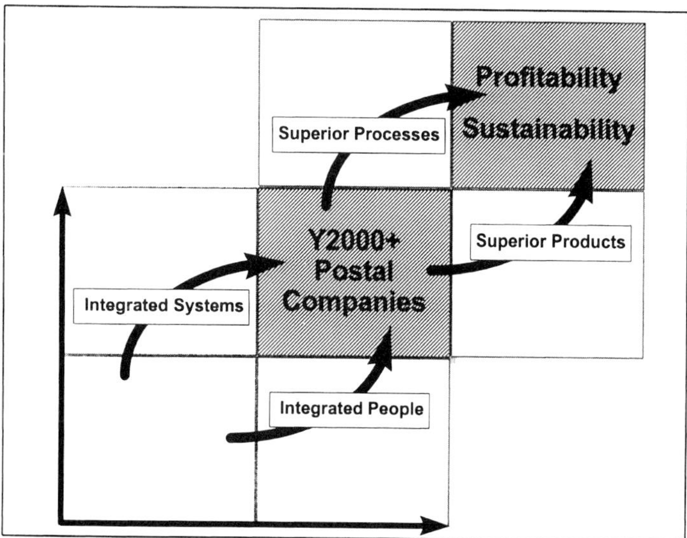

With ISCM urgent business challenges can be addressed, such as:

❶ Synchronisation of product introduction processes (i.e. postal, logistics and e-business).
❷ Realistic pricing of new service offerings (i.e. "mezzo" and "mini" contract logistics).
❸ Client relationship management (i.e. database management of addresses and response marketing with full service offering like call centres).

This will be the game among Y2000+ postal companies.

C568/045/99

The influence of information technologies in delivering successful mail organization for the twenty-first century

R F MENDHAM
CMG UK plc, London, UK

1.   SYNOPSIS

The globalisation of business has dramatically increased the volume of items carried by the international mail organisations. Components can be sourced from and end products delivered to virtually anywhere in world and the mail organisations have developed new services to meet the demand. Competition between the global courier companies and the international arms of traditional national carriers is intense and one of the key factors for success is information. This paper examines a number of the factors influencing the development of successful mail organisations for the 21st century with particular emphasis on technologies for information.

2.   INTRODUCTION

Goods and messages have been transported between destinations from the earliest days of civilisation and pace of change in how this business is conducted has accelerated dramatically over recent years. The 'Information Age' has introduced many new technologies and has, in part, enabled the globalisation of business. Whilst information can now flow around the globe at electronic speed, goods still have to be transported by other, significantly slower means and the mail organisations have had to evolve to meet new challenges.

The term 'mail' now covers a very wide range of services from pre-paid post through to specialist items requiring specific handling facilities. Whilst competition in the pre-paid postal services area tends to be restricted there are an increasing number of suppliers for parcels and specialised services and it is this area that will be concentrated on in this paper.

As we move into the new millennium the competitive market place for these mail services is becoming increasingly global and information will one of the key factors in determining which companies thrive, and which fail.

2.1   The Mail Business

In the 150 years since the introduction of the Penny Post there has been no fundamental change to the mail business, it still involves the movement of items from one location to another. The only differences are in the areas of range, scale, and transport mediums and in recent years by the arrival of serious competition. The change in scope and scale of operations has been reflected in the number of different organisations that now operate in the mail commercial market space. Traditional Postal Administrations, which frequently contain separate organisations for market segments; mail, parcels etc, have been joined by major courier companies who provide a total, end to end service, typically concentrating on the more profitable parcels and larger items sectors.

Mail organisations are now multi-faceted service suppliers, capable of meeting demanding delivery requirements between any two points, irrespective of item, distance, geography or transport schedules. The traditional single standard of service has been replaced by an array of different products, from the most basic model of delivery at an unspecified future date through to a variety of value added services covering handling, delivery guarantees and information services. The mail market area is now highly competitive with choice at virtually every level of service and this competition has driven customer expectations

The common business process is one of logistics; the transportation of items from one location to another with varying degrees of additional services being provided at each stage and the overall process can be very complex. The key technology that has influenced this explosive growth in the mail market place is information; obtaining it, managing it and leveraging it to the commercial advantage of the mail operator and the customer.

It is against this background of change and the ever-increasing importance of information, with its underpinning technology, that mail companies have to operate and evolve if they are to remain successful into the 21st century. This paper will consider a number of the factors that are driving change and some of the information based technologies that will deliver success.

2.2   The Commercial Proposition

Whilst achieving a profit is the primary goal for mail organisations, there are wider considerations, particularly for pre-paid letter post where providing an inclusive, common pricing structure results in non-economic remote deliveries being subsidised by more profitable urban deliveries. To maintain profitability in the current market place is a complex target and needs to involve year on year growth. There are two major factors in achieving improved profitability over time; improvement in internal performance and increased volume. In common with all modern commercial organisations the mail companies need to refine their processes, delivering the business in the most cost-efficient manner. The mail market may be growing rapidly but competitive pressures on margins mean that increased volumes are necessary to maintain revenue and profit levels.

3. CHALLENGES FOR THE FUTURE

From the customers perspective the market place for parcels and items requiring special handling is strong, with a number of different suppliers capable of meeting their requirements and the commercial proposition has to be right. The customer's expectation is for a unique service, delivering items to the desired destination in accordance with his needs for priority, security, handling conditions and cost. Mail organisations are faced with a series of challenges in meeting these expectations.

3.1   Challenge 1 - Competition

The international package and specialised mail business is increasingly global with a very small number of major players providing a full range of services and owning their logistic chains from end to end.   This places them in a very powerful position and, from an information perspective they can deliver their business through a single, integrated information architecture with common standards based on a single data model being used throughout the process. This gives them a potentially powerful competitive advantage, but the downside of this apparent strength is the lack of flexibility inherent in very large organisations.   Other mail companies deliver their international business through creating a logistics network from a variety of suppliers, internal and external. These components have to be assembled into a coherent structure that appears to the customer to be a single entity and which delivers the required standard of service in a cost efficient manner.

This apparently seamless chain has many internal interfaces and it is here that one of the key information challenges is faced. There is no common data model for mail organisations and to deliver an information rich logistics chain the quantity, quality and structure of the data that is passed at each interface needs to be defined and contractually specified. The absence of such arrangements will lessen the information value of the chain, degrading its overall performance.

3.2   Challenge 2 - Market Segmentation

To compete with the global companies these integrated service suppliers have to deliver competitive advantage to their customers.   This can be in terms of specialisation, the development of niche markets where a highly optimised logistics chain can provide a higher value service than the generic system, or through performance advantage. This has led to market segmentation, with companies, or separate divisions within larger organisations, specialising in specific types of business.

The market place is dynamic and the global companies continue to refine their processes. It is only by continually delivering value-added services that the other companies will remain competitive and hence survive into the next century. This places significant pressures on the companies to be very flexible, responsive to changes in the market place and to be in a process of continuous innovation. The key to this process of rapid evolution is information. Not merely information on what competitors are achieving, but the development of a highly information rich logistics chain, with total visibility of the process.

3.3   Challenge 3 – Supply Chain Complexity

From the customer's perspective the logistic chain has a limited number of features; collection, transit and delivery arrangements, timings, and cost. The overall cost of the process tends to be a function of the other variables and the customer will select the service that meets their business requirement.

From the mail organisation's perspective the logistics process is a complex, multi-component network with a high degree of variability in product, volume and routing. Whilst there is segmentation, with individual companies, or divisions within companies concentrating on specific product lines, there is no single, common logistics process. It is a network of components each of which may involve many different partners, sub-contractors and authorities. The external suppliers may be highly efficient but their processes may be optimised to a different set of criteria to those of the prime contractor. For example, an air freight company balances load factors with schedules to produce the most economic mix. This may not be synchronised with the delivery expectations of the mail company and adds an additional variable into the equation.

With multiple suppliers in the logistic chain there are two major factors that influence the overall efficiency of the logistic chain, their information systems and the process for the handover of consignments between components of the chain. All the major suppliers within the global mail market place use sophisticated information systems as the primary management tool for their logistic chain. The degree of sophistication varies and in the absence of a common data model there are significant information challenges. Identification of individual consignments can be achieved through a limited number of internationally agreed barcoding/ labelling standards but that is not a suitable basis for an end to end, multi-supplier information architecture.

This lack of commonality is particularly relevant at handovers, when consignments transfer between suppliers. Different suppliers, different information systems, different data and different management procedures all add complexity to the overall challenge of managing the logistics chain from end to end.

The challenge is to design a process with the largest number of common components, maximising the efficiencies that can be gained by optimising the process and thereby reducing the unit cost. The key to optimising the process and in delivering the business lies in information, how it is gathered, processed and utilised

The ideal situation is for total visibility of information at every stage of the process. The reality is that only partial visibility is possible, and necessary, but this requires a clear understanding of the business drivers and the underlying information needs.

3.4   Challenge 4 - Legacy systems
In common with many commercial sectors the development of information systems in the mail business area has followed a stovepipe approach. Individual components of the business were analysed and information systems developed and installed to meet that specific component's requirements. The holistic view was rarely, if ever taken. Frequently there is no common approach to information management and as a result the legacy systems tend to exist in isolation.

The established players in the mail market place have to meet the challenge of delivering integrated information architectures to enable the active management of the overall process. They also need to embrace the concepts of e-commerce, using their information systems as the basis for value-added services, order processing, consignment tracking and bill processing. This has the further potential to integrate the high value customers into the suppliers systems, allowing further services to be offered, such as sourcing and storage in transit, and each additional service adds value and promotes an enduring relationship between customer and supplier.

The investment in legacy systems has been significant and whilst they meet specific needs, they are unlikely to fit into an overall, end to end information architecture. Change is inevitable because new companies can implement the latest, inherently flexible, technologies without the costs involved with the management of legacy systems and through this capture vital market share thereby jeopardising the viability of the traditional companies.

The options are to replace the legacy systems or utilise newer technologies to integrate them into the new structure. Either route involves significant investment and may include consideration of outsourcing. This can be the quickest, and potentially the most economic way of moving to the latest technologies but has major implications. Information is the key resource in the management of the mail process and could be considered to be core business, and as such should arguably be retained in house.

Irrespective of which approach is adopted the implementation of integrated information architectures and e-commerce facilities are strong commercial drivers influencing all of the major mail organisations.

4.   MEETING THE CHALLENGES

There are a number of different technical approaches that can be taken to meet these challenges but the start point has to be internal process improvement leading to added value services for customers.

4.1   Delivering Internal Improvements

The requirements of delivering end to end efficiency and having a highly customisable logistic chain need to be balanced. Improvements in one may be at the expense of the other and the overall result could make the value proposition non-competitive. The challenge facing all mail organisations is that improvements have to be delivered in order to survive. The start point is Process Identification, defining every element of the chain from collection to delivery and including administrative processing. This is relatively simple at a high level, but needs decomposing and every additional level adds complexity.

The aim is to optimise the overall process by the individual and collective optimisation of the components. Wherever possible the processes should be automated, utilising intelligent handling equipment that can identify the consignment and apply the appropriate action without human intervention. Internal or external suppliers can provide process components and hard decisions need to be taken over which elements are core business and will continue to be delivered internally. The contractual relationship with external suppliers needs to be precise and covered by service level agreements that cover not just the standard of service but also information requirements.

The ideal process chain would be totally information rich, capable of meeting the requirements of both the customer and the mail organisation. These requirements have differences but there are common information roots to both of them.

4.2   Delivering value added services to customers

Once the components of the logistic process have been refined the elements of the chain can be costed and commercial propositions developed and delivered to customers. The process will be in the form of a network with individual components delivered by one or more suppliers. The overall process could even be delivered by a virtual organisation, where the prime contractor does not own any of the components of the chain and everything is outsourced. The sole role of the service provider would be as an information organisation, owning nothing but managing everything on the basis of information. The virtual mail organisation's business would be the marketing and selling of mail products and managing the delivery of that service.

The customer is faced with multiple choices over levels of service and costs and from the suppliers perspective the business objective has to be to move their products up the food chain; from a basic service to an information rich, added value service. The pressures to deliver the required standard of service are significantly greater, but so to are the margins. The ideal model is an integrated logistics chain that is capable of 'customisation', to deliver a unique service to each customer, but within the envelope of the overall system. Once that has been achieved additional value added services can be offered, such as procurement and sourcing, in transit storage and customer management of delivery schedules.

This approach demands a highly flexible, information rich network of components capable of delivering different levels of end to end service concurrently. Consignments will pass through the network with different priorities and decisions on the allocation of routing and freight allocations will have to be made dynamically.

This delivery of a unique service to the customer has numerous advantages for both parties. For the customer it allows them to view the mail component as an integral element of their own process and for the supplier it provides the basis for an enduring relationship. Shared objectives can be introduced moving the relationship from potentially confrontational to one of partnership.

The key to delivering these improvements is information.

5.   TECHNOLOGIES FOR THE WAY AHEAD

There are many areas where technology has the potential to deliver the goods, in terms of mail organisations fit to flourish in the next century. This paper will consider three that are highly relevant to the information component of the process, auto-ID, smart cards and internet technologies.

5.1   Auto-ID

One of the major challenges faced in managing the end to end mail system is knowing the location of consignments within the overall pipeline. There are obvious points where this information is readily available, typically the start and end points, but the pipeline is a complex integration of a number of discrete components and it is possible for consignments to travel through it under the cloak of anonymity.

Consignments can be identified in a number of ways, including hand written or machine-readable labels and bar codes. These have the ability to identify consignments, but require intervention to achieve identification. Radio frequency identification, RF ID tagging, is a technology that introduces emitting tags that can be read remotely by scanning equipment.

Each tag can have a unique 'signature' and introduces ability generate automatic identity information, auto ID. This has major advantages for the Mail Company. It is a neutral technology, independent of legacy management systems and enables 'track and trace' as well as opening the way for dynamic management of the pipeline. It also enables greater use of automated handling systems within the overall process.

Track and trace is the ability to know where a specific consignment is within the pipeline. Real time tracking can be implemented but in the majority of cases historic tracking, recording when the consignment passed a specific point, is sufficient. The consignment's progress can be tracked, its arrival at the next checkpoint can be predicted and, if necessary it can be traced. These two are complementary because tracking can be a passive management function. When a consignment is passing through the pipeline the tracking system can monitor its progress. As long as the consignment is progressing according to expectations there is no requirement to take any action. The system is performing and the consignment will reach its destination at the desired time. The added value comes when the consignment becomes delayed for any reason. At the next reporting point it will be identified as running behind schedule and two course of action are then open; it can be actively managed to speed it through the pipeline until it is back on schedule or, the customer can be notified of the revised delivery prediction. This active management of the pipeline allows the mail organisation to optimise the performance of the chain dynamically and meet customer's expectations.

Because the location information is generated automatically there are a number of additional advantages that accrue. Performance of the pipeline can be measured by taking information from a cross sample of tracked consignments, and this can then be used as the basis for the international charging regime. Another advantage is that it introduces the conditions necessary for e-commerce to be conducted within the components of the chain, and beyond.

Any chain that involves multiple suppliers generates a series of interfaces and the requirement for information as the basis for bill payment. This administrative overhead consumes resources that are not involved in the core business of mail movement and is an area of costs that needs to be optimised to the lowest possible level. The use of auto ID for consignments provides part of the solution towards automating the administrative function.

5.2   Smart cards

A complementary technology to auto ID is Smart cards, which have the potential to transform the way the overall business is conducted. Smart cards include a memory chip with the ability to hold dynamic information. This dynamic memory can be used to hold information such as 'electronic money', as demonstrate in the Mondex trial, but an area of great potential is in holding identity information. Security algorithms can embedded in the chips, providing the basis for Smart cards to be used as identity devices or as the key to accessing information and conducting business electronically.

The customer can use the Smart card as a secure entry device into the mail organisations information systems, enabling many new ways of conducting business. Orders for services can be placed on line, reducing the volume of paperwork and costly manual data entry and processing for both parties. Consignments can be handled electronically from end to end in the process, including scheduling, tracking, delivery arrangements and bill processing. This generates economies for all parties involved and has the added advantage of integrating the mail process into the customer's overall business. They have the ability to manage an end to end service through to the eventual recipient of their product, inclusive of the distribution process supplied by the mail organisation.

The multi purpose, value added services that can be achieved with smart cards is demonstrated by the Chipper project in Holland. This joint venture between ING Bank and the Dutch PTT has involved the issue of several million smart cards. These are multi-function and contain ID, ticketing and e-cash. They can communicate via public telephones, the internet, point of sales and ATMs. They can be used in a wide variety of vending machines, such as parking and ticketing terminals in public transport environments.

5.3   The Internet

The final component of this future electronic commerce based model for mail organisations is the Internet. The greatest virtues of the internet are its global availability, its low entry costs and approach to standards.   There is no requirement for expensive infrastructure to be installed to enable internet access, a simple telephone can provide the entry point into the global power of the internet. The level of sophistication of the technology behind the telephone line is not a major factor and internet access can be achieved in countries where the overall level of infrastructure development is poor. For mail companies this will enable information to be gathered from places outside the fixed infrastructure of their traditional systems architecture. This includes mobile delivery systems as well as information entry from remote sites. On line updates are equally achievable from a static installation in Plymouth England, a delivery vehicle in Plymouth Massachusetts or a small office in Plymouth Montserrat.

The internet also provides a potential basis for linking various supplies within the overall logistic process. Relatively simple sets of standards underlie the internet and provide the basis for addressing and information presentation. TCP/IP addressing is increasingly used as the basis for information routing between systems and is independent of the technology used on the system. This can provide the electronic conduits for the passage of information between the systems used by the various suppliers involved in the overall mail process. It does not, however overcome the challenge of data integration. There is no common data model in use by mail organisations and there will continue to be a requirement to translate data prior to processing. The advantage of the internet is that it enables easy access to the systems allowing data sets to be synchronised enabling greater visibility of the process irrespective of which system is responsible.

A combination of auto ID, Smart card and the internet will generate the conditions for e-commerce and enable the mail organisations to have visibility of the logistics chain from end to end. That is an essential precursor to tuning the process to deliver the integrated logistics chain with the necessary flexibility to deliver a unique service to the customer, the standard requirement for the future.

6.   SUMMARY

In order to achieve and maintain competitive advantage in the global market place, organisations must embrace the opportunities offered and created by electronic commerce. In the global market place customers will place their orders with whichever mail organisation offers them the best value proposition, and that could even be on a consignment by consignment basis. Mail organisations will embrace the concepts of e-commerce, exploiting their information systems as the basis for additional value added services; order processing, consignment tracking and delivery management. In this way costs will be reduced for both supplier and customer, paperwork will be reduced, costly manual data entry eliminated and the overall processing load reduced.

The level of business, in terms of goods to be transported will continue to grow, as organisations source items on a global basis and increasingly regard the world as their potential market place. Information will be the key to the future delivery of business and the mail organisations that integrate their customers into their logistics systems that will flourish. With internet technologies breaking the barriers between legacy systems and providing virtually global access to information the customer can be provided with on line tracking information on the progress of his consignment to its destination. This is a high value proposition, which can be provided using low cost technologies and has the added value of effectively integrating the mail function into the customer's overall process. This creates an environment of shared objectives, with mutual business benefits, which is the basis for a sustained relationship between customer and supplier.

Internally the mail organisations will refine their processes, optimising individual elements and creating an overall integrated system, irrespective of whether they own the components or not.

In the increasingly de-regulated and competitive mail market place companies will have to improve internal efficiencies as well as improving their market share – or they will probably not survive. Information is the key to success and those mail organisations that are innovative in their adoption of the latest technologies will be the ones fittest to survive and prosper into the next century.

C568/027/99

Dimensioning the right way: reliable

J SPANDOW
Cargoscan, Oslo, Norway

Dimensioning systems are becoming increasingly popular among parcel carriers and postal services around the world. Goods is becoming lighter and carriers and postal services are increasingly operating with volume based price tariffs as volume, not weight, often is the limiting carrier resource. Carriers using automatic dimensioning for re-checking shipments are experiencing very short paybacks on the investments and the market for measuring devices is growing rapidly. The market for dimensioning devices is relatively young and there are presently no clear standards for how to dimensions are to be measured and what the customer should expect from the equipment. The objective of this paper is to give the reader an overview of the benefits of using automatic dimensioning as well as highlight some issues when implementing automatic dimensioning.

A NEW DIMENSION IN DATA CAPTURE

It is becoming increasingly popular among carriers to include size and volume in their price tariffs instead of pricing according parcel weight only. Considering that carriers operate with two main limiting factors; weight and volume of the parcel load, structuring the price tariffs around both limiting factors make sense. Goods are becoming lighter and containers and trucks are more often full before becoming too heavy. Volume, not weight is increasingly becoming the limited factor. The same is true for warehouses where capacity is given in volume and conveyor systems are more often limited by volume rather than by weight.

By dimensioning the payload and computing the so-called "dimensional weight", carriers can compare dimensional to weight based tariffs. In this way, carriers can make use of the optimal pricing policy and avoid filling their fleets with light freight only to be penalized by conventional weight based revenue charges

Although the first carriers to implement volume based price tariffs were global integrators like DHL, FedEx, TNT and UPS, the concept is now spreading fast to other carriers and postal systems worldwide.

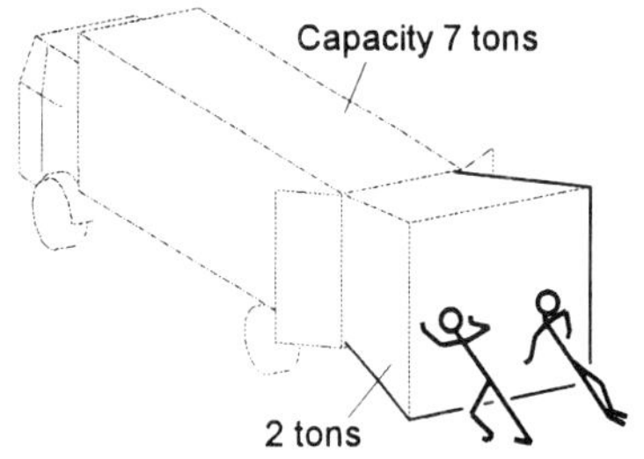

As shipments are becoming lighter, carriers more often load out on volume, not weight. As a result more and more carriers include parcel size in price tariffs.

Revenue Recovery

With pricing based on volume, checking the actual dimensions of the parcels becomes critical for billing purposes. Naturally carriers want to make sure they receive the correct revenue according the size of the shipment.

Prior to the introduction of automatic dimensioning systems, shippers were left to measure and provide goods information to the carriers as dimensioning in the sort stations where too time consuming and often interrupted the materials goods flow. Only seldom would the carrier conduct spot checks to compare actual dimensions of the goods to the data received by the shipper.

Although, shippers usually not deliberately mislead carriers when supplying data about shipments, experience has shown that manual measurements tend to be inaccurate and that the actual dimensions of the shipped goods frequently vary from the standard package measured by the shippers. By using automatic dimensioning systems in the sort stations, the carrier is able to recover revenue that otherwise would have been lost. Often carriers are astonished to see the investment in dimensioning solutions being paid back within in a few months!

Some carrier fear that re-dimensioning the shipments would create a negative reaction among shippers when charging customers for a larger volume or weight than what the shipper believed the shipments to be. Experience shows that this seldom happens. Most shippers view it as fair to pay for the actual space or weight that their goods occupy.

1.2 Improved Load Planning

It is not only billing procedures that are the driving force behind the increased use of dimensioning systems. Collecting volume and weight data also helps carriers in determining a more efficient load planning of their aircraft and truck fleets. The data collected on shipments assist in performing labor trend analysis by gathering statistics on parcel flow related to time of day or night to suitable staff the pick and place operation. Automatic dimensioning systems can therefore tabulate parcels weight to volume ratios in determining aircraft and truck freight loading, enabling a more economical fleet use and fuel consumption.

1.3 Improved Sales and Marketing Decisions

Another important benefit of capturing volume and weight data is the possibility for the carrier to determine a profile on each customer and transport route. Knowing these shipment data, customer requirements, costs and profits are made visible. The carrier can easily identify customer transport requirements and costs and profits of each customer and route. In this way, the carrier is able to tailor make customer specific products, negotiate better agreements, improve services and identify unprofitable accounts.

1.4 The Goods Check In-Station

Capturing the volume and weight data is normally done at a goods check in station, which is often located in the carrier initial receiving terminal. Alternatively the data can be captured in a major transit hub. In the check in station, parcel identity, dimensions and weight is captured. The bar code reader system is needed for linking parcel dimensions and weight to the correct shipment. The equipment is controlled by a data concentrator, which in turn feeds the information to a host computer / local area net work.

Picture 2; The Goods Check-In Station

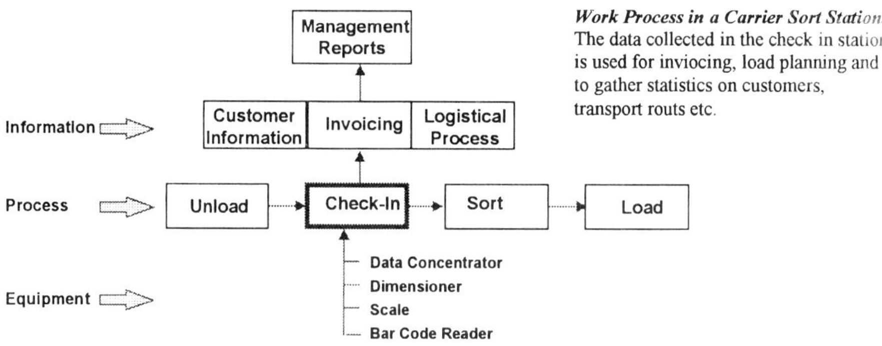

Work Process in a Carrier Sort Station. The data collected in the check in station is used for inviocing, load planning and to gather statistics on customers, transport routs etc.

The data collected is processed in the carrier management information system where it is used to identify discrepancies between dimensions and weight stated by the shipper and the actual dimensions / weight of the parcels shipped. Also with improved information on load requirements per shipment, carriers can better utilize aircraft and trucks, enabling a more economical fleet use. Statistics is gathered on each customer and transport route, which in turn becomes an important source for improving business decisions.

DIMENSIONING OBJECTS RELIABLY

With the emergency of automatic dimensioning solutions there has been a need to create a common set of standard for how objects are measured, terminology used and what the customer can expect from different dimensioning devices. New products and applications are being introduced in the markets; some with success, but also with failures. Failures could be avoided if users, manufacturers as well as legal for trade regulatory bodies could tap a common pool of accumulated knowledge about dimensioning.

Cargoscan pioneered the dimensioning technology and has been a supplier of dimensioning solutions to the transport and logistics industry for more than 10 years. With close to 1000 devices installed, approved and in operation throughout the world, the company is probably the most experienced group of people in the industry.

The focus is on *reliability*. Only when all parties are sure a dimensioning device is reliable is when it should be put into operation.

A reliable dimensioning device measures objects:

- using a clear *definition* of what is to be measured;

- with the *right precision;*

- *reproducibly,* e.g. the ability of a dimensioning device to supply the same result in repetitive measurements using the same object and independent of object location.

- *in the environment* where the device will be used.

To assist users in insuring that a dimensioning device meets certain basic requirements for reliability, many countries are starting to implement legal for trade regulatory bodies stating standards for how objects are to be measured and they define reliability and accuracy requirements for dimensioning equipment. National as well as international legal for trade bodies *(Organisation Internationale de Métrologique Legale)* also play an important role in certifying different dimensioning devices that can be used for billing purposes.

Defining Dimensions and Volume

The space occupied (volume) of an object for space utilization purposes is the most important item, when handling, transporting or storing goods. But what is the volume of an object?

Scientists define it as the "amount of space, measured in cubic units, that an object occupies". They measure it by submersing the object in water and measuring the amount of water that is displaced. For the transport and logistics industry, this is a totally impractical way of measuring space occupied.

The space occupied depends on the size, shape and the way objects are stacked together. An empty paint can has a small "scientific" volume. A filled one with a cover, may have a volume 10 times the empty one. Two identical buckets stacked side by side utilize much more space than their actual volume. When objects are not perfectly rectangular, it is difficult to utilize the space around the irregular shape or part of the object. A consistent definition and for all practical purposes the space occupied -volume- of an object is determined by:

> ***"The dimensions of the smallest rectangular box***
> ***that can enclose the object***
> ***as it is placed on a surface".***

Note that in this definition:

* The surface the object is placed on, is used as the bottom side of the smallest rectangular box.

* The dimensions of the smallest rectangular box that encloses an object may be smaller than the dimensions measured at the longest points of the object.

* For storing and handling of goods objects are usually placed with their most stable side down.

Picture 3; Dimensions Measured

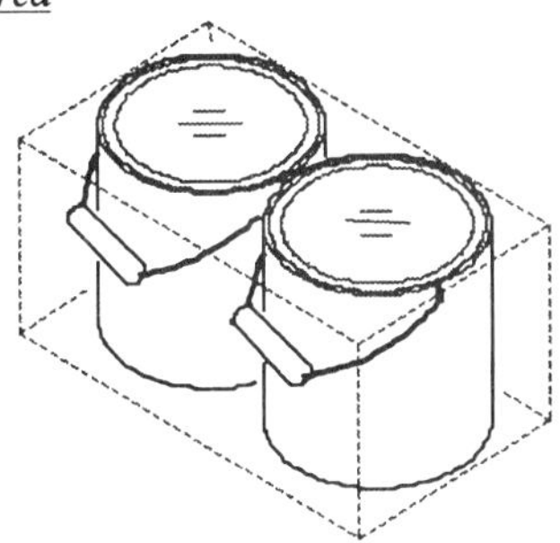

Look at Precision, *not* Resolution

Precision is how close a set of results are to the actual size of the object being measured. A device with a precision of ± 5mm (0.2") can give a range of results from 29.5 to 30.5 cm (11.6 to 12") for an object with a length of 30.0 cm (11.8").

No device is perfect. With enough repetitive measurements, it will give a result that falls outside the precision range. Cargoscan states that 99.6% of the results will fall within the precision range (in technical terms this is called a precision of 3 sigma).

Resolution should not be mistaken for precision. Resolution is the smallest measurement unit that the technology can accomplish. Dependent on a device's design, it may have high resolution but low precision in actual use.
One should always ask the vendor for performance tests that show the precision range. Be sure that test objects are similar in size, shape, texture and color, to the items to be dimensioned in the application.

Cuboidal or Rectangular Shaped Objects

Many objects are transported and stored in cardboard corrugated boxes. These appear to have a rectangular or "cuboidal" shape but they usually vary significantly from a perfect rectangular shape. A perfect rectangular shape has perfect flat sides, and perfect 90° angles. Because of overfilling and rough handling, cardboard boxes are commonly found to have:

- bulges

- skewed edges

- rounded corners

- collapsed sides

Even though cardboard boxes are not perfectly rectangular shaped, a dimensioning device still must produce a well-defined reproducible result of these items. The definition of measurement (the smallest rectangular box that will fit over the object) includes irregularities in the shape that have an effect on space utilization. A dimensioning device has no other option than to measure what is presented to it in a predictable and easily understandable manner. Otherwise users will find the device unpredictable and unreliable.

Picture 4; Not all Boxes are Square:

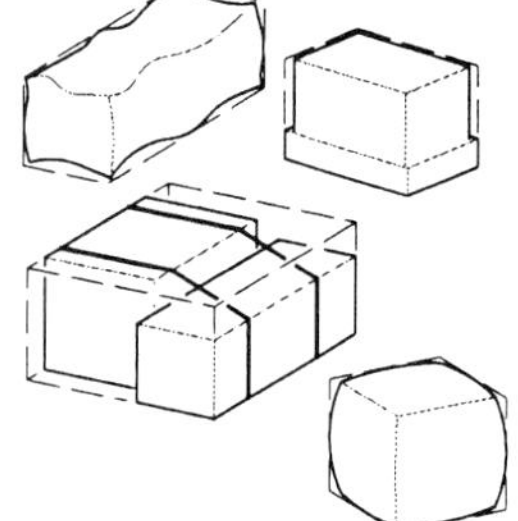

Dimensioning devices must be able to measure different shapes to produce reliable measuring results.

DIMENSIONING TECHNOLOGIES

There are a number of technologies available for dimensioning goods and parcels. Besides meeting the requirements of reliability as previously described, dimensioning devices used in the transport industry must be able to work properly in the rough environment of a typical sort station. In addition to an often dusty environment, the instruments are often exposed to heavy vibrations, changing temperature and light conditions. At least but not less important the devices must be robust and easy for station personnel to operate.

In addition, there are three important questions that should be evaluated when considering the technologies available:

1. The number of measurement points taken on the object.

2. Do the measurement points catch the objects important details and give the dimensions of the real shape?

3. Does the device use a shadowing or a reflective technique?

Measurement Points

Devices can be differentiated by the number of actual measurement points that are made on the object during the dimensioning process. For example, if only 3 points are measured on a box then those three points will determine the actual dimensions, whether those 3 points are representative of the shape or not. A device that can measure thousands of points on an object can certainly determine the dimensions on an object much more precisely than 3 points.

Measuring Points Positioned to Catch Details

Since no device can give a continuous measurement on the whole surface of an object, a device's design must try to optimize the position of the measuring points so that the important details of an object are recognized.

Some technologies use parallel light beams to characterize the object while others use angled. Parallel beams see a more representative view of an object's characteristics. Angled beams can easily be shadowed by the edges of an object and the beam may be blocked from seeing important characteristics or "details".

Picture 5; A High Number of Measurement Points Give Reliable Results

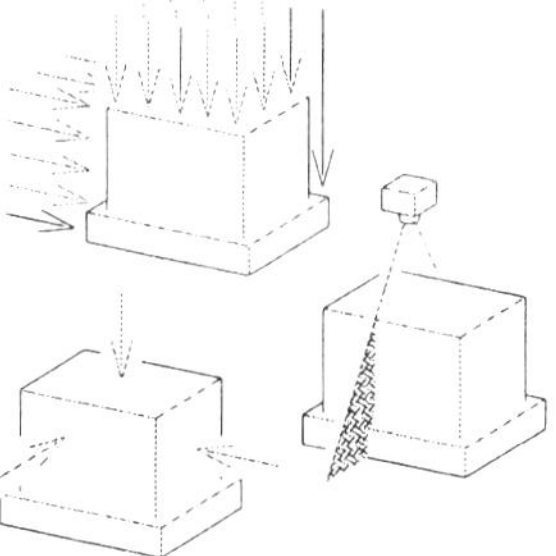

Shadowing vs. Reflective Technologies

Basically there are two different technology concepts used for measuring dimensions of objects:

- Shadow technology
- Reflective Technology

In a *shadowing* device, the object interrupts a light path and a shadow profile is observed by the device. This method works with all objects without regard to the surface characteristics of the object. Even transparent materials can be measured because of light reflects on the edges of the objects.

Reflective devices use some sort of radiation that is reflected back from the object to do its measuring. For all types of radiation there are materials and surfaces that absorb or reflect too much or too little of the radiation to allow good measurements. By modulating the radiation, a device can measure significantly more surfaces, and in addition is less sensitive to background radiation. A reliable reflective device should, be able to tell when a surface reflects to little or too much radiation, and not present false results.

The Light Curtain Technology (Shadowing Technology)

The light curtain is a shadow technology incorporated in a dimensioning frame.
Light curtains consist of hundreds of arrays of infrared emitters and receivers. Infrared transmitters are placed on one side and transmit their radiation to the receiver on the other side. One transmitter and the corresponding receiver are active simultaneously. As the object is moved through the device on conveyors, the silhouettes are stored in a computer. Movement of belts are measured with a pulse counter.

The technology is also set to focus on the edges of the objects for guaranteeing the best possible accuracy. Light curtains can measure accurately any rigid object at very high speed (200m/min, 656 f/min). Although very well suited for green field stations where the necessary space for the surrounding conveyors can be planned in advance, it can be difficult to install units in existing conveyor systems. The main disadvantage of light curtains is the requirements of precision belts to make the objects pass over a slot between the belts undisturbed.

Picture 6; The Cargoscan Light Curtain Technology

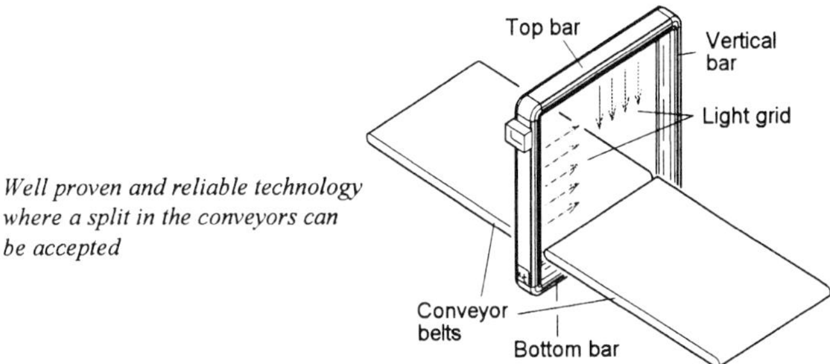

Well proven and reliable technology where a split in the conveyors can be accepted

PILAR[TM] Technology (Reflective Technology)

Some years ago Cargoscan developed and patented a technology called **PILAR[TM]** (Parallel Infrared LAser Rangefinder). The PILAR technology is a reflective technology where the device can be placed on top of conveyors. As such it is no longer dependant on the slot between the belts and can easily be mounted on top of existing conveyors. In short the PILAR is a laser radar that emits parallel eye-safe beams aimed straight down at the packages streaming below. The laser beams are reflected directly back up to the scanner, and the system measures the round trip travel of each beam from emission to return. The concept is based on a so-called phase shift measurement where the instrument compares the outgoing light wave with the reflected light wave and analyses the resulting phase shift. The system allows for more than 100.000 point measurement per second and is very accurate. The computer inside the instrument analyses the obtained data and is then able to a precise "topographic map". At the same time the instrument continuously focus on the edges of the objects; this data is stored and the dimensions and the volume of the object can be precisely calculated.

Through the high density of completely parallel beams that are being created, the PILAR[TM] technology is able to measure objects of almost any shape and it can look at several objects moving underneath on the conveyor belt simultaneously. The ability to measure overlapping objects that are not properly singulated simultaneously, is another advantage as carriers no longer are dependant of installing space consuming and expensive singulation systems in the sort stations.

*Picture 7; Cargoscan **PILAR**^{TM} Technology*

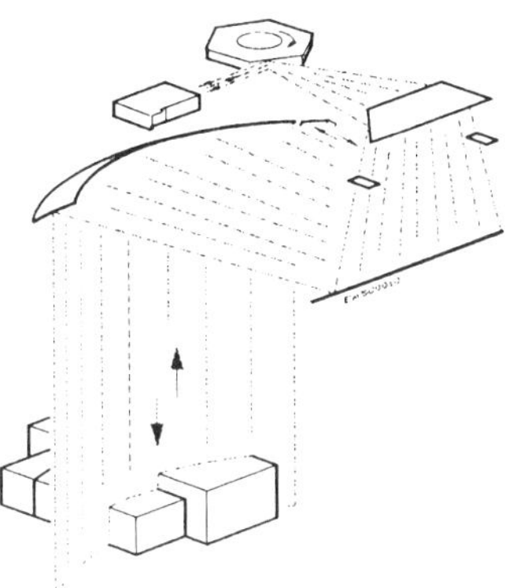

125.000 paralell laser beams per second scan the object moving underneath on the conveyor. A computor inside the instrument determines a topograpfical map and the dimensions of the object is being calculated. The technology is extremely accurate with a resolution better than 1 mm.

Dimensioning Solutions for Different Needs

There are many things to consider when evaluating a dimensioning measuring solution. Different carriers have different requirements for different sort stations. Depending on the number of parcels that need to be processed in a given time frame, degree of mechanization of conveyors, type of weighing requirement, existing IT infrastructure etc, there are a number of different solutions available.

Parcel throughput requirements on a dimensioning solution may vary from a few hundred parcels per hour to being able to measure more than 10.000 parcels in an hour.

In sort stations with low throughput requirements, a dimensioning system may be integrated in a manual conveyor system where the data capture station consisting of a dimensioner, a static scale, and a hand bar code reader. The operator controls the parcel flow manually and shots the bar code by hand.

Picture 8: Cargoscanner CS5120 at Operation in a DHL Station

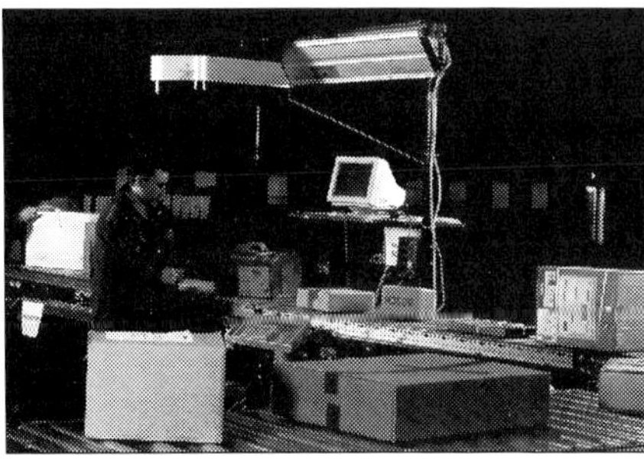

PILAR^{TM} technology in operation in a DHL sort Station. The system is designed to handle up to 500 parcels per hour

An example of a dimensioning system able to handle more than 10.000 parcels per hour is the Mass Scanning and Dimensioning System (MSDS) used by Federal Express. The complete system is developed by Adaptive Optics Associates, Boston USA, and uses Cargoscan PILAR technology for profiling the parcels. The MSDS system is a combined fully automatic dimensioning bar code reading system that will measure non-singulated parcels moving rapidly on large conveyor belts. The system is one of its kinds as it is able to dimension and read the bar codes of parcels bunched together and often with no separation between them. In addition to providing dimensional data the PILAR technology provides information on parcel shape for the bar code scanners to be pinpointed and focused on the bar codes.

Picture 9; Mass Scanning and Dimensioning System (MSDS)

The MSDS is able to scan and identify boxes and packages bunched together on a fast moving conveyor belt. The MSDS is equipped with Cargoscan PILAR™ Technology.

Independent of parcel though put, the carrier should make sure to choose a dimensioning solution that will provide the best possible *"dim rates"* (percentage of all parcels that the system is able to measure correctly). The higher the dim rates, the more revenue generated and the faster payback on the investment.

CONCLUSIONS

The market for automatic dimensioning equipment is expanding rapidly as carriers and postal services around the world are starting the realize the benefits of dimensional weight based pricing structures. At the same time there are many things to consider when evaluating a dimensioning solution. A serious vendor will consider all the items mentioned in this paper and insure the devices delivered meet the current and future application requirements.

Legal for trade officials around the world are currently working on enforcing standards of what the customer should expect from a dimensioning device. Organisation Internationale de Métrologic Legale (OIML) is attempting to make a standard code for all the legal for trade entitles in different countries to follow. In the meanwhile many individual countries have already implemented national standards and approval requirements on dimensional devices.

The introduction of these codes and standards with the approval of devices by governmental regulatory bodies, will help both users and vendors by defining what is required for automatic dimensioning devices. Until then, serious vendors and users will have to work together to define the right solutions to insure that each measurement result is a *reliably* result.

C568/010/99

Mail technology – surviving in a new market place

B W HEESTERMAN
New Zealand Post Limited, Wellington, New Zealand

SYNOPSIS

Existing postal enterprises are unlikely to remain the sole operators of letter processing machines much longer. As an increasing number of governments remove the long existing monopoly protection on letters, the functionability of the technology must change to encompass the challenges of the 21st century. Vendors and purchasers alike have an increased responsibility to ensure technology continues to develop in such a way that industry fragmentation does not occur. As a result of having to perform in a deregulated letter market, six characteristics of future technology are presented in this paper. Embracing competition and working actively to grow the letter business must remain the common goal of all forward thinking enterprises, technology vendors and industry influencers.

INTRODUCTION

It is now considered inevitable that postal companies will face competition in the standard letter market. Whether this competition is created by explicit changes to legislation, or by stealth through electronic substitution or erosion, remains a question that many Governments are at this time considering with care.

In New Zealand the Government removed all monopoly protection on Letters on 1 April 1998. In doing so New Zealand Post joined a very small group of postal enterprises fully exposed to the market forces, a move welcomed by New Zealand Post itself and long anticipated.

This paper discusses the operational impact of deregulation and presents a range of predictions of such impact on mail technology requirements in the future.

A BRIEF HISTORY OF NEW ZEALAND POST

New Zealand Post was formed as a state owned enterprise (SOE) in 1987, separating from the banking and Telecom businesses that comprised the original New Zealand Post Office. Since that time it has modernised, investing heavily in leading edge mail processing plant, production systems and mail facilities. It has also developed and deployed world best practice methodology and systems for mail operations management that has resulted in it becoming one of the worlds most consistently cost effective and high performing mail networks.

As well as developing an operationally excellent enterprise New Zealand Post has returned consistently high profits every year and in 1995 lowered the standard letter price from 45 to 40 cents in anticipation of eventual deregulation.

WHAT IS A LETTER?

For the purposes of this paper a letter will be defined as being up to C5 and up to 6mm thick. This is generally the size that most processing technology will accept and coincidentally is where most monopoly protection finishes. The letter as a product is beguilingly simple. It in-fact exists within both the communication and physical distribution markets. While typically the sender enters into the contract and pays for the service the receiver is most aware of the service (or lack of) provided.

Predictions of the letters demise during the last five decades resulting from technology such as the telephone, telegraph, telex, facsimile and more latterly email and the Internet have failed to materialise. In fact, as can be seen in figure one, New Zealand Post has experienced positive growth in the letter market over the past ten years resulting in over 50% more letter mail today than in 1989.

Figure One

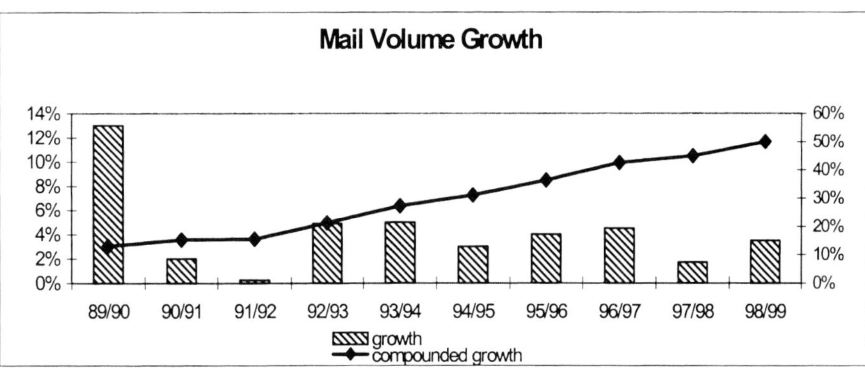

LETTER DEREGULATION

It is appropriate to discuss exactly what deregulation means to the letter market in New Zealand before discussing the operational impacts of the law changes. The main points are:

1. New Zealand Post must continue to deliver to the same number of delivery points with the same frequency (mostly six days per week) as at 1 April 1998.
2. Any organisation can deliver letter mail as a postal operator provided they are registered. Registration costs $85 and provides virtually no barrier to entry.
3. New Zealand Post must allow access to its network on terms no less favourable than customers.
4. Postal operators must have their registered identifier on mail that they handle. If the mail is handled by more than one operator then they each must have their identifier on the mail pieces.

A more detailed analysis of the impact of these issues will be presented later in the paper. While it may be expected that it will, the legislation does not provide equality between operators. Competitors to New Zealand Post have no requirements placed upon them regarding delivery frequency. Neither are they required to accept mail into their network from New Zealand Post. Finally there is no requirement to provide any national and/or rural coverage.

In practice deregulation has resulted in about 15 postal operators. All but four are individuals that pick up and deliver mail within a specific town or city at a typical retail price of around 30 cents. They will not accept out of town mail, or if they do, charge 40 cents and post with New Zealand Post.

Several larger organisations have more sophisticated strategies aimed at a greater share of the national market. To date no competitor has begun an alternative addressed letter delivery network although this is considered inevitable albeit at a lesser frequency to that offered by New Zealand Post.

The operations of these larger competitor's can be categorised into several modes:

1. Selling stamps through a retail network and establishing an "access" agreement with New Zealand Post allowing mail to be posted through its network of street boxes.
2. Collecting and consolidating small/medium business mailings which are then pre-sorted and lodged in bulk. This can achieve a discount from New Zealand Post and result in the possible lowering of the price from 40 cents to 37-38 cents.
3. Establishing a network of street posting receptacles and performing the front end collection and pre-sorting before lodgement with New Zealand Post for delivery.
4. Business to business private box service (document exchange) which operates in a niche of the letter market.

In practice some competitors fit into more than one category and it is foreseeable that existing competitors will co-operate to expand coverage in the future.

MAIL PROCESSING METHODOLOGY

In broad terms mail is processed in a semi-continuous series of processes. If there are several product offerings (first class, second class, air mail etc) then these are typically separated as early as possible and then each type is then fed into its own series of processes. The letter will typically undergo five distinct processes, more often than not separated by one or more transport sector(s).

Figure Two

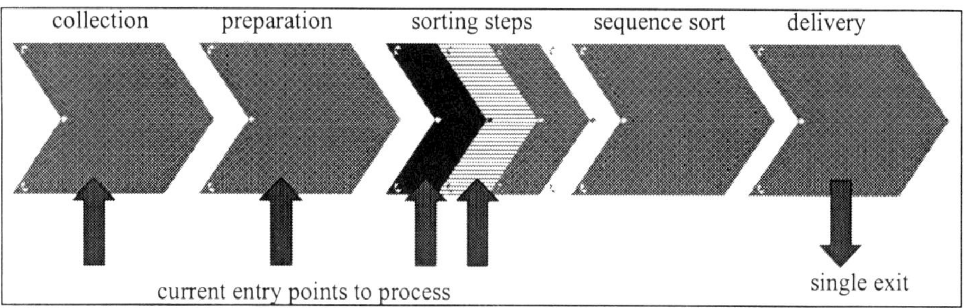

Processing technology is usually applied to steps 2, 3 or 4; that is facing up, cancelling, streaming, sorting and sequencing letters.

It is rare that mail operators need to use batch processing techniques because once streamed the processes simply continue until all mail is processed. For this reason, most technology is used as a single function machine; technology such as the integrated mail processor (IMP) is much more a more multi-tasking machine than others have been and demonstrates a trend away from the "one machine – one process" model.

PROCESSING TECHNOLOGY TODAY

The technology employed by today's postal enterprises is the result of three or more decades of development by a variety of technology vendors and of course has been aided by the rapid increase in computing power available today. The increased computer power manifests itself not just in the increased ability to read mail addresses but in machine control, accuracy, ink jet printing and database access and management techniques.

In particular the ability of the technology to recognise most machine produced as well as many hand produced cursive fonts at rates exceeding 30,000 per hour have resulted in spectacular gains in productivity, especially when applied to delivery and sequence sorting.

The orientation and cancellation of letter items has historically been achieved by the machines scanning for phosphor stamps, fluorescent franking marks or a contrast bar arrangement. The detection of the perforation and to a limited degree pure image detection systems have also been applied to orientate letter items by the machine, but to a lesser degree.

                                                 C568/010 © IMechE 1999

Significantly while there are less than ten dominant suppliers of technology nearly every postal enterprise has a different application of the technology resulting in virtually no cross-border compatibility existing for the processing of letters. That is, barcodes produced by sorting machines are in a bewildering range of formats, inks, positions and structures. While some countries can now read tag ID codes from other countries the scope in overall terms remains relatively small.

Development by the vendors has often been driven by the purchasing power of a number of large countries including USA, Japan, UK, Germany and France to name but a few; the economic power of these countries has resulted in a faster development than would have occurred without it. However parochial pressures by these countries have not necessarily allowed global considerations sufficient weight in the design stage and has certainly contributed to the lack of international compatibility.

WHAT ARE THE CHALLENGES FROM DEREGULATION?

The primary challenge facing mail operations in the deregulated market stem from managing multiple batches of mail through a variety (but not necessarily all) of the normal process steps.

For example, it is theoretically possible that a competitor will want an access contract that allows their customers to post mail in a New Zealand Post street box but will want the mail handed over once it is streamed, faced and perhaps cancelled.

Obviously the possible permutations of process steps is almost limitless. In practice we have, to date, adopted a strategy that allows entry/exit of mail from the network at points identical to customer contracts.

Thus the entry points include street posting boxes, bulk lodgements that include various states of pre-sorting. The only "exit" point is the delivery of the letter to a letter box or private box. It is considered unlikely that the above position will be completely sustainable in its entirety past the medium term.

Multiple batches of mail pose difficulty with revenue protection (counting) of mail, the possibility of different service performance requirements and the difficulty ensuring quality of competitor mail is to "our" standard.

The secondary and only marginally less vexing issue is that of the postal identifier. An example of just how far reaching this requirement may be can be seen by considering incoming international mail. Such mail will usually be date stamped by the originating country, but upon entry it requires New Zealand Post to print its own name or logo on the mail items.

Having to pass all such mail through current technology to just print the identifier adds no value to the mail itself; it delays and adds cost though.

Similar problems exist with pre-sorted mail in the domestic letter market. Having accepted it pre-sorted it must be machined to just print the identifier; often this will virtually eliminate the benefit of the customer pre-sort.

Lastly, the added complexity of the operation poses significant challenges to communicate operational instructions effectively with all employees. New and changing contracts with competitors have to date resulted in a large degree of volatility around operational procedures and the need for weekly (or even daily) change to instructions.

WILL COMPETITORS PURCHASE THEIR OWN TECHNOLOGY?

It may be tempting to assume that the answer will be no. However our analysis of the modes of operation discussed previously would indicate that consolidating mail and successfully achieving a discount from New Zealand Post could yield a positive return on investment. The market share necessary for such an investment could be as low as 3-4%, although in practice a greater share of a smaller area would be a more likely target.

However, nearly everyone here is very aware of how costly and difficult a successful technology project can be and competitors have, or soon will have, learned some of the same lessons. Our strategy is, and will continue to be, to make the option of the purchase of technology less attractive than the price of sorting through New Zealand Post's machines. If competitors do decide to purchase their own technology there will undoubtedly be significant tension to ensure mutual compatibility is assured. Also tension will arise as competitors expect to have easy access to intellectual property created by the original enterprise.

We all may feel secure in the fact that we have contractual protection for intellectual property created during the technology project(s). This may not be sufficient. This will not be, I am sure, due to any bad faith by suppliers but simply due to the fact that the specialists in their respective laboratories have "learned" how to solve the technical issues facing specific countries. So while competitors will need to spend as much for software, the difficulty to develop and lead times until delivery will benefit from our original pioneering development.

FUTURE TECHNOLOGY REQUIREMENTS

Left unchallenged there is no doubt that technology will continue to evolve. As mentioned previously the traditional driver of increased productivity is no longer appropriate as the sole input into strategic direction of technology. With a new market the time is right to consider the global direction of technology and its development.

Based on our assessment of a deregulated letter market six characteristics of mail technology have been identified as needing focus. The first three deal with more philosophical matters of design and contract while the last three relate to the machines technical and performance characteristics.

Characteristic One: Short Lead Times

A deregulated market demands short lead times. Gone are the days where 18-36 months from contract start to completion is acceptable. "First to market" will demand "fast to market" response times and as a consequence project duration's of eight to twelve months must become the norm. (It is accepted that where a large order quantity is involved the time will be longer; the first machine should still only take 8-12 months though.)

Reducing lead time is not just the responsibility of suppliers; we as purchasers need to avoid re-inventing functions that are similar to standard and we need to expect or demand less customisation; in some cases we need to swallow our pride and accept that buying the same as others is acceptable. For suppliers a greater emphasis on modularity, flexibility and "plug and play" style options is required. The pay back will be lower inventory and change control costs. More importantly will be the reduction in on site commissioning and installation costs and the ability to offer earlier "go live" dates to customers.

Shorter lead time should allow more "future proofing" of contracts. There is no reason that the decision to purchase a particular technology at a certain moment in time should freeze the performance. We need to be able to purchase technology that is not obsolete (or heading that way) by installation, so concurrent with shorter lead times, future technology contracts that include continuous access to enhancements for specified periods need to be offered as the norm.

Characteristic Two: Global Compatibility

As noted earlier very little cross border compatibility exists. A graphic example of this was pointed out to New Zealand Post by a card manufacturer who operated in the USA, the UK, Canada, Australia and New Zealand markets. All these countries use AEG or Siemens derived sorting machines and yet the maximum card size is quite different in all five countries.

Even if we, as postal enterprises, are not planning or operating globally our competition are, or will be, as are vendors and other industry influencers. If organisations such as UPS, DHL or FedEx decided to enter the letter market on a global scale it is certain that they would create compatible technology platforms and use this to create advantages in the exchange of international letters.

There are some who would argue that by continuing the technological diversity, we prevent the entry by these competitors who would find it impossible to be compatible with both local and global technology platforms. This thinking is, I believe, flawed because the economic proposition remains unchanged for the competitors who could equally consider grouping themselves quickly to form a significant block. Since competitors are more likely to be buying technology (since many of the existing operators have platforms already) then the vendors may be inclined to consider such groups very carefully; indeed they may find them impossible to ignore.

As the incumbent postal enterprise we have the opportunity to set the standard that newcomers must adopt. Global compatibility is not an option for the long term – the question is more who will do it first?

Of course this will not occur overnight but I consider that without a start now we risk being "left out" of any global technology platform. The responsibility for compatibility rests as much with postal enterprises as with technology vendors to drive forward towards the goal of maximum compatibility.

Characteristic Three: Integration

Integration of currently stand alone machine functions into one machine must be a serious contemplation for vendors. Our experience with Siemens IMP machines reinforces this view. There are several reasons to consider integration.

1. The purchase cost is usually less than the sum of the individual machines.
2. Floor space and power consumption is less with one machine than several.
3. Fewer staff are required to operate a single machine; fewer are required to maintain it too.
4. Enhanced synergies are obtained from a variety of systems running simultaneously (i.e. indicia detection and address block detection can be programmed to assist each other).
5. Total process time is vastly reduced by the elimination of multiple machines.

The disadvantages are a reduced operational flexibility in that the integrated machine is key to many processes; also a sense that if the machine fails then all process steps are jeopardised. Our experience is however, that the advantages outweigh the disadvantages considerably.

Of significance is the reduction in average process time. A theoretical modelling of the same process with three different machine combinations illustrates a dramatic variation in average process time. The three configurations modelled were:

1. IMP
2. CFC directly coupled to an OCR (CFC/OCR)
3. Stand alone CFC and stand alone OCR (CFC-OCR)

(Note all combinations included on-line video coding too).

In the simulation, queue times prior to every process were varied binominally around times from zero to 90 minutes. Every simulation included 10,000 runs and the results graphed to form figure three.

In the IMP, the average process time (the average time that a letter takes from the start to the finish of all process steps) is almost completely independent of queue time, despite the fact that rejects are processed manually and are subject to the same queue times as machines.

This theoretical view is supported by our operation of the IMP's where keeping the machine loaded with mail is the single biggest determinant to overall high efficiency. Also significant increases in indicia and address block detection occur from the synergetic application of multiple detection processes within the same machine.

Figure Three

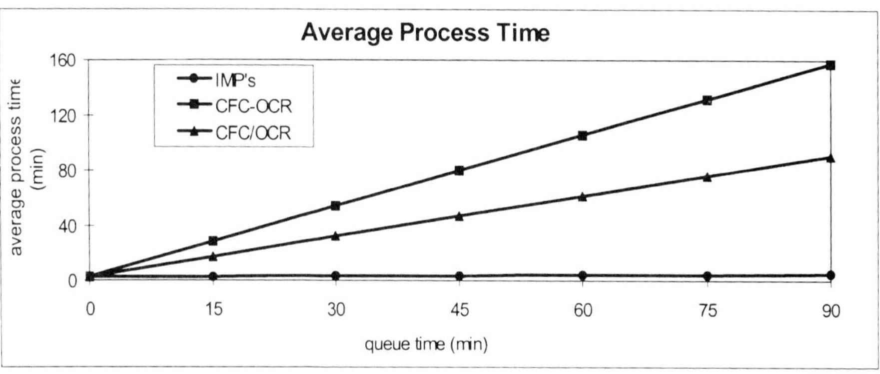

Characteristic Four: On-the-Fly Revenue Protection

The fact that the mail stream will consist of multiple batches of mail (from various postal operators) means technology will be required to play an increasing role in revenue protection.

That is, we will require machines to identify, count and probably separate mail according to the postal operator. It is foreseen that this could be achieved through new ink types, indicia barcodes, or some other detection technology such as radio frequency (RF) tags etc.

The machines will be increasingly required to have the function of invoice generation, mandating that the accuracy of the counting to be paramount. If a competitors mail volume is less than 2% of the total, we need to be confident that the counting error is not larger than the count itself.

The reverse need exists too. That is, the need to separate illicit or unregistered operators mail from the mail stream as early as possible. This is more problematic than it seems at first glance because typically technology is calibrated to recognise items even with a fairly vague resemblance to correct indicia. This creative license on the part of the unregistered operator may result in their mail being accepted by technology and remaining undiscovered until delivery, if it is discovered at all.

Characteristic Five: Smart Indicia Systems

The current methods on indicia detection have been in use for some time. Namely phosphor, fluorescent and contrast bar detection systems. While some variations such as serration detection has emerged, as has the application of more pure image recognition ideas, most enterprises have adopted a combination of the first three and changed little from the first inception.

All these systems remain a fairly clumsy tool. For the purposes of orientation of the mail item they suffice although the success rate can be reasonably variable. While the use of a green and red phosphor allows a machine to segregate mail into two streams, no significant in-roads have been made into postal tariff rate detection that gives a high rate of confidence or accuracy or as a means to differentiate multiple postal operators mail.

While some advanced systems are able to "read" the franking impression amount, until the success rate and accuracy is close to 100 percent this remains a relatively poor method of revenue collection or detection.

In the deregulated market there will be a reduced ability to enforce compliance with phosphor stamps, fluorescent franking and a myriad of other "rules". It is also possible that new entrants to the market may use the lack of such rules as a competitive advantage – certainly in the early stages of their business development.

The fact is that compliance, with the need to produce phosphor or fluorescent indicia adds to stamp production costs, introduces additional complexity and it is not attractive for competitors to comply with. Therefore mail processing technology must move forward and encompass emerging technology such as 2D barcodes or RF tags etc to produce an indicia that are:

1. Cheap to apply or produce
2. Nearly 100% successful
3. Has the ability to carry information (rate, service, postal operator etc)
4. Easy to apply and check compliance
5. Difficult to forge

To move forward we as postal enterprises must start demanding new approaches to indicia and payment systems that have been created for the new millennium not re-worked from the 70's or 80's.

Characteristic Six: Machine Performance

Lastly, but certainly not least, technology must provide increasingly superior performance. High productivity, high reliability and ease of use will remain as paramount importance to equipment selection.

The fact is that our operations are intensive and operate time sensitive environments: equipment must perform to its optimum during the hours that it is required – every time. To ensure this, technology must increasingly be equipped with high level diagnostic tools, on-line help and have at its disposal 24 hour access to expert vendor advice.

The new market will undoubtedly condense processing windows; existing postal enterprises will have a reduced ability to influence lodgement times and posting patterns. Once again, as a point of competitive difference competitors, with less volume, may well try and use later acceptance times as an advantage.

Thus the ability to process mail in the shortest possible time is increasingly vital. Integrated machines reduce the time from acceptance until dispatch, however technology must continue to have:

- High read rates
- High throughput
- Low error rate
- High reliability

These are of course not new requirements but the importance of maximising these has never been as critical. Additionally, the ability to read and sort to delivery and delivery sequence level provides huge productivity gains that cannot be ignored.

Increasingly vendors should be offering the ability to receive new developments progressively as they are developed as has been discussed previously, with the expectation that the costs of such continuous enhancements are being included in contracts from the start. As was the case with indicia, the existing postal operators will have reduced influence over mail standards; as a consequence technology must be less sensitive to variations in quality, style and presentation.

CONCLUSION

After years of development, enhancement and improvement, mail operations is continuing to become more complex. In order to survive and indeed thrive, it is increasingly likely that postal enterprises will need to decide to work with each other, vendors and increasingly competitors. Simply ignoring competition will not make it go away and doing so risks forcing competitors to create their own global standards for technology.

The future of mail processing technology will depend upon the development and deployment of new and creative ways of increasing productivity and allowing batch processing of various "types" of mail.

Vendors and users alike can actively work towards establishing and adopting global standards that strengthen links to each other adopting some or all of the six characteristics outlined in this paper.

Fundamentally choices are required to be made during the coming years that will shape the face of letter processing technology into the future. We must embrace deregulation and grow the letter business to its maximum potential, using all the intellectual capital available to postal operators, be they long established postal offices or newly emerged competitors.

C568/018/99

A process to aid product data management

C E WAINWRIGHT and **K REYNOLDS**
The CIM Institute, Cranfield University, Bedford, UK

Successful business organisations operating in today's global markets need to maximise the
strategic benefits of information systems. Time to market, improved product quality and
flexible manufacturing are common business objectives, yet communication between multifunctional teams is often hindered by the use of functionally focused information systems,
resulting in duplicate and consequently inaccurate information amongst different departments.

Product Data Management (PDM) systems provide a solution, yet implementations often
focus upon technological solutions or have a particular departmental bias. To provide an
objective solution, a process that specifies information requirements at different stages of
business processes is required. By identifying key decision points, business processes can be
simplified and stream-lined to deliver essential information at particular milestones, thereby
aiding the transition through decision risk gates.

This paper describes a process to aid PDM implementation. First the limitations of
disparate information systems are discussed, before outlining the evolution of technological
solutions and critiquing the use of Business Process Re-engineering (BPR) techniques to
improve business performance. A process is then outlined involving the use of business
modelling techniques to specify information requirements, thus encouraging information
sharing and re-use throughout the process.

1 INTRODUCTION

Product Data Management (PDM) solutions vary in scale and domain according to an
organisation's business requirements and the existing technologies and processes in place.
This paper describes a generic process framework based on observation and participation in

two different organisations, Matra BAe Dynamics and Pitney Bowes Limited. Consequently this reflects the author's opinion and not the participating organisations.

Matra BAe Dynamics is an established producer of guided weapon systems and is a partly owned subsidiary of British Aerospace plc (BAe). Global political changes during the last decade resulted in an investigation of Dynamics' current situation with a view to formulating an approach to meet future business needs. The fragmented growth of Information Technology (IT) within the company presented a possible barrier to business integration, as the historical accumulation of computer hardware and software had resulted in a high degree of incompatibility, with different packages co-existing to serve the same purpose. Data maintenance is an important contractual requirement for a defence supplier, with data being maintained for the life-cycle of the product. With a guided weapon system this can be for a 20-30 year duration. Yet due to the number of information systems in place, data was held in a variety of formats resulting in time consuming practices when attempting to retrieve or merge data. Consequently a process to achieve information system integration was required and the requirements for a PDM system specified.

Pitney Bowes Ltd. is a satellite organisation of the American owned global manufacturer Pitney Bowes Inc. Historically engineering design was American based, with European customisation and development undertaken in the United Kingdom (UK). As one of the largest office equipment suppliers, Pitney Bowes enjoys a large market share of the postage meter market. Changes in market requirements have demanded the migration of mechanical postage meters to mechatronic ones customised using software. Such a shift has allowed other organisations with similar design capabilities to compete in this lucrative market. To maintain its competitive standing with time to market, improved product quality, customer service and flexible manufacturing as generic goals. Information sharing and global design are paramount to their achievement of strategic advantage from IT systems. Subsequently information systems needed to be aligned to business objectives and product requirements. This was to be achieved through the use of a PDM system.

2 PDM SYSTEM EVOLUTION

2.1 Initial development

The need for an information system to integrate engineering systems was first illustrated when the effect of technological advancement on business performance was dampened by the necessary paper based systems to connect localised technical solutions. Computer applications in manufacturing organisations focused on direct functions. Paranby (1) identified these functions as: design, purchasing, production planning, production control and manufacturing. Typical applications were built for functional requirements. With localised technological investment justification, benefits focused on departmental efficiency. The integration of disparate technologies was complicated by the number of linkages required to interface discrete computer applications across different platforms and their respective repositories, holding data in different formats and consequently structures. The number of applications also increased the likelihood of duplicate data, where different departments used different versions of the same data. Early PDM systems evolved from the application of computer based systems to engineering. A PDM system was used within an enterprise to organise, access and control data related to products and to manage the life-cycle of those products (2) (3). PDM solutions at this time were developed in-house, typically by larger organisations that contained their own software development departments.

2.2 First generation PDM solutions
By the late 1980's, the first generation of commercial PDM solutions were available. Some were developed by independent software houses, however PDM vendors already provided Computer Aided Design/Manufacturing/Engineering (CAD/CAM/CAE) application and used their market awareness to identify customer requirements and consequently provide an additional module to increase sales of their existing product suite. The mechanical CAD market is testimony to this trend, especially as the current demand for PDM solutions represents a large proportion of the revenue generated in manufacturing computer applications.

Typical first generation PDM solutions consisted of a relational database to record parts and their related files. These tended to have a manufacturing focus by concentrating on the downstream use of design data. Key features included the release of design data to manufacturing and engineering change management. Given the frequency of changes to unreleased design data and manufacturing's preference for static and stable product design, PDM systems provided a buffer to aid the data transition from design to manufacturing and hence supplied version control and BOM management capabilities.

2.3 Second generation PDM solutions
The move from relational to object oriented database technologies allowed greater flexibility in PDM systems. First generation PDM systems were governed by functional boundaries and the transformation of data across functional barriers. The successful use of Japanese manufacturing techniques (4) and their lean and collaborating philosophies achieved dramatic cost and quality benefits thus improving business performance. Collaborative principles used to improve manufacturing were introduced across the design/manufacturing interface and were fundamental to the implementation of Concurrent Engineering (CE). Advances in technology and the decentralisation of decision making required computer applications to support CE practices to improve strategic objectives such as time-to market. PDM systems were required to support the product life-cycle from initial concept to product obsolescence and CE practices with the conceptual design phase, allowing individuals to work with consistent data and managing the release of product data across intermediate decision and risk gates between initial concept to final release.

The process improvement principles from Total Quality Management (TQM) resulted in the introduction of Business Process Re-engineering (BPR) as a customer focused and 'value adding' initiative. This was associated as a pre-requisite for IT implementation. Hammer (5) suggested reviewing operational practices from a customer's perspective, in particular organising processes around customers and the removal of 'non-value' adding activities. His work was based on research undertaken by Massachusetts Institute of Technology (MIT) (6), Porter's proposal to align IT solutions to competitive advantage (7) and the construction of customer focused processes and the development of value chains (8) using process improvement techniques detailed by Harrington (9). BPR has received mixed reactions (10) as instead of continuously improved incremental changes associated with TQM, BPR sought step changes with high risk and failure rates. Some of the difficulties associated with the BPR concept can be attributed to the early pioneers. Hammer became more radical between articles (5) (11) after which the emphasis seemed to be on dramatic change (12), whilst most implementations were often associated with incremental changes (10). Similarly Davenport (13) proposed information technology as a method to innovate processes, criticised hammer and Champy's book (14) for being too dramatic (15), and subsequently suggested the best method to implement information technology was through changes in an organisation's information culture (16). Likewise BPR has been criticised for the association with redundancies (17), as a fashionable trend (18), the lack of a comprehensive definition (19), the emphasis on major change (20) and that the technology has ignored the message of integration through people (21).

Despite the criticisms levelled at BPR, the approach has had an important contribution to the re-alignment of business processes to corporate strategy. However, Majchrzak and Wang (22) found that restructuring around processes was not sufficient to integrate functions, as many organisations neglected the essential training to remove the functional mind set and values from employees. Consequently the misunderstanding associated with BPR implementation and the manufacturing focus of second generation PDM led to the implementation of functional focused IT systems which failed to achieve true cross-functional integration.

2.4 Future developments

The move to Internet technologies has generated a mechanism to package data in Java Applets, allowing the transfer of data across different processes, providing the software mechanism to access the data with the data necessary to make decisions, across different technological platforms. Powerful search engines make information retrieval easier and screens can display information from a variety of sources. This move to distributed applications with their own databases is less reliant on the data vaults evident in traditional PDM systems. Instead the emphasis is on the ability to index a variety of applications. However issues such as data integrity and commercial security have made third generation PDM systems more of a concept than a reality.

Similarly the growth of Enterprise Resource Planning (ERP) vendors based on the replacement of non-millennium compliant legacy systems and the need for process based IT solutions, as defined by the BPR initiatives, have influenced the PDM market Large CAD and ERP vendors are now competing for the lucrative PDM market. Consequently the selection of a suitable PDM solution is dependent on an organisation's beliefs and business requirements, for example the use of an indexing system versus a data vault and the preference between an open system and storage in neutral formats or the integration of a process using proprietary enterprise-wide software. Either way true cross-functional integration is only achieved through tight linkages in time and in communication between individuals and groups working on closely related problems (23). The misunderstanding associated with BPR implementation and the lack of an objective framework to follow makes successful PDM implementations difficult to achieve. Therefore a process framework needs to be defined to aid the implementation of a PDM system.

3 PROCESS FRAMEWORK

There are five phases necessary in any information system modelling and specification (24):
- Process requirements
- Process description
- Information requirements
- Software specification
- Software development.

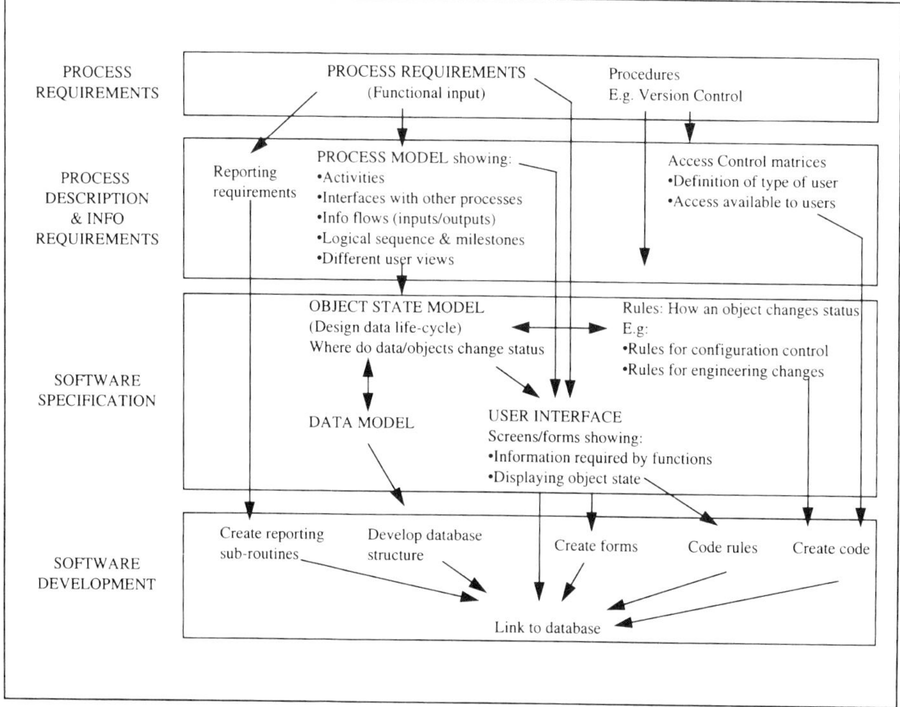

Figure 1: A process framework for PDM implementation

3.1 Process requirements

The requirements of any business process relates directly to business performance and the contribution to competitive advantage. Processes usually have operational, strategic or supporting roles. Efficiency is the main aim for non-strategic processes as they provide little or no competitive strategic advantage, whilst effectiveness is the objective of strategic processes. In both participating organisations' the engineering process was strategic in nature and hence had to be effective in meeting dependent business objectives such as timely product development. Related business requirements were imposed on the engineering's process requirements. Similarly business and technological constraints had to be incorporated, for example government legislation and quality standards and the data model underlying the commercial PDM system.

3.2 Process Model

A model is a simplification of a real world problem which is used to deduce conclusions which can be applied to the problem the model attempts to emulate. Baines *et al*, (25) found within manufacturing organisations where modelling techniques were used, the most common technique was SADT/IDEF$_0$ (26) (27). This technique was used to describe the process in the form of hierarchical activities, where the information inputs and outputs were illustrated together with activity resources and constraints. This provided a high level view of the process, however IDEF$_0$ failed to describe the timing of information flows through the process. Consequently a secondary flow chart was constructed to aid the specification of information requirements.

3.3 Information requirements

Using the IDEF$_0$ model as a basis, activities were mapped according to a logical sequence in a flow chart form. This evaluated the flow of information through the process by mapping a critical path of information flows. Decision risk gates were used to describe the completeness of information required before commencement of consecutive activities. Once the existing process and its corresponding information flows were modelled and documented, this was used to highlight deficiencies in both the process and the existing systems used to supply information. Using a team consisting of all relevant functional representatives, the proposed process and the information flows was be specified. The level of information completeness was also specified by listing the relevant attributes and whether these were mandatory or optional.

The information required before passing a decision gate transition was also defined and the necessary sign off procedure followed. This helped define the workflow of the process, which when programmed, forced users to follow a set or variant of a defined process. The use of reporting and queries allowed the notification of obstacles in the information flow and that essential verification had been made and by whom. This removed the reliance on paper based systems, indicated duplication in multiple systems and improved product development visibility.

A tool associated with workflow is IDEF$_3$ (28), one of the more recent additions to the IDEF modelling suite. This overcomes some of the timing limitations associated with IDEF$_0$. IDEF$_3$ models the process in a series of objects and Boolean algebra, and can be used to define alternative routes or variant processes and the corresponding objects' status through the defined process. Object state modelling such as IDEF$_3$ can be used to serve process definition or aid software specification. Alternatively models such as GRAI grids (29) can be used to highlight deficiencies in information flows by mapping decision points to the availability of information within a given time period. Each decision point depicts the time frame individuals have to make a decision, whilst mapping information flows against time periods illustrates the relevant information available.

3.4 Software specification

The majority of commercially available PDM solutions are built on object oriented databases. These provide a set of pre-defined objects which can be modified according to an organisation's preferences and requirements. Object state modelling (IDEF$_3$) was used to describe different characteristics of a pre-defined object through the process and its relationship with other objects and their relationship with an object's attributes. This allowed the modelling of the data life-cycle through the PDM system. Given this object definition, rules were then specified documenting how and when an object changes status. This technique is particularly useful in Release Management when for example an engineering change is proposed, accepted and then introduced into production. Using this specification, corresponding menu displays and forms were designed to illustrate the various stages of the objects status. This allowed an operational PDM system to locate an instance of this object in the engineering process.

3.5 Software development

The underlying data model behind any commercial PDM system governs the database structure and consequently the software development required. Development involved linking these objects to the database structure and developing the user interface. Forms were constructed to develop a user interface. Commonality of screens was important, both to software users and developers as this reduced the level of software development and users training required. Similarly the rules which determined the workflow and the procedure for product configuration and version control were coded. These were then linked to the database. Information outputs such as the defined reporting requirements need to be coded such that the sub-routines to find and retrieve data do so in the most efficient manner. By following all stages in the process framework, the most effective business process was defined, the information requirements to support the process documented and the corresponding technological solution specified and developed. In doing so, the process framework provided an outline to successful PDM implementation.

4 CONCLUSION

The evolution of commercial PDM systems has resulted in a functional bias concentrating on the transfer of product data form design to manufacturing. Instead PDM systems should be considered as a mechanism to improve business performance by evaluating process requirements before proposing a technological solution. Given the limitations and misunderstanding surrounding BPR philosophy, a five stage process framework was proposed to incorporate requirements and their translation into a PDM system to improve strategic advantage.

REFERENCES

1. PARANBY, J. (1979). "Concept of a Manufacturing System", International Journal of Production Research, **17**, (2), pp 123-135.
2. GASCOIGNE, B. (1995). "PDM: The Essential Technology for Concurrent Engineering" World Class to Manufacture, **2**, (1), pp 38-42.
3. CIMdata (1993). PDM Buyer's Guide, CIMdata, Ann Arbor, Massachusetts.
4. SCHONBERGER, R.J. (1982). "*Japanese Manufacturing Techniques: Nine Lessons in Simplicity*", The Free Press, New York, USA., ISBN 0-02-929100-3.
5. HAMMER, M. (1990). "Re-engineering Work: Don't Automate-Obliterate", Harvard Business Review, **68**, (4), pp 104-112.
6. SCOTT MORTON, M S. (1991). "*The Corporation of the 1990's: Information Technology and Organisational Transformation*", Oxford University Press Inc., New York, USA., ISBN 0-19-506358-9.

7. PORTER, M. and MILLAR, V.E. (1985). "How Information Gives you Competitive Advantage", <u>Harvard Business Review</u>, **63**, (4), pp 149-160.

8. PORTER, M. (1985). *"Competitive Advantage: Creating and Sustaining Superior Performance"*, Free Press, New York, USA., ISBN 0-02-925090-0.

9. HARRINGTON, H.J., (1991). *"Business Process Improvement: The Breakthrough Strategy for Total Quality, Productivity and Competitiveness"*, McGraw-Hill Inc., New York, USA, ISBN 0-07-026768-5.

10. CURRIE, W.L. and WILLCOCKS, L., (1996). "The new Branch Columbus Project at Royal Bank of Scotland: The Implementation of Large-Scale Business Process Re-engineering", <u>Journal of Strategic Information Systems</u>, **5**, pp 213-236.

11. HAMMER, M. and CHAMPY, J. (1995). *"Re-engineering the Corporation: A Manifesto for Business Revolution"*, Nicholas Brealey Publishing, London, UK., ISBN 1-85788-029-3.

12. HAMMER, M. and STANTON, S.A., (1995). *"The Re-engineering Revolution"*, Harper Collins, London, UK., ISBN 0-00-255657-X.

13. DAVENPORT, T.H. (1993a). *"Process Innovation: Re-engineering Work through Information Technology"*, Harvard Business Press, Boston, USA., ISBN 0-87584-366-2.

14. CHAMPY, J. (1995). *"Re-engineering Management: The Mandate for New Leadership"*, Harper Collins, London, UK., ISBN 0-00-255521-2.

15. DAVENPORT, T.H. (1993b). Book review of HAMMER, M. and CHAMPY, J. (1993)., "Re-engineering the Corporation", <u>Sloan Management Review</u>, **37**, (4), pp 103-104.

16. DAVENPORT, T.H. (1994). "Saving IT's Soul: Human-Centred Information Management", <u>Harvard Business Review</u>, **72**, (2), pp 119-131.

17. KAVANGH, J. (1994). "Business Process Re-engineering", A-Z Computing, <u>Financial Times</u>, 26 April 1994, p3.

18. EARL, M., (1994). "The New and Old of Business Process Redesign", <u>Journal of Strategic Information Systems</u>, **3**, (1), pp 5-22.

19. COOMBES, R. and HILL, R., (1995). "BPR as 'IT-enabled Change': An Assessment", New Technology and Employment, **10**, (2), pp 121-132.

20. WILLMOT, H. (1995). "The Odd Couple? Re-engineering Business Processes, Managing Human Resources", New Technology, Work and Employment, **10**, (2), pp 89-98.

21. SCHONBERGER, R.J. (1994). "Human Resource Management: Lessons from a Decade of Total Quality Management and Re-engineering", <u>California Management Review</u>, **36**, (4), pp 109-123.

22. MAJCHRZAK, A. and WANG, Q., (1996). "Breaking the Functional Mind-set in Process Organisations", <u>Harvard Business Review</u>, **74**, (5), pp 93-99.

23. CLARK, K.B. and FUJIMOTO, T., (1991). *"Product Development Performance: Strategy, Organisation and Management in the World Automobile Industry"*, Harvard Business School Press, Boston, USA., ISBN 0-87584-245-3.

24. OLLE, T.W., HALGELSTEIN, J., MACDONALD, I.G., ROLLAND, C., SOL, H.G., VAN ASSCHE, F.J.M. and VERRIJN-STUART, A.A., (1991). *"Information System Methodologies: A Framework for Understanding"*, Second Edition, Addison-Wesley Publishing Company, Wokingham, UK., ISBN 0-201-54443-1.

25. BAINES, R.W., SMALL, C.A. and COLQUHOUN, G.J. (1996). "Modelling Methods in the Manufacturing Change Process" IN: "Advances in Manufacturing Technology", <u>Proceedings of the Twelfth National Conference on Manufacturing Research,</u> University of Bath, September 1996. pp 91-95.

26. ROSS, D.T and SCHOMAN, K.E. (1977). "Structured Analysis for requirements Definition", <u>IEEE Transactions on Software Engineering</u>, **SE-3**, (1), pp 6-15.
27. ICAM. (1981). "Integrated Computer Aided Manufacturing (ICAM) Architecture Part II, Volume IV - Functional Modelling Manual (IDEF0)", Air Force Material Laboratory, Wright-Patterson Air Force Base, Ohio 45433, USA., AFWAL-TR-86-4006.
28. MAYER, R.J., CULLINANE, T.P., KNAPPENBERGER, P.S., PERAKATH, W.B. and Wells, M.S., "Information Integration for Concurrent Engineering (IICE), IDEF3 Process Description and Capture Method Report", AL-TR-1992-0057, Wright Patterson United States Air Force Base, Ohio, USA, 1992.
29. PUN, L., DOUMEINGTS, G. and BOURELY, A., (1988). "The GRAI Approach to the Structural Design of Flexible Manufacturing Systems", <u>International Journal of Production Research</u>, **23**, (6), pp 1197-1215.

Postal information framework

D J MILLER
IBM Global Postal Business, Feltham, UK

SYNOPSIS

Around the world, almost without exception, Posts are planning and implementing strategies for reinventing themselves and for re-engineering their processes. Huge investments are being made in mail processing systems, distribution networks, track-and-trace technologies, and electronic alternatives to traditional products and services.

Posts can increase the value of investments in information technology by adopting a systematic approach to identifying required information from an overall perspective, rather than a project-specific one. This can guide the development and prioritisation of projects by identifying information requirements and gaps, and ensure integration of mail processing equipment and systems with more general support systems and IT infrastructures.

1 BACKGROUND

When posts want to improve their productivity, or offer new services to an ever demanding customer set, like most businesses in the world today they turn to information technology for a solution to the problem in hand. However, like most businesses, they are sometimes disappointed with the end result. Why is this? It often comes down to one of three factors.

First of all, there is a communications barrier between the business users and the information technology providers. Assumptions are made (on both sides) about the knowledge and understanding of the other party. The business users assume the information technology providers understand the problem they are trying to solve, and the information technology providers assume the business users are *au fait* with the current systems, the opportunities and constraints of current and imminent technology, and jargon abounds (again on both sides).

The language used to communicate is imprecise; sometimes individual words have several meanings even within the business community - who is the "customer" in a customer information system for example? It is not surprising, therefore, if the end result of communication based on incorrect assumptions, unclear definitions and imprecise wording is not what the users wanted.

Secondly, there has been a lack of reuse. In other words, developments have been made which do not build on previous business systems or technical infrastructure. Whilst this is to some extent driven by the relentless advances in information technology, there are signs that enough common standards are coming together, through the emergence of the internet, that reuse will become a much easier proposition. This is more than just using Commercial Off-The-Shelf (COTS) products, although that has a role to play. If systems are designed with reuse in mind, based on a common or even universal set of standards, then it is more likely that a system implemented today can be usefully extended tomorrow. This would save costs and time, and reduce the risks in information technology projects.

Finally, not enough thought is given to future uses of the function or information within a proposed system. Business users and information technology providers reach for their blinkers and only look at the system in hand, because they believe looking at the "big picture" is too hard and takes too long when they are (naturally) looking for quick results, This is why posts find it difficult to implement track and trace systems, for example. Not because the information isn't there, but because it is spread across many systems designed without this future requirement in mind. Then, having gone though the pain of integrating all the necessary data, they suddenly realise that they also need a container tracking system, when some structured planning would have enabled them to come to that conclusion earlier. This lack of integration has created "islands of computing", which serve their own purposes but don't provide the enterprise wide benefits that they could.

In summary, the business users don't get what they want, get it late and over budget, and miss the opportunity of providing new services quickly. This is not a happy position to be in.

The Postal Information FrameWork is an approach which attempts to get to the bottom of these problems, and to offer a practical solution to improving communications, reuse and integration of information systems. So what is this magic cure?

2 INFORMATION FRAMEWORK

At the end of the 80s, IBM Research and Development staff examined this problem and proposed a concept called "a framework for information systems architecture". It used the analogy of an architect designing and building a house. In it, consistent information about a business could be portrayed in a framework providing different "views" (such as what to build, how to build and when to build) in increasing levels of detail for the owner, the architect and the artisans (such as builders, plumbers and electricians). Each view and level would be linked so that, for example, a change in requirements (such as an extra bedroom) would be reflected in a design change and appropriate instructions to the builders.

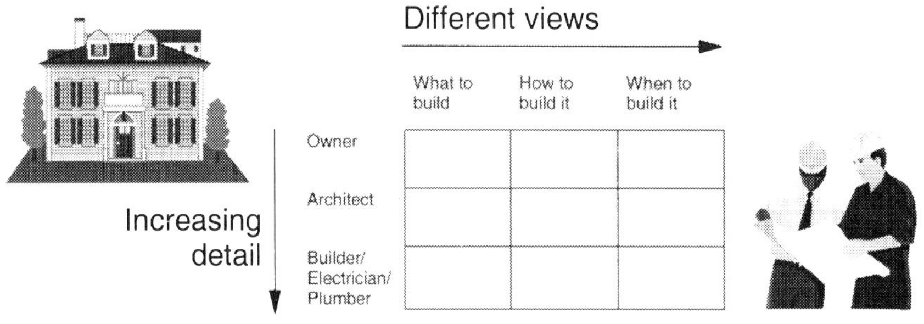

Figure 1 Analogy with building a house

This concept was taken up during the early 90s by IBM's Banking and Financial Services sector, and investments were made to create what is now called the Information FrameWork (1). Further investment has been made to add postal knowledge to it to create the Postal Information FrameWork.

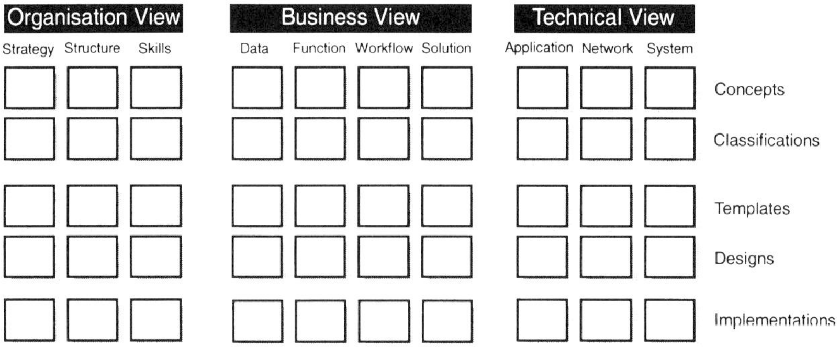

Figure 2 Architectural framework

The business version of the earlier picture shows a ten by five matrix of cells, which are linked on many levels. Across the matrix, the columns are divided into three sets of information:

- Organisational information about strategies, organisation structures and skills
- Business information about data, functions, work flows and solutions
- Technical information about applications, networks and systems

Down the columns, the information becomes more constrained, starting at the conceptual level and moving through logical and physical designs to actual implementations.

This architectural model is the first element of the framework. The second element of the framework is the content of the cells, for example a function model or a technical architecture. Models are structured specifications or classifications of information, usually held in textual form in some kind of knowledge management tool, where they can be easily searched, selected, modified and arranged to assist in the task at hand.

This is where the postal content features. A comprehensive function model and data model have been created which can be used for various planning and design activities, of which more later. In addition a solution model containing various business solutions applicable to posts has been devised, and finally a technical architecture based on e-business standards.

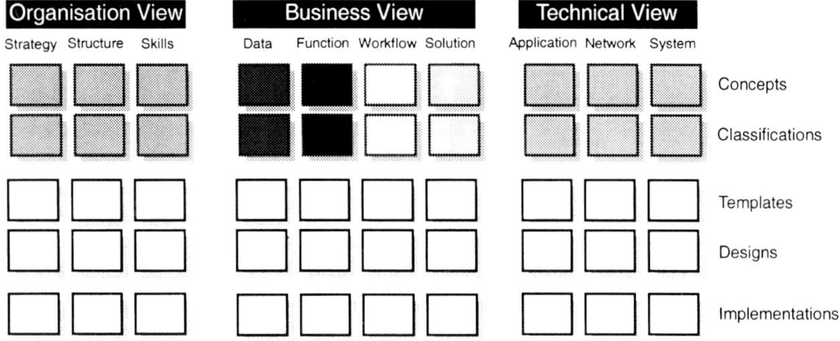

Figure 3 Business models

3 FRAMEWORK APPLICATION

The value of the framework approach is not just having the information held in a consistent manner, but also that these individual information items can be linked. Let's take an example of a strategy change, such as to become more customer focused.

Adopting a new customer focus strategy has structural implications. The post is likely to be organised on a "line of business" basis -- letter mail, parcels, and so on. Such a strategy would also require new customer facing skills, and enhanced written and verbal communications from staff. Thus there are organisational and training issues to be addressed.

The post may choose to provide customer access to information about its services through a telephone call centre and the Internet. This may require new data to be held in the customer database (contact points, contact history) and associated functions to maintain that data. This would mean system changes.

Offering a telephone service means some processes might have to be re-engineered. There's only a few minutes in which to satisfactorily handle a telephone call. Improving the productivity of the call centre staff, a new call centre solution would provide an interface between the customer database and the telephone system to automate caller recognition or dialling. This would involve a project to introduce new technology. This ability to co-ordinate changes from business strategy to technology illustrates the power of a linked information base such as that provided by the framework.

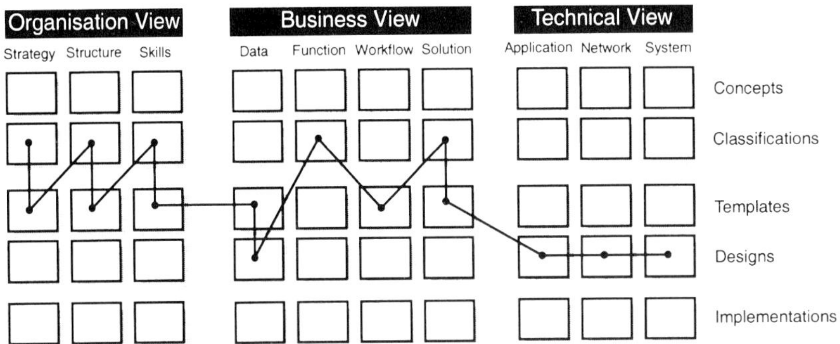

Figure 4 Linking between models

The framework has standard "route maps" which show how to use the framework. These include projects addressing the following areas:

- application development
- data warehouse
- enterprise technical architecture

So how does the framework address the problems raised earlier?

Firstly, it facilitates communication by providing clear definitions of business function and information and unambiguous meaning of names and definitions. These are created and jointly owned between business users and IT staff.

Secondly, it encourages reuse. The same function and data definitions are used for all projects. Information collected at a high level for planning purposes can be reused for design purposes. Identifying IT infrastructure standards makes future integration easier.

Finally, it helps avoid the "islands of automation". By revealing the "big picture", it helps everyone to understand the scope of a single project in respect of other systems. The tool allows everyone easily to view the gaps and overlaps between systems. The "big picture" avoids systems development in isolation.

4 EXAMPLES OF USE

Let's see some examples of how this helps, by using the function and data models to help with three project examples:

1. Project scoping and package selection for a call centre application
2. Defining management information for parcels processing
3. Creating an e-business strategy

Before we start, here's an example of the kind of information held in the function model.

NAME: Relationship Monitoring

ID: FMP394

DEFINITION: The management of activities which are required to monitor relationships between the Post and its customers and other organisations.

EXAMPLE: Activities supporting this function are, for example, to become aware of all the involved party roles of a particular customer, to monitor which products are used by that customer and how often, and to formulate a profile of all the business relationships an organisation may have with the Post.

PURPOSE: To support the buying and selling functions.

Figure 5 Relationship Monitoring definition

Each cell has a standard format showing the number, definition and purpose, along with examples. There are over 500 functions defined in the Postal Information FrameWork. These models have been created using a software tool called *m1*, which provides:

- navigation of all the content models,
- full support for customising and extending the content models,
- project views that allow users to select the items that are relevant to a particular project,
- export facilities to use the information in the models in CASE tools, and
- model management including model comparison, model merge, audit and security.

4.1 Project Scoping

Now let's look at the project scoping example. By using the provided function model outlines, functions can be easily marked for inclusion, perhaps with some indication of priority. The definitions can be reviewed and changed during the exercise, so that by the end of the exercise a high level, complete, agreed project scope can be produced. The result is an unambiguous definition of the project with requirements defined at the same time.

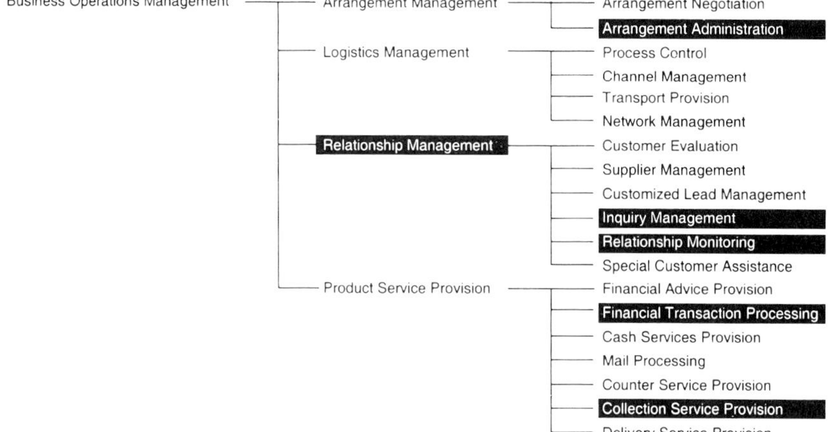

Figure 6 Project scoping

In this example of a customer collection call centre, the high level functions selected include those required to manage the relationship, handle inquiries and to process mail collection requests. An equivalent exercise could be conducted on the data model to identify what information is required by the agent in order to perform these functions.

The package selection is similarly conducted. Each package being evaluated can be matched against the agreed scope, and highlighted on the model to show where the gaps and overlaps are. This provides a neutral and clear picture of what each package provides, enabling an informed choice to be made.

4.2 Management Information for Parcels

There is much information the business would like to know about the parcel mail it processes, such as information about the mail piece (who from, who to, size and weight), the equipment used in the processing (availability, throughput rate and service history), the people (video coding keying rate, error rate, facing quality), not to mention information the post would like to provide to its customers as part of an improved service, such as track and trace information, proof of delivery and service level information. The list is endless.

To manage this "endless" list a systematic approach to data requirements is needed. The data model of the framework can provide this inventory of possible data requirements. Once identified, the data items selected can be further modelled to produce a database design, reusing the structured information already gathered. In addition, the data requirements can be mapped against existing data sources, so that a system design can be done.

We can look at a few examples. The framework has nine fundamental data concepts; some of the key ones are Product, Resource Item and Event. These fundamental data concepts can be further classified into types, descriptors and relationships. To give an example of each of these we can take parcel processing and look at three aspects of it:

• a product definition that can provide delivery to a specific address or to a PO Box (perhaps with some discount scheme) - this is shown as a type of Product,
• selected descriptors of a parcel (a type of Resource Item), in this case size, shape and weight
• a machine monitoring facility whereby faults (a type of Event) are monitored and service scheduled based on a threshold number of faults being reported.

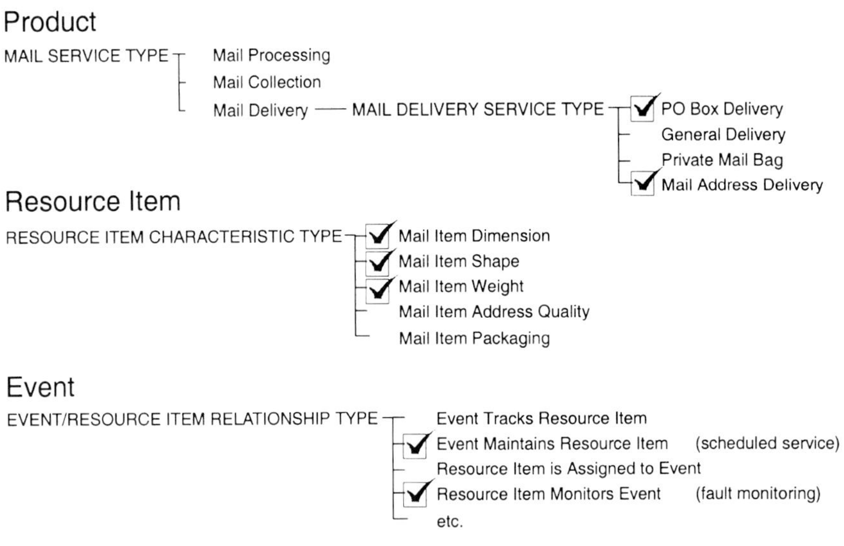

Figure 7 Management Information

Using this approach the "endless" list can be defined and agreed, and a database and system design put in place based on this agreed list. What's more, because this is based on a common repository of information, there is no excuse for building databases that duplicate information held elsewhere, that only have to be interfaced to or merged at a later date.

4.3 e-business Strategy

Finally we can use a similar approach to help define an e-business strategy. Potential areas for e-business can be analysed and selected, providing a basis for discussion across lines of business and to enable understanding of the implications on existing systems.

This example could be taken much further. Methods and tools are available to help their customers move to the "e-future". The following diagram shows the overall approach.

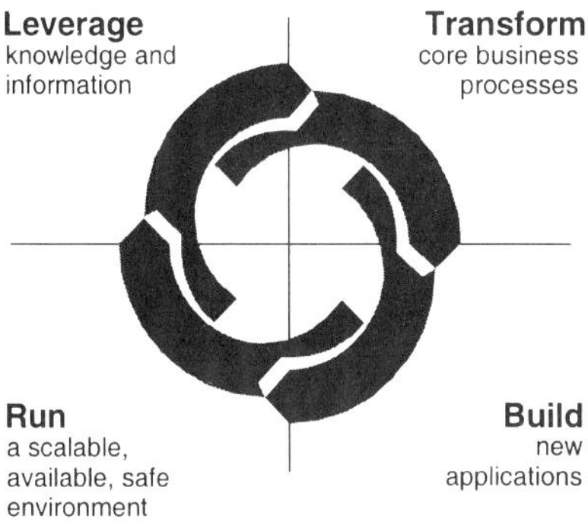

Figure 8 e-business cycle

Posts, along with most enterprises, have lots of data that remains to be exploited as information, and knowledge within their people that is used to provide the functional capability of the post. The framework is an excellent way of capturing the extent of that knowledge and information, and to examine ways it can be better used - "leveraged" to use the American!

To transform the business is an undertaking that requires both organisational and process change. The framework allows the definition of strategy, structure and skills, which can be linked to workflow definitions, to help understand the impact on the organisation of any potential change to a process, and *vice versa*.

Building an application to support the new processes can be simplified by using information already gathered in the planning and design stages, and the framework can be used to support this information reuse. Also there may be solutions already available and as described earlier the framework is a great tool for project scoping and package selection.

Finally, these applications need an environment in which to run, a technical architecture which is secure, scaleable and highly available, characteristics which are essential in the 24 by 7, universal access "e-future" we are moving to.

5 SUMMARY

In summary, the framework contains information, methods, tools and solutions which can help a post take full advantage of information technology, to help them understand where they are today, where they want to go and how to get there. Other industries, particularly banking and financial services, have faced many of the same issues of lack of communication, reuse and integration, and have used the self same approach to address these problems.

The Postal Information FrameWork is a map, essential if you want to know where you are and how to get somewhere - all you have to do is know where you want to go!

REFERENCES

(1) Information FrameWork, Roger Evernden, IBM Systems Journal, Vol 35, No. 1, 1996

C568/039/99

The evolution of message processing and communications

W S CONARD, R J WOJDYLA, and **M YOST**
Bell & Howell Mail and Messaging Technologies, Chicago, USA

Information distribution in the next millennium will undergo many changes that will embrace new electronic-based communication while evolving the quality and effectiveness of more traditional direct mail. The key to effective communication between information distributors (advertisers, financial organisations, etc.) and their customers is to provide a completely integrated level of service that starts with the message to be distributed and ends with customer action. This process is broken down into seven steps that govern the circle of communication to the customer and back. This circle of communication is shown in the IMPACT™ Model below. "IMPACT" stands for "Integrated Message Processing And Communications Technology". Section One of this paper examines each element and its significance to information distribution, while Section Two focuses on three elements: message finishing, message distribution, and process management. The process management associated with these elements is the area that has the most potential for positive impact on postal authorities and their customers.

To stay competitive, businesses must adopt flexible, upgradeable solutions that improve information gathering, processing, distribution, and receipt. This paper first splits this process into seven components and examines each component and its relationship to the other components. We then drill down to examine where the unique relationship that exists between mailers and posts fits into this model and why posts must either adopt new distribution methods or improve upon their current methods to remain competitive. Finally, suggestions are offered that could help both posts and their major mailing customers improve their efficiency and profitability.

Figure 1. IMPACT™ Model

To survive and prosper, a business must adapt to changes in its environment. While that maxim has been true for centuries, the recent convergence of communication technology has taken it to the extreme. Newer, better ways of communicating are developed so quickly that many systems run the risk of being obsolete before they can be fully implemented.

It is happening in company mailrooms all over the world. In fact, one day soon, the "mailroom" will not be a room, and it will not just deliver mail. It will evolve into a set of integrated functions and networked systems located throughout a corporation. And it will handle postal mail, hybrid mail (the combination of electronic and physical delivery of information), the Internet, CD-ROM—any of a dozen messaging channels that will connect businesses to their customers.

                                          C568/039 © IMechE 1999

In order to stay competitive, businesses must understand this new world order and position themselves to adjust to it. This involves finding flexible, upgradeable solutions that improve almost every point in the process of gathering, processing, distributing and receiving information. This process can be split into seven components, as shown in the IMPACT™ Model above: 1) customer information management, 2) media selection and message formatting, 3) message finishing, 4) message distribution, 5) response management, 6) process management, and 7) customer care. All of these components are necessary to close the communication loop that connects companies to their customers and provide integrated, end-to-end management of business messaging systems.

1.1   Customer Information Management

Customer information management includes processes and technologies that manage the format of customer messages and the information in underlying customer databases. Using this information, businesses can create highly targeted one-to-one marketing programs to strengthen customer relationships and increase revenues.

Creating a one-to-one marketing program requires customer-profiling tools that help target messages, the ability to segment customers by demographic characteristics and buying patterns, and a software-driven infrastructure that enables fast, reliable and selective insertion of marketing materials.

1.2   Media Selection/Message Formatting

Media selection and message formatting technologies match and format customer messages to the most appropriate medium in terms of customer preference, cost, speed, integrity, and impact. To compete most effectively in this arena, businesses must be able to do the following:

- Merge and combine documents

- Create, reformat, and redesign electronic and paper-based output

- Manipulate customer data for print or electronic transmission without the time, cost, and risk of changing legacy systems

- Cleanse address databases

- Secure the largest postal discounts and highest levels of distribution efficiency

- Convert information to media such as Web, e-mail, fax, print, or personal software applications

- Dynamically insert personalised content and targeted marketing material into the electronic statement process

- Integrate physical, electronic, and hybrid mail to obtain the most effective method for information distribution

Media selection and message formatting take on even more significance with the rising popularity of electronic bill presentment and payment (EBPP), and the ability of businesses to provide customers with a choice of formats. Electronic delivery of consumer bills is expected to grow rapidly during the next five years, fuelled by the expanded use of the Internet and home computers. To remain competitive, billers will need to offer EBPP to their customers, even though most of their bills will still be delivered via traditional print and mail operations.

Until now, billers have implemented EBPP as pilot projects, separate from their paper billing operations. As the volume of electronic bills grows, however, the need to integrate with paper-based operations will also grow. This integration will simplify the movement of customers from one delivery method to another, improve marketing efforts, ensure seamless customer service, streamline data processing, and measure the performance of the entire billing operation.

Regardless of the delivery method, data from billing systems must be reformatted before reaching customers. For electronic delivery, the data must be parsed and converted into a format suitable for presentation on the Internet, using HyperText Mark-up Language (HTML).

For paper delivery, this print stream also must be manipulated before printing to add finishing equipment instructions, verify addresses for postal discounts, and improve the appearance of the bill. This data preparation process can be integrated for electronic or paper delivery using common tools for print stream manipulation. A single tool set for both delivery methods can reduce training and support costs.

1.3   Message Finishing

Message finishing systems process and package messages for delivery to customers and prepare them for distribution, either as traditional mailpieces, e-mail, CD-ROMs, or some other form of electronic message. The ultimate goal of a successful finishing operation is to achieve the lowest cost per mailpiece and highest document integrity available, while also ensuring efficient, cost-effective, and high-impact communications to customers. Requirements for achieving this goal include the following:

- Systems for accumulating, folding, inserting, evidencing payment, and sorting filled envelopes
- Electronic message image creation and transfer to e-mail, CD-ROM, or other media
- System durability and modularity
- System scalability for upgrades to pace business growth
- System flexibility to support traditional mailpieces and cost-effective hybrid mail; this flexibility also provides customers with optional methods of receiving their mail

Nowhere are the changes wrought by technological advances more obvious than in the mailpiece itself. Historically, the document sent to a customer was simply formatted, transaction-oriented, and measured largely by its cost. The new document is personalised, improves cash flow, markets new products and services, and differentiates the sender from the competition.

Greater use of software in the inserting side of the business has significantly changed what a mailer is able to do with a mailpiece. Host-based software, for example, has enabled the mailer to make changes without changing legacy applications. The advent of spot colour printing at high speeds has enabled better personalization of the mailpiece, and improvements in the mail delivery infrastructure have ensured quicker response times.

New levels of customisation, however, mean more elements that must be co-ordinated. That need for greater co-ordination requires higher levels of integrity in the print/finish environment. Because more integrity usually means more money, managers must balance their desire for 100 percent mailpiece integrity against the realities of their budgets.

Some mailers merely want to track the completion of a mailing by job. Others want to send information to other departments showing the time each mailpiece was sent and what it included. Other mailers want to track postage costs by job, department, machine, and operator. Barcode technologies available today enable all of these capabilities. Businesses must decide which technologies are best suited to their organisation.

1.4   Message Distribution

Message distribution processes improve the speed, efficiency, and accuracy with which messages reach their recipient. They also enable businesses to track the progress of outgoing and incoming messages. For outgoing mail, businesses can verify whether a document has been sent, ensuring that billing statements and marketing opportunities are not missed and that customers are not inconvenienced by duplications. An automatic document reconciliation feature allows businesses to detect duplicate documents, produce accurate manifests, customise reports and queries, and make last minute address changes.

Detailed shipment documentation speeds permit or postage mail into the distribution system for faster processing and final delivery. With message distribution products and processes in place, businesses can provide accurate, comprehensive manifests of mailings by individual mailpiece, tray, and job. Businesses also can expand control over back-end operations with carrier-acceptance tracking and reporting systems.

1.5   Response Management

Response management systems make sure customer responses—whether paper or electronic—are processed in a timely and cost-effective way, expediting information flow to individuals, application processing systems, and databases. For example, high-speed, intelligent sorters can read, interpret, and group paper-based incoming messages by destination, while imaging solutions convert physical documents into electronic data.

Additional products and processes would allow businesses to do the following:

- Rapidly identify incoming checks for quicker processing and better cash flow

- Integrate enterprise-wide directory for instantaneous look-up and routing

- Use imaging systems to make archiving and retrieval more efficient

- Interface seamlessly with application processing systems

- Streamline multi-step processes while reducing errors, manual labour, and hand-offs

1.6   Process Management

Process management borrows from the tools used in the manufacturing systems sector. These technologies give businesses effective management of the customer messaging life cycle. Within the mail processing environment, this includes closing the loop between mainframe systems and the mailroom floor with detailed tracking of individual jobs, documents, pages, and inserts across the entire message processing environment. Elements of an effective process management program include the following:

- Advanced scanning devices for complete audit tracking by document

- Centralised management control software for supervision of local and remote operations, regardless of vendor

- Real-time status reports on equipment, jobs, and operators to alert managers to diminished production or equipment problems

- Reporting tools that provide comprehensive trending data and analysis, leading to improved workflow and optimised resources

Advanced scanning devices that can read all major types of barcodes, including DataGlyphs, help process items quickly and accurately by matching the barcodes on each piece of a document set. With this technology, complex statements can be processed quickly and with the correct number of pages, in the correct order, on the correct letterhead, with the correct page orientation—virtually every time.

Centralised management control software can allow supervisors to monitor multiple facilities—even if they are located in different parts of the same country. Similar centralised monitoring of operations in foreign postal operations will be possible even through control will not be authorised, and information collection will be made available in real time. It also can provide real-time status reports that allow managers to find potential problems and take corrective action without leaving their offices. By fixing small problems before they become large and expensive barriers to customer communication, this process management tool can increase the likelihood that bills, marketing materials, and other customer communications will be sent on time and boost the potential for positive cash flow.

A closely monitored messaging operation generates a cornucopia of performance data. Using automated analysis and reporting tools, managers can organise performance data into trends that influence future decisions about resource allocation and processes. All of this can be accomplished without losing production time now spent filling out forms or consolidating data into spreadsheets.

1.7   Customer Care

Customer care in the mail processing industry traditionally has been limited to hardware and software maintenance. Today, it must go well beyond those boundaries. Companies must embrace a complete service philosophy that encompasses everything from remanufacture and warranty of old systems, to integration of leading-edge technology, to immediate and informed responses to every inquiry. As the messaging industry changes, the availability of consulting services and financial services becomes more important. Consulting services, supplied by industry experts, ensure that businesses stay on the cutting-edge of technology, and make the most effective use of this technology to meet their business goals. Financial services ensure that businesses make the best use of their investment dollars with cost-effective software, hardware, and technology replacement programs. In addition, effective customer care programs include the following:

- Training in areas such as postal knowledge and equipment operation

- Consulting services, customised solutions, and project management to increase productivity and implement new technology

- Recovery services with contingency plans to maintain work flow during times of natural disaster or crisis

- Financial services to minimise up-front outlay, bring solutions and new technology within financial reach

2. ADDING WIN-WIN VALUE TO THE UNIQUE MAILER-POST RELATIONSHIP

This part of the paper will examine the relationship that exists between major mailers and posts. According to the IMPACT™ model shown in Section One, major mailers can be defined as "message finishers" while posts typically fall into the role of "message distributors". Process management can provide connectivity within and between the message finishing and message distributing processes.

While every postal authority has a goal to increase mail volumes, every message finisher wants to effectively get the message to the customer at the lowest cost and maintain the highest level of product value in an increasing competitive messaging market. It is these overlapping goals that will drive processing changes, and that will result in new levels of performance. Section Two will first examine what is needed to maintain mail volumes with increasing competition from the electronic messaging market (e-mail) and then review the opportunity for increased data communications between the organisations that will yield new services and new levels of delivery performance.

2.1   The Way It Is

Over the last 20 years, posts world-wide have been changing their methods for sorting and distributing mail from manual and mechanised methods to mostly automated methods. The cost of processing mail has been increasing with regularity, making automation the best choice for posts if they are to improve productivity and reduce expenses. Automated sorting of 1,000 letters saves the United States Postal Service (USPS), for example, $36 over older and less efficient methods.

For posts to stay competitive and increase value of physical mail messaging, cost and quality of delivery are important drivers when competing with electronic or non-physical messaging competitors. Transaction confirmation and the physical everyday presence of a message are areas with which electronic messaging have and will have difficulty competing. Even though there will be an increasing level of transactions, payments, and purchases occurring via the Internet, proof of the transaction occurrence and the security against data manipulation will most certainly continue to be questioned and challenged in the future. Confirmation of these types of occurrences can only be assured via a physical document. Physical proof will always hold up when challenged by any legal system. Therefore, speed and reliability of physical mail delivery and confirmation of delivery will play a very significant role in the future in support of electronic transaction messaging markets.

Track-and-trace, capacity planning, and direct interjection into downstream processing will play an increasing role in the future as the cost of entering the electronic messaging market arena becomes less and less costly, and more people become comfortable in its use. For posts to leverage this new market, investing in these technologies will assure a competitive place in the future for physical messaging.

In the advertising markets where revenue-driven motives are the main reasons for using drive mail messaging, customer response to a mass advertising mailing must exceed the cost of the mailing. Internet advertising is becoming a less expensive way to reach potential customers, but it cannot target the market effectively. Direct mailing campaigns can be tuned to customer needs. The mailings can be selective, whereas the Internet or e-mail advertising cannot selectively market, thus reducing advertising impact. Saturation mailings can also be timed to produce availability within a particular region. Electronic interactive communication between the direct mailer and the post can provide feedback to the mailer as to when a mailing is being dropped to the customer via sampling track and trace methods. Feedback to the mailer would allow the mailer to make a follow-up call.

To maximise the efficiency of their automated processing methods, posts use sorting equipment that processes only "machinable" and "readable" mail. The term "machinable" refers to mail that is of the proper size and shape and made of the proper material for high-speed transport through automated systems. "Readable" mail generally contains a machine-printed address or barcode for scanning and sorting by optical character readers and barcode sorters. With the advancements of technology, the pool of "readable" mail is increasing to include non-machine printed mail (i.e., hand-written addresses). However, automation compatibility will continue to be higher for mailpieces with stricter addressing standards.
Some posts have created address standardisation and mailpiece design regulations to increase the amount of "machinable" and "readable" mail. Those posts have published these regulations and made them available to their customers. It is important that "major mailers" have these regulations and practice them, as well. Major mailers (hereafter referred to simply as "mailers") can be defined as those postal customers who make large mailings of more than a defined number of pieces of letter mail per mailing (e.g., more than 200,000 pieces of letter

mail per mailing in the United States). Typically, mailers who make such large mailings could be eligible for postage discounts.

A few posts offer a variety of discounts for correctly prepared, automation-compatible mailpieces. Mailers can qualify for automation rates for delivery point barcoded mailpieces submitted according to postal standards. As mentioned above, some posts publish their regulations and standards for mailers. In addition, mailpiece design analysts are often available locally to assist mailers in designing automation-compatible mail. It is obvious that both mailers and posts benefit by increasing the amount of automation-compatible mail. Mailers receive postage discounts and improved service due to automation, while posts enjoy increased processing efficiency and minimise the cost of automation. Whereas these practices should continue, both mailers and posts can do even more to maximise their gains. First of all, posts that have not yet published their addressing standards should do so to increase the likelihood of receiving automation-compatible mail from their customers. In addition, those posts that can offer postage discounts as described above will give their customers incentive for submitting only machinable and readable mail. These are the first two steps that posts can take to increase their own efficiency and profitability. Major mailers can achieve the same results, as well. The following section examines several opportunities for data mining and sharing that represent even more opportunities for significantly enhancing the unique mailer-post relationship.

2.2   The Way It Could Be

Mailers and posts both stand to gain by mining and sharing data that are in some cases already available.

Figure 2 below identifies some of the data that can be mined and shared to support all players in this process improvement effort.

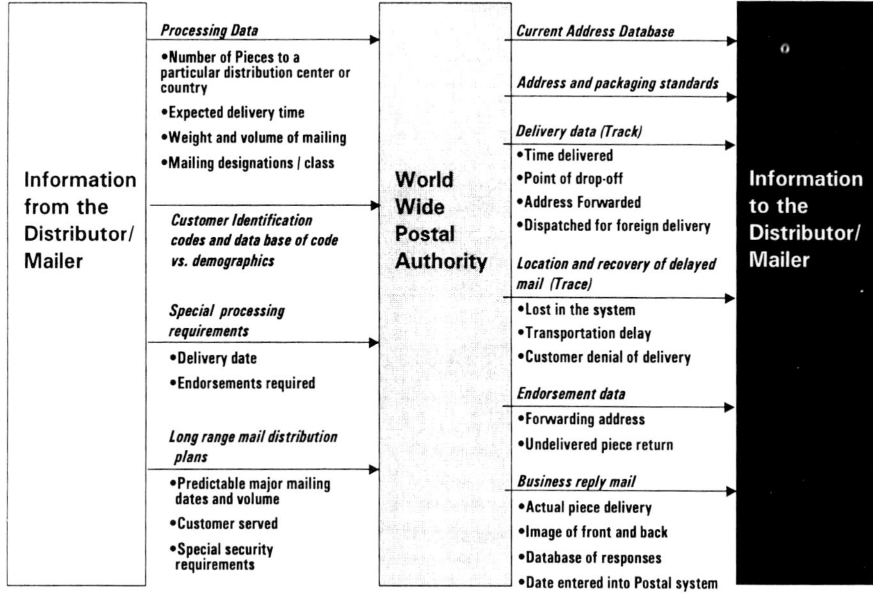

Figure 2. Value Add Information Flow

By investigating and implementing the suggestions for improved data utilisation described in this paper, mailers could do the following:

- Receive additional postage discounts for sharing data with posts

- Receive improved speed-of-delivery and accuracy-of-delivery results

- Receive proof-of-delivery results

- Increase the rate/cycle of their mailings

- Protect their investments/revenue

- Reduce the likelihood of problem mailings (i.e., rejects)

- Broaden the overall lines of communication with their respective posts

At the same time, posts could do the following:

- Utilise mailer data to improve efficiency
- Realise labour, transportation, and automation savings
- Increase revenue through increased volume due to higher levels of customer satisfaction
- Increase revenue by providing track-and-trace services
- Protect revenue through electronic analysis
- Broaden the overall lines of communication with their respective customers

We have separated the methods by which the above benefits could be realised into two categories:

- Mailer to Post Data Transfers
- Post to Mailer Data Transfers

Both categories are examined in more detail below.

2.2.1 *Mailer to Post Data Transfers*

Some commercial mailers create mailings with piece counts in the millions. Unfortunately for both mailer and post, such mailings are often delivered to the docks of a post's mail processing facility without warning. Imagine the logistical and staffing nightmare that this substantial, additional volume must create for a facility manager who has already devoted most or all of his/her available equipment and personnel to processing volumes already introduced into the automation stream. Surely, no facility manager considers this situation to be ideal for achieving maximum throughput.

Mailers would do better to electronically alert individual mail processing facilities at least several days in advance of impending drop-offs, including the following data in their transmissions:

- Date and point of drop-off
- Quantity of each mailing by destination ZIP or region
- Mailing designations (e.g., first class, standard class, etc.)
- Foreign destination data
- Delivery-due requirements

With this data in hand, mail processing facility managers could more adequately prepare their staff and equipment to process the mail in the most efficient manner possible. In return for this critical logistical data, posts could offer additional postage discounts to those mailers who choose to supply it. Mailers might gain by receiving more efficient delivery, as well.

Mailers would receive the best possible service for their money by implementing a transfer of such data from mailers to posts, as well. Posts would also increase the likelihood of maximising the reductions to processing costs (e.g., labour, transportation, equipment) that automation brings.

2.2.2   Post to Mailer Data Transfers

Mailers must not only design, but also maintain automation-compatible mailings to maximise their postage discounts and level of service. Proper mailpiece design is the first step toward fully realising the postage discounts available to them from posts for providing automation compatible mail.

Those posts that publish postal regulations and mailpiece design documents, make them available to mailers. Typically, for a nominal fee, a mailer can obtain these publications from their local post, or they can order them. Most regular mailers are aware of these regulatory documents; they have them in their possession and use them. Some mailers, however, particularly new mailers, might not be aware of their existence, let alone the importance of mailpiece design standards. Many posts also offer the services of local mailpiece design analysts, who can work with mailers to ensure the creation of automation-compatible mail.

While it is agreed that *all* posts should publish their standards and *all* mailers should use them and consult mailpiece design analysts, it is difficult to gauge how many mailers take advantage of these documents and services, and whether mailers re-examine their design practices for conformance with any regularity. An incentive program would motivate mailers to use such documents. For instance, if mailers meet the standards to improve mailpiece delivery, then the postage would be less than the normal postage rate. The benefits to the mailer would then be faster and accurate delivery of their mail.

Further data transfer beyond the published regulations/standards from posts to mailers could help to maintain and, when necessary, refine addressing practices or standards.

Posts could supply mailers with addressing data. This would allow mailers to identify isolated instances of non-compliance or general trends in mailpiece design that are resulting in rejected mail. Mailers could then use this data to correct addressing problems themselves. To properly and adequately resolve any problems, mailers should discuss the reject data and their corrective actions with local postal officials.

Additionally, as mailers become more interactive with the post, mailing lists could be transferred electronically to the post, printed, and then inserted at a local delivery mail processing centre. Sorting or mailing handling would be eliminated upstream, reducing the need for automated processing equipment. Customised mailpiece design would increase the value of mailing and this new kind of service.

Conversely, as mailpieces enter the post (e.g., surveys, payments, and products), order placement forms can be sorted or scanned and processed for the mailer. These types of services could add significant value to physical mail messaging.

Posts could regularly study reject data from individual mailers to identify "problem" trends experienced by mailers in conforming to mailpiece design standards. Communication of these problems back to the mailers would eliminate the continued inefficiencies associated with improperly designed mail.

The overall benefits of posts sharing addressing data with mailers are "win-win" for each party. Mailers who analyse reject data to develop corrective action would be ensuring the integrity of their addressing practices, thereby reducing their rejects and maximising their postage discounts. Posts could proactively alert mailers about how to avoid specific problems, thereby reducing the number of rejects and the overall efficiency and cost benefits of automated processing.

Posts could also supply mailers with online address forwarding data. If mailers had online access to this data, they could regularly update their directories, thereby reducing the amount of automation necessary, and increasing their chances of yielding the fastest delivery results possible. Posts would also benefit. By keeping customers updated of changes to their address forwarding directories, posts could reduce the amount of mail in their forwarding automation streams, thus lowering the cost of automation per mailing.

2.2.3   Track-and-Trace Barcoding

Wide area networking within a post and linking a customer's mailing list within the network would allow the mailer and the post to track and trace the mailpiece within the distribution network.

These track-and-trace type barcodes can be read by postal sorting equipment, which, in turn, makes the encoded mailpiece information available for mailers to access electronically over a centralised network server. Mailers can learn, for example, when outgoing mail is nearing delivery or when they will receive customer reply mail.

Such information could allow mailers to improve customer service, identify delivery trends with regard to specific mailings, refine their "due date" accordingly, and co-ordinate more efficient staffing.

Posts that develop barcodes that track mailings electronically and provide additional information about each mailpiece could reassure mailers that they are receiving the best possible service available for their postage investment.

Along the same lines, pre-printed reply envelopes such as those for bill payment could be tagged so that the mailer would be pre-notified that the payment was coming in from a particular customer. Also, pre-printed ordering forms could be marked or printed with a special barcode that would notify the mailer that a certain product was being ordered. The benefit to the mailer is improved inventory and cash management.

Lastly, track-and-trace would allow the mailer to determine address quality. Mailing could be mailed by regional groupings that could be tracked by individual mailpiece for barcoding accuracy.

2.2.4 *Peak Processing Data*

Some posts collect and analyse "peak processing" data from mail processing facilities and bulk mail centres. They then use that data to identify those facilities/centres that can most efficiently process specific customer mailings, offering additional discounts to those mailers who choose to take advantage of this strategic data. Such analysis is key to a post's ability to operate at peak efficiency.

For example, let us assume that a fictional bank in New York, New York is known by the United States Postal Service (USPS) to make regular mailings to bank customers in Los Angeles, California. Through its analysis of data from its mail processing facilities in both New York and Los Angeles, the USPS has determined that mail volumes in the New York facility are at their peak during the period that the bank regularly makes its large customer mailing to Los Angeles. The data also reveals that, conversely, the Los Angeles mail processing facility's mail volumes are "light" at the time the bank regularly makes its mailing to Los Angeles. The USPS could offer additional postage discounts to the bank if it agreed to work with a print shop and a pre-sort house in Los Angeles to prepare, sort, and drop off the bank's mailing at the Los Angeles mail processing facility. For agreeing to comply with the USPS's request, the New York mailer would not only be more likely to receive the optimal rate of automation processing, but also additional postage discounts. Likewise, the USPS benefits through eliminating the need for initial processing of the mailing in its New York facility.

3. CONCLUSION

Methods of message processing and communications are constantly evolving. Posts and mailers who share the types of data described in this paper can enhance their operational efficiency as well as their revenue. Furthermore, the mere act of analysing and discussing such "out-of-the-box" concepts can only serve to open wide the lines of communication between customer and service provider—the cornerstone of any good business relationship.

E-services and Hybrid Mail

C568/011/99

Re-engineering the mailer–post interface

K P SMITH, R D G SPENCELEY, and J L WELLS
Technology Centre, Royal Mail, Swindon, UK

SYNOPSIS

REMPI (**Re-E**ngineering the **M**ailer-**P**ost Interface), an electronic commerce pilot project, is the successor to TIMBRE (Technology In Mail Business Re-Engineering) which established the basis for re-engineering and process integration in the postal business.

This paper discusses both projects, indicating how TIMBRE provided the foundations for REMPI and other pilot projects, and how these could revolutionise postal operations by effective use of recent developments in information technology.

REMPI objectives are explained, achievements to date are listed and potential opportunities for improving the mailer-post interface are explored. The use of the Internet as a cost-effective vehicle for electronic data interchange is also outlined.

The paper indicates how the REMPI pilot will demonstrate the benefits for both the Posts and their customers, and how these could be exploited in the future.

1. INTRODUCTION

Despite many prophesies that it would be overtaken by technological advances such as telegraph and facsimile, the postal industry has survived and flourished. It is one of the world's largest and most international service industries; the universal service providers ("the Posts") have 662,000 outlets in more than 178 countries, and annual volumes of over 420 billion items, generating around euro 120 billion pa turnover, of which around 7% is between countries. In Europe volumes exceed 80 billion items, resulting in a turnover of euro 30 billion pa, providing employment to more than 1.4 million people and serving a customer base which includes virtually all businesses and private consumers.

However, today the Post is under threat more than ever before, from substitution of physical mail by electronic alternatives, through revenue loss due to fraud & payment-service mismatch, and because market liberalisation is bringing increased competition.

Technological developments, however, also create the opportunity for major improvements in the efficiency with which the Posts operate. Royal Mail has for several years invested in these new technologies, particularly in the fields of mechanisation, optical character recognition and the computerisation of tasks which have historically been carried out manually. Recently, there have also been significant advancements in item identification & tracking, printing technology and multi-media communications.

If the postal industry is to meet the new challenges presented by market liberalisation and electronic alternatives to traditional mail, it must respond by offering new and improved services, with greater efficiency of operation and higher reliability. Tinkering with existing procedures or piecemeal adoption of new technologies are unlikely to achieve the necessary results. Effective use of Information Technology must be made, in order to control the physical handling process by integrating it into the postal system, and thereby re-engineering the whole postal process.(1)

TIMBRE, an acronym for Technology In Mail Business Re-Engineering, aimed to address this issue by bringing together a consortium of interested parties, both Posts and major suppliers, to establish the requirements for re-engineering the postal process.

2. TIMBRE

2.1. Background

In October 1995, Royal Mail joined forces with An Post (Ireland), Post Danmark (Denmark), Siemens ElectroCom[1], Elsag[2], Pitney Bowes and IPC Technology, in the first phase of this process integration and re-engineering project. TIMBRE sought to make significant improvements in the productivity and quality of postal services throughout Europe.

TIMBRE was supported by the European Commission under its ESPRIT programme for Research and Technical Development (RTD)[3]. The project, which lasted only seven months, investigated the requirements for completely re-engineering the whole postal process by embracing recent advances in information technologies and integrating them into the postal system.

2.2. The Postal Process

The mail handling process is composed of three major functional areas:

- Acceptance and Payment
- Sorting and Transportation
- Delivery and Delivery Services

---

[1] formerly AEG ElectroCom
[2] formerly Elsag Bailey
[3] ESPRIT - Task 7.1 of the RTD in Information Technologies Workprogramme.

Figure 1 illustrates these major areas and indicates how each one comprises several processes which link the physical mail flows together.

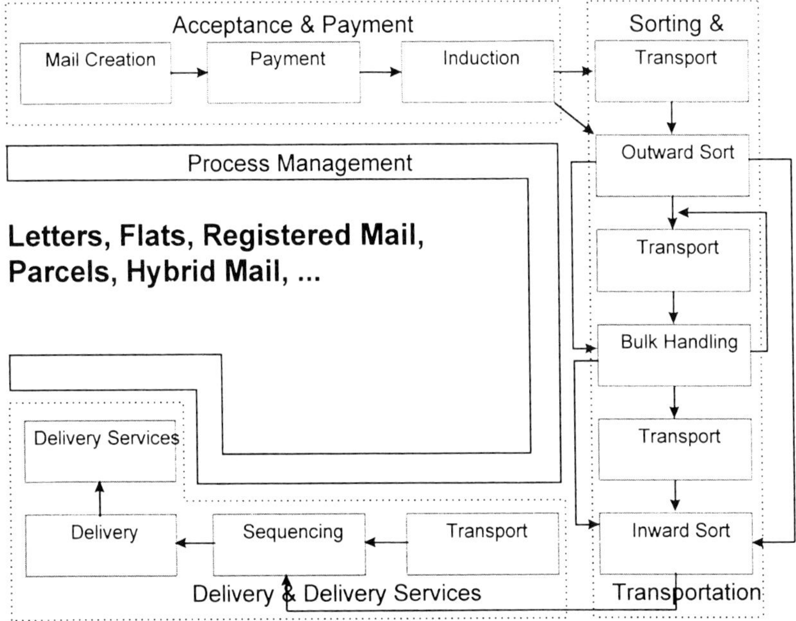

Figure 1. Mail Processing Diagram

The diagram also emphasises the chained sequence of physical mail processing, where the process begins with mail creation by the customer and ends with delivery to the consumer. The diagram also shows that physical mail processing is inextricably linked to process management, the information required to ensure successful processing and delivery of mail items. It is in this area that improvements in the richness of information can be used to enhance the quality of service provided to customers, improve the efficiency of the physical process itself and provide the opportunity for new services, such as day-certain delivery and track and trace of normal mail.

Before any progress could be made to improve the quality of this information it was necessary to understand the current circumstances surrounding mail processing. This involved seeking information from postal administrations about what automation they currently employed and what plans they had for the immediate future. Equipment manufacturers were also canvassed to establish what plans they had for improving mail processing equipment. These investigations were carried out during the initial stages of the TIMBRE project.

2.3. Postal Operators State of the Art Review

Royal Mail's first TIMBRE task consisted of managing a review of the current status of and short to medium term plans for postal automation worldwide. Questionnaires were distributed to several postal administrations whose representatives attended, or indicated an interest in, the TIMBRE Postal Users' Group. Although the timescale for completion was extremely tight, six questionnaires (from the US, Canada, Portugal, Ireland, Denmark and the UK) were returned in time for the information to be recorded in the TIMBRE Postal Operators' State of the Art Report.

The report contained a synopsis, which compared the status of all six countries, and more detailed information on a country by country basis. These provided a *Background* to the regulatory environment, an outline of the mail processing *Infrastructure*, information on *Mail Traffic and Services*, and a report on the *Mail Processing Automation Status*, including the status of *Quality Management and Management Information Systems*. The report also examined the *Automation Strategy for the Future, Posting Trends*, arrangements for *Transportation* and procedures for handling *Redirected, Returned and Wrongly Addressed Mail*.

As would be expected, investment into postal automation was generally based upon the size of the operator and the volumes of mail traffic handled. However, the report revealed a marked difference in the degree of automation between administrations and various approaches towards future developments. Royal Mail for instance, which processes 17.5 billion items annually, was, at the time, heavily involved in completing its automation programme for letters up to C5. This included re-equipping its 76 automated mail centres around the UK with newly developed systems, employing state of the art technology both in materials handling and information technology.

It was also clear that many Posts had already identified the need to invest in information technology and were planning to improve their infrastructures accordingly, whilst others were concentrating on improvements to their core business.

2.4. Postal Automation State of the Art Review

The *Postal Automation State of the Art Report*, which was co-ordinated by IPC Technology, contained suppliers current and future plans for mail processing equipment. The report covered:

- Business Mail Preparation and Payment
- Mail Induction, Transport and Delivery
- Coding, Sorting, Mechanised Handling, In-Plant Monitoring and Control Systems
- Communications
- Tracking & Tracing
- Hybrid Mail

The report covered a diverse range of technologies associated with the Postal industry, but the proposals concerning electronic data interchange between the mailer and the Post for improving meter mail & PPI[4] management and controlling mail induction, are of particular importance. This area of the postal process, the mailer-post interface, has hitherto been susceptible to intentional fraud and unintentional miscalculation of postage, both of which inevitably result in loss of revenue.

2.5. Postal Process Information Model

The *Postal Process Information Model*, which was led by Post Danmark, addressed the decomposition of postal process functions into 31 data entities and 103 data attributes. The model indicated how postal processing activities, from mail production to delivery, could be supported by information technology.

2.6. TIMBRE Architecture Model

Royal Mail's second lead role in TIMBRE was to manage the production of the *Architecture Model*. This documented the options for data capture, processing and management of the data identified in the Information Model. The report concentrated on the information supplied by Posts and suppliers, and took particular account of any limitations identified in "State of the Art" reports.

The Model established six high level data processes which related to the processing functions of the *Information Model*. It also identified areas within the postal system where data could be captured, stored and transmitted. It was evident, however, that because both mailers and Posts would need to make substantial investments to migrate from existing equipment, a single architecture would not provide the solution for improving the exchange of information within the postal process. It was proposed, therefore, that by defining & promoting standards associated with system interfaces, it would be possible to create the basis of a system which could evolve progressively towards an open architecture and ultimately lead to exploitation opportunities for electronic data interchange.

2.7. Standardisation Requirements

The *Standardisation Requirements Report* described a total of 14 items ripe for standardisation. These included mail item identification tagging by Posts and customer item identification and encoding. The standards would be used internally within the Posts, externally at the mailer-post interface and for postage payment and induction procedures. The report was issued to the CEN TC331 Post[5] and to the UPU Technical Standards Board[6] to provide the foundations for future standards definition.

---

[4] Printed Postage Impression

[5] Comité Européen de Normalisation, the European standards body. Its Technical Committee TC331 deals with standards for Postal Services.

[6] Universal Postal Union, the United Nations body which regulates the world-wide postal service.

2.8. Pilot Demonstrator Projects

The project partners concluded that there were five major areas which should be considered for exploitation in future phases of the project. These were:

- revenue management
- extending the mailer-post interface
- improving mail handling through item tagging
- quality of service improvements through logistics management
- business reply analysis & information delivery

It was generally felt that the first three in the list were by far the most important of the pilots. In view of its importance for associating processing data with an individual mailpiece and the potential benefits of cross-border co-operation, Royal Mail, the United States Postal Service and Canada Post decided to develop item level tagging. Revenue management and extending the mailer-post interface pilots were combined to form REMPI - Re-Engineering the Mailer-Post Interface, which ultimately attracted funding support from the European Commission.

3. REMPI

3.1. Background

REMPI is an electronic commerce pilot project aimed at revolutionising the customer interface with the Post. It will result in a totally new form of access to, and payment for, postal services, and will provide the foundation for the development of new customer orientated services, such as proof of acceptance, day-certain delivery and tracking & tracing of normal mail.

Historically the Posts have invested heavily in automation and, as a result, a large proportion of mail is processed by machines which are at the forefront of advanced materials handling and information technology. Mailers have also invested in advanced systems which generally employ computer systems to control mail finishing procedures. This should enable postal information to be both captured and exchanged automatically.

REMPI is being undertaken by a consortium comprising Deutsche Post, Royal Mail, Neopost and Pitney Bowes, co-ordinated by IPC Technology. It is scheduled in three main phases:

- Specification
- Design & Component Development
- Operation & Evaluation

running over a period of 27 months, commencing in January 1998 and due to complete in April 2000.(2)

3.2. Objectives

REMPI will develop and practically evaluate a new, electronic interface to postal services. through :

- Customer Applied Item Identifiers

- Data Capture, Exchange and Use

- Electronic Accounting & Payment Evidencing

- Proof of Posting & Customer Information Access

REMPI will contribute to, test and operationally evaluate:

- standards for the customer identification of mail items and for the electronic communication of mail item data

- open, internet based, electronic accounting for mail services, linked to computer generated printing of payment evidence on mailpieces

- the use of encryption for privacy, data integrity and authentication purposes

- reconciliation of electronic payment data with physical mail services provided

- electronic provision of proof of acceptance and tracking services for customers.

3.3. Achievements

Key achievements to date include:

- completion of the Customer and Postal Operators' Requirements Specification

- creation of Customer Focus and Postal Users' Groups

- selection of three pilot customers, two in the UK and one in Germany

- Mailers' and Posts' Conference, held 1-2 February 1999 in Rugby, hosted by Royal Mail, with guest presentations by IPC Technology, Deutsche Post & Neopost, with 122 participants from 60 organisations representing 14 countries

- agreement on an overall System Architecture

- completion of the Functional and Architecture Design Specification

- agreement on the outline design of two pilot implementations and on the design of the digital postage marks to be used for postage evidencing

- the adoption of draft standards for digital indicia by the UPU Technical Standards Board and their acceptance, by a parallel CEN sub-group, on the basis of European standardisation

- completion of the project's initial Standards Contributions, including a proposal for customer-applied item identifiers, for an EDIFACT Statement of Mailing Submission and draft inputs to the joint CEN/UPU working group on Digital Postage Marks

- presentation of the project to a Public Workshop, held on 14 October 1998

- definition of electronic data interchange messaging for mail submission and induction

3.4. The REMPI Pilot Project

Inevitably, REMPI will not deliver all of the system functionality during the pilot phase; what it aims to do is demonstrate and evaluate concepts, prior to investment in infrastructure improvements & integration. Royal Mail has forged a healthy working relationship with Deutsche Post, and this has led to the production of a system design. This design could readily be transformed into a viable, robust system, which would ultimately satisfy both the Posts' own demanding requirements and those of their customers.

The REMPI pilot will demonstrate the benefits which could be reaped from automated information exchange between the mailer and the Post, using customer applied item identification and the capture of payment and address information at mail creation. Standards have been proposed, in accordance with both CEN and the UPU Technical Standards, for customer applied item identification which can be used either in the address block or in the payment evidencing mark.

It will be possible to exchange mail item information, such as address and service required, in advance of the actual submission of the physical mail items. The Post could immediately use this information for order handling, collections scheduling and sorting & process management.

The mailer will benefit from proof of postage, the ability to reconcile electronic accounting data with the physical mail services provided and access to computer-generated postage evidencing. Customers would also obtain access to postal service information, such as postage rates and address databases, and would be able to make enquires into the status of its mailings.

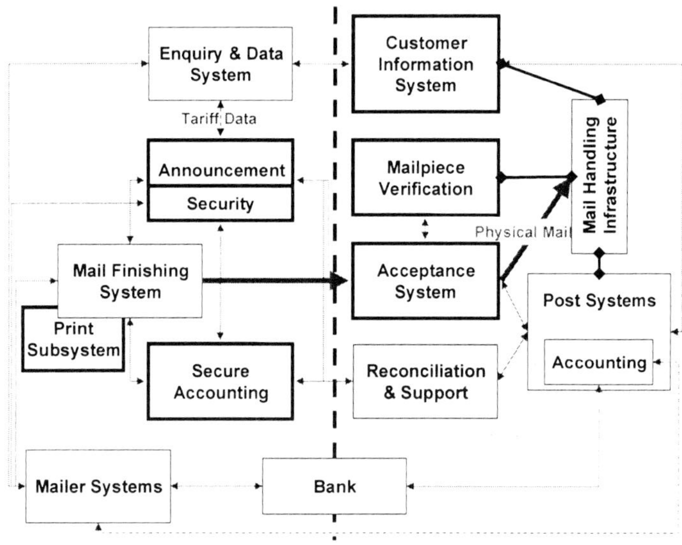

Figure 2 Overall REMPI Component Achitecture

The REMPI components architecture is shown in Figure 2 above. To the left of the dashed vertical line lies the mailer's system, comprising general mail finishing components and includes the REMPI "Announcement" facility. The Post's equipment is situated to the right of the vertical line and this will provide the components for "Acceptance" and "Verification".

Royal Mail and Neopost will build the UK pilot system. Neopost is producing the "Announcement" system, which will be used to apply the digital indicium and generate the unique customer applied item identifiers; Royal Mail will construct the "Acceptance" system.

Royal Mail has acquired the support of two of its key customers (Reader's Digest and Motivation Marketing) for trialling operations in the UK.. In Germany, Deutsche Post will be working with Lufthansa Airplus and Pitney Bowes.

The "Acceptance" component will be installed at two Mail Centres in the UK and will receive Statements of Mailing Submissions. It will be linked to the "Verification" system being built by Deutsche Post. This will process and authenticate payments and verify the payment evidence printed by the "Announcement" component and included in the indicium printed on each mailpiece

Deutsche Post also intends to link with hybrid mail operations in Germany, in order to establish the benefits of electronic data interchange with its hybrid mailers. In the UK, Royal Mail has designed and will construct a Customer Enquiry System to provide mailers with access to data associated with mailing submissions. Royal Mail is particularly interested in the benefits of providing this information over the Internet; benefits include minimising implementation costs, standardisation of customer access, flexibility of operation and reliability. The opportunities created by the use of the Internet are described below in the section entitled "REMPINET".

4. REMPINET

4.1. Internet Technologies

4.1.1. Messaging

Postal administrations deal with a wide range of different types of business customers, from large corporations to small enterprises. In order to reflect the differing needs of these customers, the Posts are required to implement a wide variety of ways in which operational and business data can be exchanged with mailers. Issues of punctuality, availability, reliability, security, confidentiality and existing infrastructure all play a part in selecting the most appropriate means of communication to suit a particular need.

Detailed data about mailings are contained in messages that follow UN/EDIFACT standards. Typically, in order to describe all the mail items in a large mailing submission, the messages may reach several megabytes in size. Medium and small mailings would produce correspondingly smaller messages, but these are still significant if items are described individually.

4.1.2. Connectivity

For the REMPI pilot, practical considerations of infrastructure and security have ruled out the use of the Internet as a satisfactory means of communications. Messages about mailings will be exchanged between customers and Royal Mail using conventional data transmission methods over the public telephone network (see Figure 3).

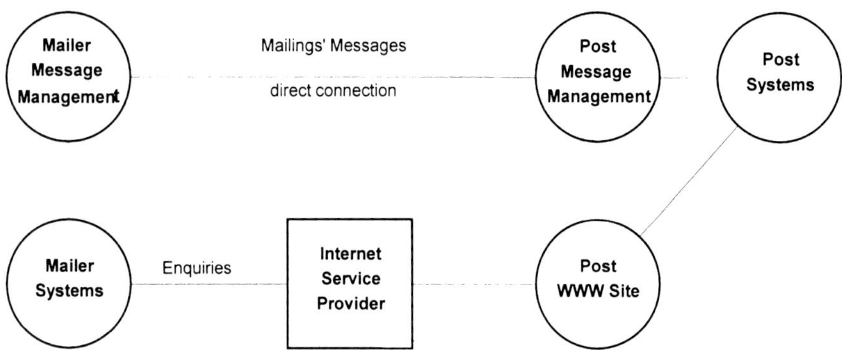

Figure 3 Typical REMPI connectivity

However, the Internet will be used for enquiries by customers about the status of their mailing submissions and for more general enquiries about products and services. This will give both parties experience of using the Internet in an operational environment, and will help to build confidence in using it for electronic commerce in the future.

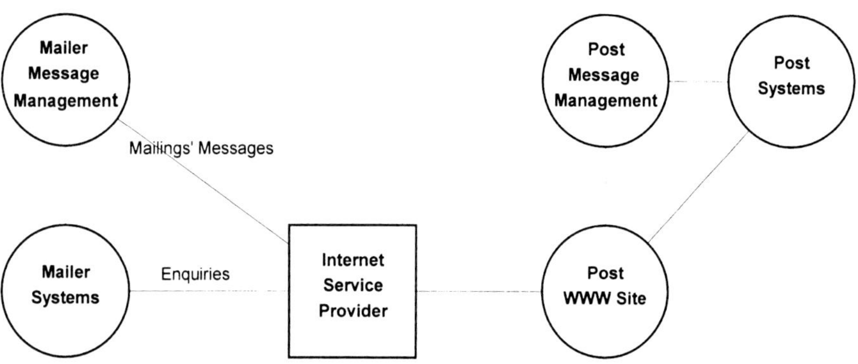

Figure 4 Internet-based connectivity

It is expected that full implementation of Internet and WWW[7]-based communication between customers and the Posts will ultimately be offered (see Figure 4).

For customers with limited resources, but who already subscribe to Internet services, the attraction of the WWW is that it provides a cost-effective, single communications channel for all their electronic commerce needs; larger customers might also adopt the same approach. However, many mailers currently use EDI[8] for electronic commerce, and they may wish to continue with this method. The Posts will cater for such customers through evolving transparent interfaces.

4.1.3. Scripting Language

In the future, mailers who wish to use the Internet to communicate with the Posts may be provided with templates for the entry of management data about a particular mailing submission. These templates could be based on HTML[9], Javascript, XML[10] or other scripting languages. Java has a possible applications role in automating the interaction between the customer and the Post, as well as providing enhanced security. Templates could also be provided for the automated collection of individual item-level data from electronic mail finishing equipment.

Once created, the pages would be transmitted through the customer's Internet Service Provider to the Post in the form of an electronic mail message; the large file of item-level data would be transmitted as an attachment. Future WWW browsers will allow for automation of this process. Automated response and warning features in the application will advise mailers of successful transmission or alert them to delays. Such delays can occur through network congestion in the store-and-forward approach of Internet email. Contingency plans need to be put into place to respond to transmission failures or delays. For example, following an advisory telephone call, the mailing message could be sent by facsimile or even accompany the physical mail being submitted. The message could be scanned and interpreted through optical character recognition systems, an acceptable interim measure.

4.2. Future Development Opportunities

4.2.1. Services

Future access via the Internet to a dedicated mailing site could provide customers with a variety of information links. For example:

- submission of mailing messages

- advice of mailing amendments

- tracking of individual items

- enquiries about the processing and delivery status of particular mailings

- outstanding acknowledgements and response messages

- new accounts creation

---

[7] World-Wide-Web
[8] Electronic Data Interchange.
[9] HyperText Markup Language
[10] eXtensible Markup Language

- templates for message creation

- links to sources of product, service and tariff data on the Post's public WWW site

Looking further ahead, services such as address cleansing might also be offered through the WWW.

4.2.2. Security

The sensitive nature of the data involved demands a high level of security in its transmission. As a minimum, 40-bit SSL[11] will have to be used, coupled with Java encryption or ActiveX[12] controls. The most sensitive messages (from either the customer's or the Post's point of view) will incorporate an additional digital signature to authenticate the source.

The basis for the digital signature will be an encryption key, managed by the Post, and also used in creating Digital Postage Marks for printing on envelopes. A practical alternative approach would be for customers to dial into a secure server which uses Java-based software for managing accounts and message exchanges. While this detracts from the ease-of-use and general availability of Internet communications, it would meet the need for the specific interaction involved.

Each time customers contact the Post's mailings site, they would be required to follow a password-protected log-on procedure. A satisfactory response would allow the interactive process for exchanging mailing messages to begin, allowing the mailer to select from several available options.

One of the issues still under consideration is the impact of privacy legislation (e.g. Data Protection Acts) on the storage and manipulation of mailing data. This is another reason for demanding a high level of transmission security, involving encryption to ensure that data cannot easily be read by unauthorised personnel.

4.2.3. Functionality

As the frequency and intensity of access to the Post's dedicated mailing Web site increases, enhancements to much of its underlying functionality would be required. Site architecture would be improved to minimise the number of pages customers must pass through to access the function they require. Enhanced search engines would offer greater flexibility in the inputting of data and an improved response to enquiries. Performance would be improved through faster connection speeds and easier access. Data file reception would be fully automated to enable message transfer, through a firewall, into the Posts' systems as required; state-of-the-art anti-virus protection measures would be implemented.

There are already many examples of uniquely identifiable items being tracked in the postal field. The banking and travel industries also provide good examples of WWW-based interactive account management. It should be possible to build on these as best practice.

Even without WWW browser capability, the adoption of Internet technologies could benefit both mailers and Posts, through the use of hypertext-linked documentation and the gathering & formatting of data. The addition of WWW opens up the possibility of greatly improved interactive communication for example:

[11] Secure Sockets Layer
[12] A proprietary system for providing control functions

- better dissemination of message generation applications

- simpler enquiry facilities

- visually appealing information presentation

- simple search procedures

- simplified electronic commerce communications channels

5. EXPLOITATION

The REMPI pilot will demonstrate the benefits of electronic data interchange between the mailer and the Posts. It will utilise defined standards for operating a more efficient postal network and identify improvements to those standards. REMPI will also pave the way for future exploitation of electronic commerce within the postal industry. This will include new products and services to be enjoyed by all customers of the Posts, starting with large volume mailers, but ultimately including small and home office customers.

Utilisation of customer applied item identifiers will improve handling, increase efficiency of operation, provide detailed information for the customer, reduce fraud and thereby reduce costs for both the Post and its customers.

The Internet and the World-Wide-Web afford exciting opportunities for the Posts and their customers. Royal Mail already provides detailed information about its products & services and address & postcode data over the Internet, as do several other major postal administrations. Future developments in this area will help to realise REMPI's objectives and provide a transparent interface between the mailer and the Post.

ACKNOWLEDGEMENTS

The assistance of Deutsche Post, IPC Technology, Neopost, Pitney Bowes and Royal Mail is gratefully acknowledged.

REFERENCES

(1)   WELLS, J.L. TIMBRE: Publishable Final Report - Publicly Available Version of the Final Report for the TIMBRE Project Requirements and Definition Phase, 1996, (AEG ElectroCom, An Post, Elsag Bailey, IPC Technology, Pitney Bowes, Post Danmark and The Post Office-Royal Mail)

(2)   WELLS, J.L. REMPI: RE-ENGINEERING THE MAILER-POST INTERFACE Annex I (Project Programme) for an Electronic Commerce Pilot under RTD in Information Technologies section of the Fourth Framework Programme of the European Union, Part I: Financial and Administrative Data - Publishable Summary, ppii, 1997, (IPC Technology)

C568/031/99

Mail piece data capture and exchange systems

R A LAW and **R P SANSONE**
Pitney Bowes Incorporated, Connecticut, USA

SYNOPSIS

The ever-increasing usage of computers in the mail generation process has lead to both posts and the mail originators demanding greater flexibility, more control and better information from the equipment used to physically produce mail. The facilities added to satisfying the mail originator's needs may also be used to collect, collate and report details of the mailing process to the mail carrier.

This paper will highlight the purposes, capabilities, values and implications associated with several types of mail piece Data Capture solutions. This paper will base some of the findings on the series of postal pilot projects conducted by Pitney Bowes, Posts and Mailers over the past ten years (1) (2). Figure 1 summarizes the diverse data flows associated with the above solutions.

1. DEFINITION OF DATA CAPTURE

"Data Capture" is the capability to identify, record, store, summarize and report the postage payment characteristics (i.e. type of mail, class, destination sub-class, weight, rate and special fees) and addressing information from a mail piece. This product functionality is becoming a vital marketing and administrative concern, and increasingly a postage meter approval requirement, of posts worldwide.

2. DIVERSE DATA CAPTURE SOLUTIONS ALREADY EXIST

Over the past 10 years, several different classes of mail-data capture devices have evolved. The differences would seem to have evolved from the specific information requirements of the ultimate mail-data user. The four broad categories are as follows:

2.1 Mailer based solutions

These are IT solutions which owe their existence to a mission critical need for timely information by the Financial and Operations Management of large to midsize businesses.

The Mailer based IT solutions include:

- Postage meters (data capture equipped) connected and reporting to a central internal Enterprise mainframe.

- Enterprise mainframe applications (open loop) managing either local or remote printing applications that produce the mail content as well as the addressing and rating information, or produce labels for parcels containing goods being delivered to customers.

- Enterprise mainframe applications (closed loop) managing both the printing and the mail finishing application that produces the mail piece in the mailroom.

- Inserter based system solutions (closed loop) that use Enterprise mainframe information to produce, finish, document and report to mainframe and or Post directly.

- Stand alone mailroom manifest systems that capture information and periodically send it to an Enterprise mainframe or a compatible Postal Application.

2.2 Meter based solutions

This class of data systems consists of data capture equipped postage meters connected and reporting directly to a central internal (IT) processor or external postal data center or a meter manufacturer managed data center.

2. 3 Automated acceptance device solutions

These are a class of mail processing devices consisting of highly automated, accurate mail verification hardware. In one pass, these devices perform a broad range of mail acceptance tasks that today are done manually. These devices will automatically read indicia, permits, pre-canceled stamps, names and addresses. They can read and analyze ZIP codes and determine mailer presort accuracy. They can analyze the accuracy and quality of all mailer produced bar codes. They can differentiate and count the number and type of each piece in a mailing. They can check mail pieces for "short paid" postage and verify the qualification status for automation of other rates.

A proposed benefit for both the Post and business mailers is ensuring that each mailing is verified using a set of consistent and repeatable criteria applied objectively to all mailings at all entry sites. Additionally, highly detailed reports generated by such devices would provide a consistent data set that could be used by a Postal Operations Group to continually improve mail acceptance through sophisticated process management techniques.

2.4 Postal counter network solutions

This is a class of message-oriented middle-ware software product that provides posts with both end-to-end solutions for networking and for ongoing application development. The software automates many mail services and customer services, such as retail, agency, and banking transactions when a customer is at a postal walkup counter.

3   THE POSTAL VALUE PROPOSITION

Postal interest in Data Capture is emerging and expanding. Several Posts have made their requirements known through new meter specifications and discussions with the meter manufacturers around the approval and content of Remote Meter Resetting Systems. These Posts include Ireland, Finland, Australia, France, the Netherlands, Norway, Switzerland and Malaysia.

Postal interest and direction is being manifested in different ways. Some Posts are mandating Data Capture functionality to the meter manufacturers.   Others are encouraging the manufacturers to provide the capability by providing postage incentives and discounts to the mailing customers that are intended to stimulate purchase demand for meters that have Data Capture functionality.

Based upon what is known to date, there is a diversity of interest in data - it varies by Posts, by functional area with the Post, by customer mail segment, and by detail. The functional areas within the Posts that have an identified need for data are Marketing, Finance, Security, Mail Processing and Special Services. The distributions of expressed needs are as follows:

- **Marketing** is interested in understanding the customer usage of the rate structure -types of mail (letters, parcels) and the class/rate (weight, destination, type of service). This information would be used to evaluate the effectiveness of the rate structure, target new rates and services, and economically allow discounting for use of special rates (such as depended upon volume of mail or time of day franked and delivered to the Post).

- **Finance** is seeking to use the information to provide automated invoicing for postage usage. This could be targeted to valuable and high revenue customers who object to pre-payment or dislike the manual effort involved in post-payment. Another use for the data is to monitor contract station usage of metering and to provide accurate discounting for postage sold. A further use of the data is for the proper application of required taxes (on selected services, rates, fees and types of mail).

- **Funds Security** is interested in the capability to detect the mis-rating and under-rating by the customers. This would be a more effective method than today's sampling detection schemes in the processing plants.

- **Mail Processing** is looking for ways to improve mail acceptance and level the mail processing workload throughout the day. The efforts are focused upon identifying the details of the mail when it is created through variations of electronic manifesting. Postal interest is in using the information to alter the delivery timing, point and method. These include picking up the mail when it is created, scheduling the delivery and providing alternate acceptance points.

- **Special Services** is interested in improving the availability and usage of their products to customers. This includes the automation of necessary forms and documentation, postal notification from the customer location, and non-counter mail deposit.

Postal interest is primarily focused in mid to high mail volume customer segments. This is because the highest percent of revenue is gained from these customers. This interest appears to be across the board by functional areas within the Post. To every rule, of course, there is an exception, such as franchise contract monitoring and access to Special Services.

Finally, within the mid to high mail volume segments, the amount and type of detail to be captured varies with postal functional area. Marketing is looking for finite detail of individual rates, while mail processing is looking for summarization. But even here, some Posts are asking for more detail than others.

4   Implications of Postal Interest

Postal infrastructures are being changed by the Post's desire to improve their operations and the availability of new technology. This extends beyond new sorting equipment and bar code readers into new Information Systems. This is especially true as it relates to Data Capture. No matter what the functional area of the Post, new Information Systems need to be developed to accept, process, analyze, store and use the information from the mailing products. These new systems will connect with the other postal systems to send and receive information. They will report results to postal management.

The Posts that have set Data Capture requirements are creating new information systems to accept and use the new information. The pace of the systems development has been slow and the effort has been primarily outsourced.

The magnitude of these Information Systems investments is not trivial. The risk to the impact and revenue potential of Data Capture is that the Posts will marginalize these investments to only create *pockets of use* i.e. Marketing Information, instead of multifunctional uses of the captured data. The probable approach that the Posts will follow is that additional systems would be introduced over time, as additional data use opportunities are understood.

5   THE CUSTOMER VALUE PROPOSITION

Customers will be both affected and benefited by Data Capture through the multiple impacts of meter migration, postal incentives/discounts, increased information on their usage of postal products, revised payments methodologies for postage, improved mail acceptance, processing and delivery, and convenient access to postal special services. The magnitudes of these impacts will vary or only be present for customers depending upon their mail volume and the Post whose services they use. Customers in the mail volume segments will benefit and be effected in proportion to the size and frequency of their mail volume. This is directly related to the amount, detail and timing of the data to be captured from the mail processing products.

New mail products, capable of Data Capture, will be mandated and/or incentiveized by the Posts. This will apply to both rental and purchased markets. As the higher mail volume segments usually involve system meters, the migration will extend to entire mail products. This is especially true when scale functionality and control are required to accurately and securely (by deactivating the standard meter keyboard) determine the correct postal service. Most Posts, recognizing this meter product upgrade requirement on the customers, are proposing postal purchase incentives and/or discounts that are intended, over time, to offset the cost of the migration to the customers.

All customers will benefit by improved access to postal special services. These include those that require the use of unique postal notification and processing.

Many customers in the higher mail volume segments currently use the departmental accounting features of the products. This is usually where the cost of the mail has to be allocated or charged to a user of the mail product. Today, these products are limited to reporting the total volume and value of the postage used. True Data Capture will provide the capability to report the mail usage by individual postal service. This would allow review of postage spending by types of delivery, special fees, etc. With pattern analysis, this would improve the customer ability to fully manage postal spending.

Depending upon the market and country, customers either pre or post pay for the postage used in their meter. In post-pay markets (usually utilizing ascending register only meters), this involves monthly manual reporting by the mailer of the postage used. Data Capture will eliminate this activity, and allow automated invoicing or direct debiting by the Post.

For those mailers in the top two mail volume segments, who represent a high proportion of the Post's revenue, Posts are looking for ways to improve the services that they are provided. These can include, special rates and services for mail, agreements covering acceptance of the mail and improved delivery performance. This translates into faster through put of the mailer's mail, which results in earlier end-customer receipt of invoices, offers and notifications. This in turn could result in improved cash flow from quicker bill payment (1), influenced purchase decision and improved end customer compliance with recalls, returns, etc.

6 THE METER MANUFACTURER VALUE PROPOSITION

The inclusion of Data Capture in mail processing products will provide significant value to the meter manufacturers because Data Capture:

- Provides renewed and increased worth (value) of metering as a postage payment alternative to posts and mailers

- Facilitates market expansion of remote reset meters in both quantity and share into new countries, where the value of the current service alone is insufficient to convince the Post to approve the conversion.

7 CONCLUSION

Pitney Bowes sees an opportunity for both mailers and mailroom equipment providers to meet the emerging Postal data capture requirements by re-purposing their already existing mail-data capture processes. This is because of the existing investment in a broad array of products that are being used by mailers to link their mailrooms to the enterprise's information technology group as well as to Posts and other carriers. There exists a mailer perception that their "end to end" processes includes the carrier delivery to a mailer's customer. Some current "Mailer" production processes include mail/postal data capture. In addition, existing data flow is most often backward toward an internal IT application such as departmental postal cost allocation or QA or QC processes.

Mailroom equipment manufacturers currently offer mailers these data capture means. Introducing new Postal data capture requirements to mailers or even re-purposing an existing data capture process might impact the in-place internal IT data flow. Understanding the cost of new hardware, upgrading existing hardware, or changes to the IT process software and hardware, could strongly influence the rate of implementation of data capture Postal solutions.

Further, Information Privacy Issues, driven by either mailer willingness or the legality to share other than basic mailer data with posts or carriers may prove to be another significant barrier to timely implementation.

REFERENCES

1 C472/024 ImechE 1994 Piloting of Advanced Mailer-Post Mail Process Data Exchange Systems ; J L R Jones, R. P. Sansone, pages 129-134

2 TelePOST*98 May 1998, Washington DC, USA
Postal process re-engineering through technology; Joseph E Wall, Ronald P. Sansone

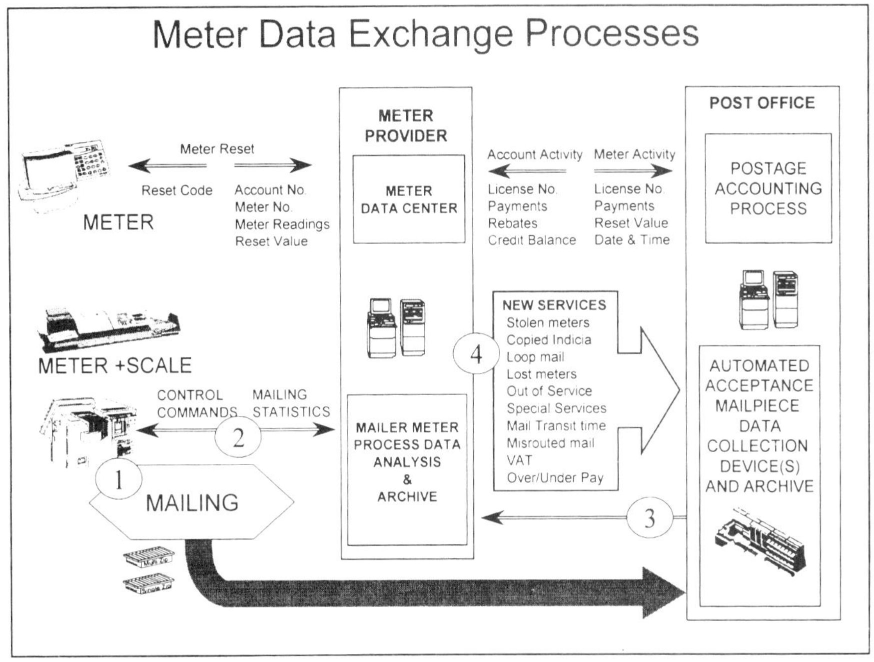

Figure 1. Existing and Emerging Data Capture Processes

Business reply mail – comprehend the message

N BARTNECK and **B ROSENBERGER**
Siemens ElectroCom GmbH & Co, Konstanz, Germany

Hybrid mail is a rather young postal service, which came up during the recent years combining the advantages of physical and electronic information transport and processing. One direction of hybrid mail, including the conversion from electronic mail to paper is mainly used between business and private households and has a strong growing rate. The opposite direction of hybrid mail including the conversion from paper to electronic presentation is most appropriate for the analysis of business reply mail (BRM) a communication form between private households and business with increasing extent. This paper shows the current state of BRM services, describes a technical approach for automatic information extraction, and discusses the benefits of automation.

INTRODUCTION

Against a background of increasing competition postal services around the world are looking at the feasibility of introducing new services and products as well as reducing operating costs. One potential area identified by postal services is the added value service of processing hybrid mail.

While the information exchange in business world more and more takes place in electronic form, paper-based mail still plays a very important role in the communication between business and households, even showing reasonable growth rates.

The general goal of the "hybrid mail" service is to use the most appropriate media for information transmittance between business and households on its way from the sender to the receiver. This service requires appropriate transformations between the electronic and the physical (paper) world and vice versa.

For the information transport between business and households the computer generated information is send electronically to special local distribution centers, which typically are equipped with highly efficient print and enveloping devices and from which the letters have to be transported only a short way to the receiver. The worldwide number of letters send and distributed in this hybrid form has a high growth rate today.

In the direction from household to business, hybrid mail aims at the automatic capturing of data and information from business reply mail. Processing the physical mail piece today is labor intensive. The data input task is performed by the business mailer himself or a contracted third party. The central goal of future BRM service is to support an automatic data capturing and information extraction process, bringing the content of BRM into an electronic presentation form, which is an appropriate base for an efficient further processing.

In the following chapters we will first describe some more details of BRM, before we outline a technical approach for an automatic BRM processing, with special emphasis to automatic information extraction. Finally we will summarize the economic advantages of the automation of BRM services.

1   BUSINESS REPLY MAIL TODAY

Business reply mail is a special form of mail (some examples are shown in figure 1), distributed by companies and public authorities, which typically allows customers to reply to an advertisement or survey by filling in appropriate data on a pre-printed mail piece and sending it back to the originator. BRMs are distributed via junk mail (personalized or anonymous) or newspaper enclosures. The pre-printed mail piece usually contains the business mailer address on one side, often together with special optical markings, and a structured format on the reverse side intended to allow the customer to enter specific data.

Certain postal organizations are providing a specific service to business mailers, the possibility of reply mail being mailed by a customer without the need to apply postage. The revenue is collected then from the business mailer based on the quantity of reply mail handled.

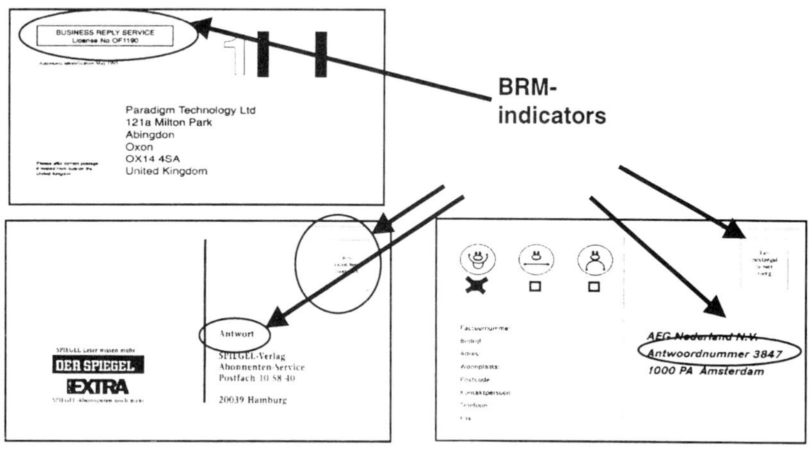

Fig. 1 Examples for BRM with different indicators

The BRMs are separated from the normal mail stream through recognition of special indicators, like

- ➤ a pre-printed facing indicia mark (e.g. at USPS, CPC and Royal Mail)
- ➤ a non geographical post code (Deutsche Post AG)
- ➤ or license boxes.

The postal services deliver the physical mail pieces back to the business mailer. At this point the BRM service of most postal organizations today ends. The business mailer must process the reply mail in order to extract the relevant data.

Typical users of BRM are

- ➤ Insurance companies: to recruit new customers insurance companies often send application forms for membership by junk mail or distributed as inserts of newspapers. These cards are very similar but not identical and have a form-like design. Typical information units of these cards are name, address, birthday, license number, claims, make and size of a car to be insured.

- ➤ Mail order companies, who have a huge number of order forms to be registered and processed every day. These mail order forms typically contain the customer's address, the number and names of the article the customer wants to order and special information boxes or labels.

- ➤ A third user for BRM services are large publishers, who have similar requirements as the mail order companies, combined which extensive advertising campaign.

2   AUTOMATION OF BUSINESS REPLY MAIL SERVICES

To release the business mailers from the labor intensive data extraction process different levels of automation data capturing are under development or already in practice.

2.1 General overview

Automatic BRM processing starts normally during the mail sorting process in the inward mail center. It typically consists of the physical BRM processing , the image processing and the information extraction .

During the automated mail processing, the physical BRM has to be identified. Identification criteria are described in chapter 2.

BRMs however could be processed directly during the automated mail processing.

- The lowest level of data capturing is simply scanning the images (front and/or rear) and storing the images in an appropriate image data base. The database could then be used as electronic document archive for further processing in the company.

- The first step of automatic analysis is to determine the regions of interest on the BRMs and to store only those parts from the image with a relevant content in a database for subsequent coding and processing.

- The next level of data capturing is to recognize special marks on the BRMs, which contains relevant information, like
  - check boxes
  - circled numerals
  - labels

  These marks could be evaluated, used as hint for different kinds of necessary processes and stored in a database.

- The highest level of capturing is to extract all the information contained in marks and the written text. This capturing process requires besides the pure recognition process application specific knowledge. The extracted information of the reply mail is stored in an suitable data base and could be used to start and control subsequent processes, being related to the BRM-content.

Data, which could not be read by the automatic system with sufficient confidence will either be captured and transferred as image or will be extracted by interactive video data entry systems.

In the next step all data will be stored and prepared for the data transfer with the help of the Electronic Data Interchange (EDI), which allows an easy data exchange (images and/or extracted data) between the user(s) of BRM System and the customer.

As additional service the processed BRM items are counted and could be automatic charged directly to the business mailer account.

2.2 Automatic information extraction

The most demanding way of transformation the BRM into electronic form is to extract the relevant information from the document. In the following, a technical approach will be described, which considers the task as a free form analysis and assumes that the image is acquired by a gray value scanner as applied usually in today's address reading machines.

The general architecture of the system as shown in figure 2 consists of a tool box of preprocessing and segmentation tools, high performance recognition methods, a frame-based modeling language and a powerful parsing algorithm. The efficiency of the system has been proven already in different free form applications for clearing houses and insurance companies.

- Model description of the BRM

The performance of BRM information extraction strongly depends on the availability of knowledge about the BRM-form and the expected content, which both strongly relates to the application and the mailer, who has designed the BRM. Therefore a sophisticated model description is necessary which holds this knowledge in an appropriate form. This model describe all the available general and mailer-specific knowledge about the BRMs:

➢ Layout knowledge

The layout description contains the possible positions, attributes and geometrical relations of the different units of a BRM. In addition to most of the commercial available form processing systems, the considered system doesn't need an exact description of the layout positions, but uses techniques from Fuzzy-logic to cope with non-exact knowledge. Further parts of this layout description are e.g. available features about the expected writing styles (e.g. machine printed or handwritten) in the different BRM-units.

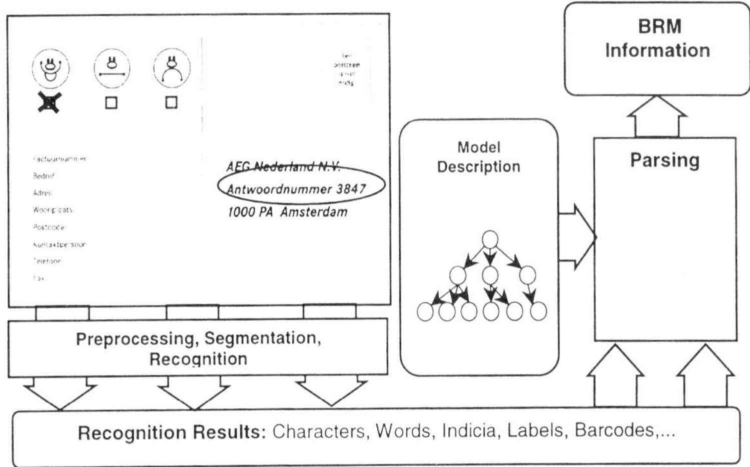

Fig. 2:   General architecture of an information extraction system

➢ Content knowledge

The model description about the content of the BRM includes knowledge about possible alpha-numeric characters in the BRM-fields, lexical knowledge about allowed words, reference lists of possible marks and labels, syntax descriptions of complex BRM-entries, like date or license number.

• Model generation

These described models are BRM-specific, requiring the identification of the BRM-type in a previous stage. Today this model description can be generated very efficiently with the help of a modeling language. For future applications we are aiming at graphical design systems, which enables the postal service or the business mailer as postal customer to create this BRM-model automatically during the design phase of his action-specific BRM.

• Preprocessing and analysis of the BRM-image

Depending on the appearance of a specific BRM different preprocessing steps are necessary to separate the relevant information from the background. High sophisticated binarization operators, as applied in address reading, are used, completed by a additional operations to reduce disturbing noise or auxiliary boxes and lines. Applying well-proven segmentation operators the result is a set of the relevant region of interest (ROI), segmented in lines, words, characters or label images.

• Recognition of marks and text

The next step after segmentation is to recognize the different information units,
➢ to determine the meanings of characters and words
➢ to find out the specific label category

For a successful recognition process it is necessary to have information about the possible content for a specific unit. Useful knowledge sources are
➢ writing style (handwritten or machine printed)
➢ dictionary about possible entries
➢ reference list of possible marks
➢ relations between different BRM items (like article number and article name)
which are included in the model description.

The first step of the recognition process is to find out the writing style of the specific BRM item: Machine printed, hand printed or cursive script. If there is no confident information in the model a classification algorithm can be used to extract this feature from the ROI-image. This classifier also determines text specific features like font-size or line thickness, which helps to select the appropriate classifier.

In the case of machine printed or handprint text high-sophisticated classifiers are applied, which are working very successfully in the address reading machines. The recognition results are set of alternatives for each character and a confidence value. With the help of dictionaries the alternatives are used to find out the best recognition result on word level. The confidence value helps to decide, whether a recognition result can be accepted or has to be rejected and given to further processing, like interactive videocoding.

For the recognition of cursive script statistical methods are used, working very successful in address reading machine. These recognition methods typically are dictionary-based, requiring again a list of possible words for the specific item. Today's recognition rate on word level is above 95% if the dictionary contains about 100 word entries. If this lexical knowledge does not exist recognition methods are available, which segment the written word into characters and deliver best estimations for the sequence of characters without a context check. Again a confidence value helps to decide whether a recognition result has to be rejected.

The mark and label recognition is a typical classification process, to find out which of the possible marks are placed on the BRM. This requires that the system knows all possible marks. This task is very similar to the stamp value recognition which is included in   address reader machines.

2.3 Data Capturing versus Information extraction

For the defined goal of automatic information extraction it is necessary to have as much as possible knowledge about the possible content - mostly in the form of dictionaries - . Only this knowledge enables the system to really get the information. A character sequence like l-i-v-e-r-p-o-o-l is only a set of meaningless data as long no list of English city names could be used, where 'Liverpool' is among the entries and it can be verified that 'Liverpool' is a legal city name. Without this knowledge the result of the recognition processes can only be integrated as sequence of characters to a database.

3   Economic chances and benefits

There are two segments in letter mail showing reasonable growth rates, which are Business – to – Households (Advertising Mail) and Household – to – Business (Business Reply Mail).

The idea of seeing these growing segments of communication between business and households as a complete Business Messaging Cycle (BMC - see figure 3) could give benefits both to business and to consumers.

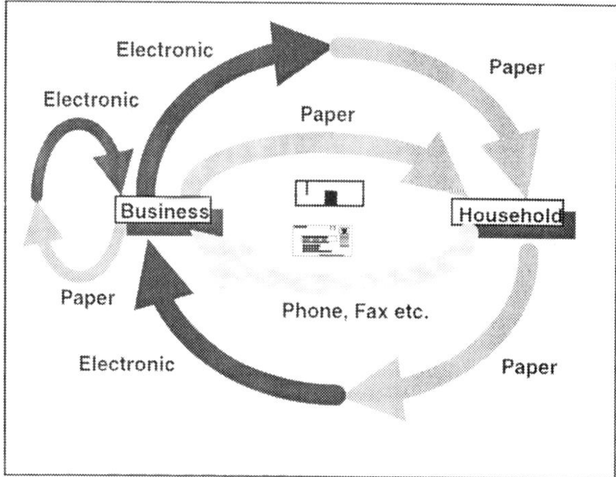

Fig. 3: Business Messaging Cycle

The shorter the response time the better it would be for business cycle. A message prompts a response.

Because most of the mail is generated with electronic data processing, the basic idea is a one-stop-shopping point for the business mailer:

- Sending data
- Receiving data
- Taking cycle control over the mailing activities
- Getting statistics

from a single point.

The idea of processing BRM automatically is forming part of the Household - to - Business Messaging Cycle and having a high potential for rationalization which some postal services want to realize.

The substitution of data entry from the physical mail piece by automatic data entry and information extraction from the mail piece image makes the data entry task more efficient and therefore less costly.

The result of an automated and accelerated BMC would give benefit to the consumer, the business mailer and also to the postal service.

The consumer will benefit from:
- Expedited receipt and response

The business mailer will benefit from:
- Reduction in mailroom costs for the incoming processing
- Reduction in costs of data entry / extraction
- More sales cycles because of reduced response time
- Automatic determination of marketing data
- Integration of mail applications into business process
- Better consumer relations
- More satisfied consumers due to the shortening of time between order and fulfilment and increase of cash flow
- Competitive advantage because of shorter order time and delivery costs
- Transparency of costs of his mailing activities.
- No capital investments, no capital costs
- Automatic integration of data into the workflow

The benefits from this new offered added value service for the postal service are:
- Better customer integration
- More mail volume because of increased number of sales cycles
- Extended revenues :
  Revenue protection because of automatic invoicing (at CPC about $Can 6 Mio./year)
- Saving of personnel resources for processing and delivering the mail pieces

- ➢ Automated billing and accounting:
  Invoicing the Mail Order Company for unpaid reply mail (at DP AG about 75
  Mio. DM/year)
- ➢ Get mail volume back from telephone services (e.g. 0130- number in Germany or 800-
  number in USA) . The USPS, e.g., predicted that the new service could be up to 60
  percent cheaper then maintaining an 800 number.

4   Conclusion

Automation of the BRM service is a challenging task with large economic benefits.
Technical solutions to automate this process with the help of sophisticated document analysis
and information extraction systems are under development. The task of processing BRM is
predestined for automation. The Integrated Mail Processor (IMP) of Siemens Electrocom
builds the ideal platform to provide this added value service. The Twin Scan System of the
IMP offers the possibility to capture the information data both on the front and on the rear
side of a mail piece. The competency of SEC in the area of mail handling, forms processing,
handwriting recognition and video coding is the ideal prerequisite for targeting this
challenging application.

Materials Handling

C568/063/99

The new Hong Kong air mail centre

P ROMANI and **D W WHITER**
British Postal Consultancy Service, London, UK

SYNOPSIS The new Hong Kong International airport was officially opened on 6 July 1998. As an integral part of the facilities provided at the new airport BPCS has designed a new air mail centre, embodying the latest in postal mechanisation technology, to serve Hongkong Post well into the next millennium.

OVERVIEW

The tiny island of Chek Lap Kok just off Lantau island - the largest island in Hong Kong waters, formed the basis for the 1248 hectare platform of reclaimed land (roughly the size of London's Heathrow) on which is built the new Hong Kong International Airport.

The £13.2bn project, including the new road and rail links to the central district of Hong Kong some 35km away, vast areas of reclamation along the west coast of Kowloon, the Tsing Ma bridge (the longest single span road/rail cable stayed suspension bridge in the world), two new cross harbour tunnels and new rail terminals, began operations on 6 July 1998. It is capable of full 24hr, round the clock, operation and will ultimately handle some 35million passengers and 3 million Tonnes of cargo a year.

With the enforced move to Chek Lap Kok Hongkong Post had the unique opportunity to build a new air mail centre which would not only cater for the growth in mail volumes over the next 12-15 years but could also include additional processing capacity to relieve the pressure on the two existing offices at the General Post Office (GPO), which serves Hong Kong, and the International Mail Centre (IMC) serving Kowloon and the New Territories.

DESIGN STUDY

As a prerequisite to determining the key factors to be included in a design brief for the new air mail centre, BPCS was commissioned to carry out a feasibility study to determine the precise needs of Hongkong Post well into the next millennium. This produced a report with a number of recommendations which was presented to Hongkong Post in September 1993. The design brief was then prepared based on these recommendations, which were:

- Single floor working - sized for the next 12 years

- Office with both Air side and Land side interfaces

- All items into the office X-rayed

- 30 min (max) processing period

- Containerisation

- Track & Trace of all Speedpost (EMS), Parcels, LC & AO

- 'State of the art' automation

BUILDING

The site for the airmail centre is approximately 2 hectares. On this has been provided a single story, light weight industrial building with a distinctive, wing-like roof. This provides a net operational area of approximately 12,500 sq. m.

The air mail centre operates as a Tenant Restricted Area (TRA) which means that the whole building is technically 'Airside' and all mail and personnel are security screened, using the latest image enhancement X-ray equipment and metal detectors, as they enter the building. This also applies to transit air mail coming from airside.

EQUIPMENT PROVISION

State-of-the-art, to quote a popular phrase, is evident in the range of equipment installed under the HK$208 million (about £16.6 million) contract awarded to Siemens ElectroCom GmbH in April 1995. The equipment line up includes:-

- One IMP (Integrated Mail Processor - as currently being installed by Royal Mail) with 128 stackers.

- One IMPEX (an IMP without the initial segregation facility) with 128 stackers.

- A shrink wrap plastic wrapping machine for letter bundles.

- Two channel type packet sorting machines (PKTSM - Universal Sorter System 2200) with 100 selections each.

- A Tilt Tray parcel sorting machine (TTPSM) with 44 selections. This latest version (Crisplant Sorter 2000) of the popular tilt tray sorter utilises a linear motor drive system.

- A fully computerised, three aisle, high bay, Container Storage and Retrieval System (CSRS) provides 504 cells for temporary storage of towable wheeled containers holding mail in bags or trays awaiting dispatch.

- A semi-automatic Elevating Transfer Vehicle (ETV) system incorporating 20 adjustable height workstations and 50 storage bays for handling the majority of standardised airline containers (usually known as Unit Load Devices - ULDs). The workstations are built in pairs so that each pair can be locked together to handle large size containers.

- A materials handling system to assist in the movement and handling of the principal mail streams within the AMC comprises parcel and packet conveyor systems, telescopic boom conveyors, scissors lifts and vacuum lifting aids. The whole system is modeled onto a computer graphics display system to facilitate monitoring and diagnostic functions.

- Towable wheeled containers (TWCs) and electrically powered tractors are provided for movement of bagged/trayed mail within the mail centre and outside to the loading area for transfer to the airline handlers.

- Manual sorting fittings, DBFs, stamp canceling machines, a bag tip conveyor for the IMP and a mechanised meter table.

- A comprehensive CCTV system provides total coverage of the site, both inside the mail hall and outside, and a PA system provides for building wide or zoned alarms or announcements but can also play background music etc. if desired.

- A fully integrated Management Data Acquisition (MDAC) system which is networked with the control computers for all individual machines comprising the AMC automation system.

PRODUCT STREAMS

Mail arrives at the AMC by van and is unloaded directly onto telescopic boom conveyors to transfer the mail into the building via the X-ray screening equipment. The boom conveyors incorporate rise and fall actions as well as 'droop snoots' (shown in Figure 1) to facilitate unloading from a wide range of vehicles down to ground level.

Figure 1 - Landside Boom Conveyor

The various product streams - Speedpost (EMS); LC & AO; parcels and local collections are then segregated into their generic types i.e. letters, packets and parcels to facilitate processing on the specialist automation equipment provided. The mail flow through the centre is shown in Figure 2.

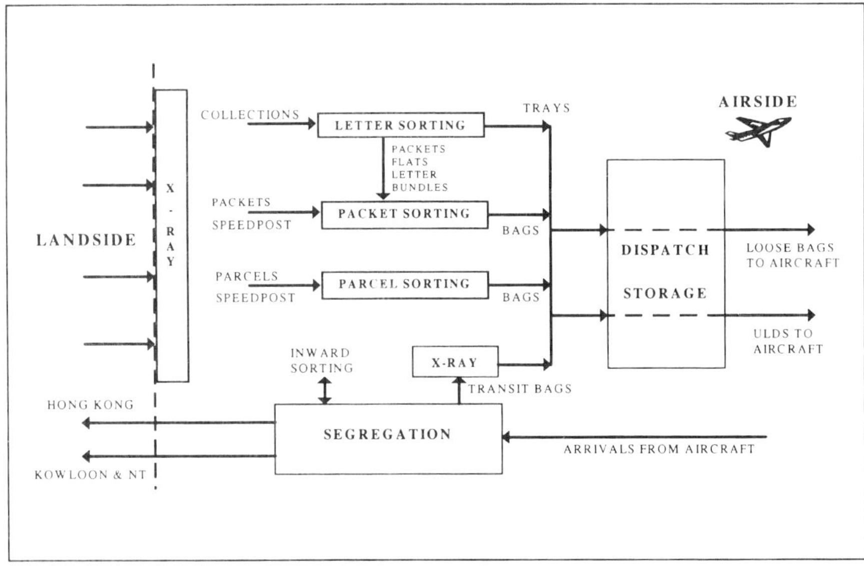

Figure 2 - AMC Mail Flow

Letter Processing

The Integrated Mail Processor (IMP) automates the processing of machinable letter mail into a single continuous stream, eliminating the need for manual transfer between successive stages of the process, so that the mixed mail from the street collection boxes etc. that enters the machine is taken from the output stackers faced and fully sorted and needs only to be directed to the dispatch bag or tray. Total time through the machine is < 1min at a processing rate of approximately 33,000 iph.

Machinable mail is in the range:

Length:	138 - 250mm
Thickness:	≤ 6mm
Height:	87 - 167mm

Collection mail i.e. mail taken from street collection boxes is first tipped into the hopper of a pre-culling conveyor. This allows for the removal of any small items e.g. loose keys, rubber bands or other debris before passing through the segregation drum which removes items that are too thick or letters that have been folded and cannot pass through the measuring slats. Large flats (≤ 6mm thick) are extracted at a three stage flats extractor. Other checks for overlength, stiffness and metal content are made in the culling contents module (CCM) before the letters are passed into the detection module. A direct feed into the CCM is provided for bulk mail from large volume posters.

Continuous high speed performance demands an accurately monitored and controlled mail stream. This means that any items that are detected to be skew, overlapping or have too small a gap to the next item must be rejected to avoid problems in processing.

Figure 3 - Collection Mail Input to IMP

The detection module checks for the presence of facing indicia marks (FIMs) and incorporates a 'Philtec' reader which contains three separate detectors to check for contrast, fluorescent or phosphorescent marks on the letter or the stamps. The results are entered into an 'Attribute table' for each mail piece and determine the precise actions within the machine for that item e.g. should a cancellation mark be printed or a routing code applied, before it is finally sorted to a stacker

Both sides of the mail piece are scanned to ensure that all available information has been captured and the OCR electronics tries to identify and read the address block. If the OCR reader cannot find a reliable match the images are presented to human operators at video coding desks to input the correct decision. The video coding desks operate as a common pool serving both the IMP and the IMPEX.

The manual attribute allocation (MAA) task is independent of the mechanical transport of the letter through the machine so that there is no time constraint on the operator to reach a correct decision. If a decision is made within the machine delay time allowed - about 10secs, the item is coded 'on-line' and is finally sorted. If a decision is made after the delay time the item is printed with a 4-state bar code ID tag and directed to a 'refeed' or 'video coding' stacker. The item can then be passed through the machine again, after a suitable time delay, to allow the sorting decision to be matched to the item.

The AMC processes both inward and outward mail. The manual coding requirements for these two streams of mail are very different and have been resolved by utilising a two keyboard system as shown in Figure 4:-

- **Inward** mail coding uses the same keyboard design (an ergonomically designed split qwerty) as is currently in use in the Hongkong Post MLSS (Mechanised Letter Sorting System) and requires the entry of a complex extract code with an on screen prompt system to guide the operator when ambiguities arise (1).

- **Outward** mail coding is based on a simple 3 digit code and can be input more efficiently by means of an ergonomically designed numeric keyboard. This numeric code is based on the already well established IDD country codes and hence is easily memorised (to minimise training requirements and improve staff flexibility, the same code is used for packet and parcel sorting).

Figure 4 - Video Coding

The working mode i.e. Inward or Outward, of each video coding desk is set by the supervisor at a control terminal and the appropriate keyboard is then enabled when the operator logs on.

Ink jet cancellation allows postal services a new freedom in the design and application of cancellation and quality marks. The printers are software driven so the cancellation impression can be changed at short notice - and even for successive mail pieces, on recognition of a preprogrammed FIM, and as the machine is computer controlled the date/time impression can be in real time.

Routing code or ID tag mark printing is performed after the cancellation function and before passing to the final sorting modules. This allows the mail piece to be read by other machines when a finer sortation is required.

The IMPEX performs the same functions as the IMP but without the initial segregation functions that are required for collection mail.

Bundle Wrapping

Most letters are taken from the stackers in bundles and by pressing a button associated with each stacker a bar coded destination label is printed. The bundle and label are then placed in a compartment on a conveyor which transfers them to the plastic shrink wrapping enclosure at one end of the system as shown in Figure 5. The wrapped bundle is then output to a conveyor which transfers to the packet sorting machine for final sorting into the dispatch bag.

For those countries that prefer their letters in trays e.g. USA, the letters are transferred manually from the stackers to tables where the trays are made up.

Figure 5 - Bundle plastic shrink wrapping

Packet Sorting

Two channel type sorters (Universal Sorter System 2220) are provided to sort packets and letter bundles. The range of packet sizes is:

Length:	100 - 400mm
Width:	0.2 - 220mm
Height:	80 - 300mm
Weight:	0.01 - 9kg

Mail is presented to each of 4 input stations by a tilted band conveyor which discharges through sliding gates feeding each position. The gates are plc controlled to ensure even filling of the operational feed lanes.

All lanes are provided with both a numeric keypad and a bar-code reader. Coding is achieved by keying a 3 digit code (the same code as used on outward letter mail) or, in the case of letter bundles, by scanning the bar code label.

A particular feature of the Hongkong Post system is the ability to read and sort to the bar-code identification labels on Speedpost items. This is achieved by downloading the destination files from the track and trace system computer to the PKTSM and TTPSM computers via the MDAC.

Each PKTSM (see Figure 6) has 100 outputs which can be either bags or wheeled containers; the latter are used in particular if a secondary sort is required or when the output consists of Speedpost items which must be tracked into the bag.

Figure 6 - Packet Sorting Machine

Parcel Sorting

Items which are too large or heavy to be classed as packets are directed into the parcel stream. The parcel system comprises twin uplift conveyors to two pairs of induction lanes on the Crisplant sorter 2000 design tilt tray parcel sorting machine. This allows Speedpost items to be given priority treatment.

Parcel stream separators, basically a system of moveable flaps in the bed of the feed chute, control the flow of parcels to the induction lanes and as with the PKTSMs each input station comprises a numeric keypad and a hand held bar-code scanner to process Speedpost items.

Figure 7 - TTPSM (showing feed conveyors to ETV chutes)

The machine handles parcels within the UPU specified range of 1.5m long and 3m length plus girth (800mm maximum width) and up to a maximum weight of 25kg.

The machine runs at high level, as shown in Figure 7, to provide adequate storage in the destination spiral chutes; these are the Royal Mail 'Safeglide' design to provide constant velocity motion with minimal impact force. To provide added convenience four outputs for heavy selections are located adjacent to the ETV.

Containerisation

In order to move the mail quickly and easily around the office and at the same time reduce the manual effort involved, one of the key recommendations at the design stage was to implement some form of containerisation.

This has been accomplished at the AMC by providing 6 battery driven tractor units together with 750 towable wheeled containers (TWCs). The TWCs can be safely towed in trains of up to 4 containers but can also be moved easily by one person; a manually applied brake system for parking is incorporated.

The containers are constructed from round section tubing with 50mm mesh infill on the side panels and doors and mounted on a MS base. The size is based on the well established Royal Mail 'York' container i.e. 800 long x 600 wide x 1500 high. In order to store the containers in the CSRS it was necessary to ensure that the tow bar could be securely stowed and that the braking mechanisms were contained within the overall size envelope but allowed sufficient space for the CSRS stacker crane lifting forks to engage on the underside of the chassis.

Each container carries a unique bar code identifier.

Container Storage and Retrieval System (CSRS)

Mail awaiting flights, particularly when the flights are infrequent, must be stored in a way that ensures easy accessibility at all times while occupying the smallest possible space in the building. A unique solution to this perennial problem has been adopted at the AMC (as far as we are aware this is the first application of high bay container storage principles to a postal operation).

Initially bags of mail for any particular destination are stored in a standard TWC in the dispatch segregation area; as containers become full they can be moved to the CSRS for storage.

The container storage and retrieval system is a fully automated high bay facility which occupies the full available height (10.5m) of the building. The storage cells are arranged on both sides of 3 stacker crane aisles in a 14 long x 6 high matrix providing a total of 504 cells. Each aisle includes one input and one output station so that storage and retrieval functions can run during the same operating cycle for improved throughput. The cycle time is typically one minute.

The whole process is computer controlled and the system status is fully detailed on a storage image. Briefly the main operations are:-

— **Storage -** the operator scans the container ID bar code at the main control terminal and enters the flight number together with any supplementary information e.g. class of mail. The computer allocates a storage cell and advises the operator at which aisle to load the container. After loading the container at the input station the container ID is again scanned to confirm that it is in the correct location; closing the safety gate initiates the storage action.

- **Retrieval** - the operator enters a flight number and all containers stored against that number are then displayed in a table form for the operator to select the particular containers to retrieve. The retrieval process is fully automatic and interleaved with any storage actions in process at the same time. When the container is set into an output station the operator must remove it before another can be brought to the same station, although the retrieval cycle will continue to run up to the point of delivery to the output station.

To ensure minimal disruption to the operation in the event of system failure or maintenance work, each aisle is constructed as a separately enclosed area and full safety interlocking is provided. An overview of the CSRS is shown in Figure 8.

Figure 8 - CSRS

Elevating Transfer Vehicle (ETV)

The bulk of air mail is carried in bags which are loaded into airline containers - called ULDs (unit load devices). These come in a variety of shapes and sizes to suite the aircraft types in service but can broadly be classified as:

- Standard or half belly hold - typically LD3

- Large or full belly hold, structural igloo etc. - typically AQ6

The mail handling operation becomes much more secure, efficient and flexible when the bags can be loaded directly to the ULDs at the mail centre. The AMC incorporates the unique facility of an ETV (as far as we know the only other mail centre to incorporate this type of equipment is at London Heathrow operated by Royal Mail International).

The work area is arranged with 10 work stations on each side of the ETV track and two further tiers of paired storage bays above, providing a total capacity of 70 standard size containers.

The work stations incorporate scissors lift beds so that the container height can be adjusted above and below floor level during the loading operation and thus assist the loading task. They are arranged in pairs so that each pair can be locked together to handle full size containers. The arrangement is shown in Figure 9.

Figure 9 - ETV

The ETV is controlled by a driver although the actual movement and alignment functions are usually performed automatically - the driver enters the number of the required bay and presses the start button. The transfer of the ULDs between the ETV, work stations and roller decks is controlled by the driver. A separate control panel at the end of the roller decks controls the transfer of ULDs between the ETV system and the airline handling agents transport dollies.

Container details - flight, container number and status i.e. empty, partly full or full, can be entered into a computer terminal in the driver's cabin as each container is accepted into the ETV and updated as the container is moved between the vehicle and the workstations or storage bays etc. A second terminal is provided for the dispatch supervisor which replicates the information in the driver's cabin and allows messages to be passed between them.

MDAC

A Management Data Acquisition Computer (MDAC) networks with all of the control computers for the various systems which comprise the AMC automation equipment and also provides an interface to the Hongkong Post T &T system for the transfer of data relating to Speedpost items.

The MDAC provides the interface to the letter, packet and parcel sorting systems for the transfer of all address database, sort plans, printer cancellation plans, etc. and gathers all machine operating and fault statistics from every item of automation equipment throughout the AMC. It can also hold the complete stores database for the spare parts held on site to maintain the automation equipment and provide a comprehensive stores management function.

The MDAC database can be accessed from a number of key locations in the AMC e.g. the Plant manager's office, to monitor the equipment function within the AMC or gather operating or fault statistics, but any function which will change the operational parameters of any equipment is password protected.

Closed Circuit Television (CCTV)

A comprehensive colour CCTV system has been installed to provide both security surveillance and equipment monitoring functions.

External cameras are mounted in environmental housings and provide complete coverage of the outside of the building, car park, landside and airside loading areas. Cameras within the sorting hall are mounted in tinted domed housings at high level so that the cameras inside are hidden from view. The cameras used have low light sensitivity and hence it is unnecessary to provide supplementary infra-red lighting for night time use. Most of the cameras incorporate full pan, tilt and zoom facilities but a small number of fixed cameras are provided in smaller rooms e.g. registered letter enclosures.

Twin control rooms have been provided so that the security requirement can be operated independently; in the event of a conflict between the two control panels the security panel has priority. The control consoles each incorporate 6 x 14" colour monitors which display sequential pictures from groups of cameras together with 2 x 21" colour monitors which can be switched to individual cameras. The security console also includes a time lapse video recorder.

Public Address (PA) System

A fully featured PA system has been installed on a zoned basis so that messages can be directed to specific areas of the building or the external loading areas if required. Radio, tape and CD facilities allow staff to enjoy music while they work!

TRAINING

While many of the staff employed at the AMC were familiar with the equipment provided for the MLSS the new technology equipment and extensive materials handling systems introduced a wide range of new processes and procedures to be learned. A lengthy and detailed programme of training for both operators and technicians was incorporated into the overall implementation plan. In the case of the video coding training an existing computer based training system had to be updated to include lessons for the numeric keyboards.

A particular feature of the Siemens training facilities provided for the IMP equipment was a computer based training (CBT) system which provided interactive tuition on an individual basis. The system allows trainees to select between 'Operator' and 'Engineer' so that the particular needs of each group can be fully met.

PROJECT MANAGEMENT

The contract for the whole of the equipment supply was awarded to Siemens ElectroCom Gmbh who are based in Konstanz, Germany. They specialise in the design of the letter and packet handling machines and hence subcontracted the materials handling systems (MHS) - conveyor systems, parcel sorting, postal and airline container handling to CEGELEC - AEG based in Frankfurt. Their partners in the realization of the materials handling system were:-

– Conveyor systems - FAM (Germany)

- Parcel sorting - Crisplant (Denmark)

- CSRS and ETV - Mannesmann Dematic (Germany)

- Site work - AEL (Hong Kong)

Overall project control was maintained by detailing each task as an item in MS project. Each subcontractor then ran their own project control system within the constraints imposed by the overall system.

BPCS managed the whole of the work on behalf of Hongkong Post.

The new Hongkong Post AMC is currently the most technically advanced air mail centre in the Asia region and will set a new standard for postal administrations around the world to follow.

ACKNOWLEDGMENTS

We gratefully acknowledge the assistance so freely given by Hongkong Post together with their permission to reproduce the photographs used to illustrate this paper.

© The Post Office 1999

REFERENCES

(1) Powell, R. W. and Whiter, D. W. - The Hong Kong mechanised letter sorting system (Proceedings of the Institution of Mechanical Engineers 1994-4 - Mail Systems 2000)

C568/065/99

Commercial materials handling technology in postal operations

W R GOODE Jr
United States Postal Service, Tray Management System Program Office, Virginia, USA

In 1995, the United States Postal Service initiated a program for the development of commercially based material handling systems to support postal mail processing operations. This program, known as Tray Management Systems or TMS, was the Postal Services initial venture into the development and adaptation of commercial material handling technology to support the material handling requirements within our processing and distribution centers. The strategic objective of the TMS program was to implement a technology solution addressing the increasing costs of allied material handling labor supporting our automated mail processing operations. For the Postal Service these costs were greater than $3 Billion per year and continuing to show positive growth. This scenario created a tremendous opportunity for the Postal Service to invest in material handling technology to offset these costs. There were additional issues to be addressed, beyond the escalating labor costs. Market place competition and customer demands were forcing the Postal Service to look into the needs for real-time work in process information. If we could capture this information, we could utilize it to more deterministically operate our plants as well as provide added value to our customers by providing them with accurate status as to where their mailings were in our operations. Combining the need to invest to offset increasing labor costs and the desire to develop an information rich mail stream defined the macro level technical objectives of the TMS program.

Understanding where we wanted to go, we now had to focus on how we were going to make TMS happen. Answering the "how" question brought about another concern. In order for the strategic objectives to be realized, we had to develop an acquisition strategy that would enable the Postal Service to develop and deploy TMS into greater than 250 plants to create the information infrastructure that would be part of this new standard plant material handling system. Traditionally the Postal Service strategy for plant material handling systems was to only deploy them into new facilities as the buildings were being built. This meant that system deployments were limited to perhaps 4-6 systems per year, depending solely on the amount of new construction being undertaken. A secondary concern was a lack of standardization born of the acquisition philosophy. When we installed a system into a new facility, we had a standard set of guidelines for the installation, but no standardization in system architecture, functionality and component selection. Finally, we had to deal with the unique problems associated with our dimensionally unstable containers. While their current design is driven by cost and the shear quantities required to support our mail processing needs, the lack of rigidity is a design challenge for material handling technology. To solve all of these concerns of standardization, system architecture/functionality, production deployment and achieve our strategic objectives, we made a business decision to take our problem to industry for a solution. The tremendous economies of scale that an organization the size of the Postal Service offered, made it attractive to leverage this position with the commercial material handling industry to work a solution. Concurrently the Postal Service was investigating some changes in our purchasing philosophy that would allow for more partnering with our supplier base, which lent itself to the needs of TMS. The combination of the strategic material handling needs of the Postal Service and our willingness to take the requirements to industry for a solution created the opportunity for the TMS program.

1. RESPONSE OF THE MATERIAL HANDLING INDUSTRY

When the material handling industry responded to the needs of the Postal Service with regards to the TMS program, they looked to apply technologies that had a proven record of success in other industrial applications. Research was done to investigate how current state of the art material handling technology was being applied to industries such as warehousing, distribution and manufacturing, and how that technology may make sense in application to the Postal Service. All of this research and development activity lead to the Postal Service awarding three contracts to three separate system integrators to solve the TMS problem. The integrators task was to select the appropriate technology to service the operational needs of the Postal Service, and integrate that technology into a intelligent material handling system which could enable the networking of our processing and distribution centers for data/information collection and utilization. Selection of appropriate commercial technology, proven in other industrial applications, was a starting point for the system design. Once selected, this commercial technology had to prove itself capable of meeting the operational demands as well as functioning in the unique environment of the processing and distribution centers. Additionally it had to prove itself to be an economic winner, to the extent that the system acquisition costs were affordable for the Postal Service. A final hurdle for this technology was the ability of the Postal Services maintenance organization to assimilate and support the equipment. For the Postal Service, outsourcing this function was not an option, which made it critical that the technology be supportable by our traditional means.

Since the Postal Service would want a competitive environment for the ultimate production deployment of the preferred technology, having three competing integrators/technologies was beneficial in supporting our strategic objectives. While we had three different integrators competing in the TMS program, there was an amazing amount of functional similarity to the designs, which one might expect since they are each solving the same operational problem. Fundamentally the operational needs of the Postal service for TMS broke the individual system designs down into four primary functional areas, which are discrete, random access work in process staging, batch staging, transportation and system controls. The balance of this discussion will focus on these four primary functional areas and how solving the practical needs of postal material handling operations affected the selection, design and application of this technology.

2. DISCRETE RANDOM ACCESS WORK IN PROCESS STAGING

Operationally, the Postal Service mail processing requirements drive a need for discrete, random access staging for mail in certain areas of the distribution operations. These operations, such as mail processed in the remote bar-coding system (RBCS), require the mail to be presented in a logical fashion to subsequent processing operations, driven by timing functions for elements of the process. From a TMS perspective, this mail requires intermediate handling, or staging between operations. This operational challenge is met by utilizing material handling technology that can store and randomly access any unique tray of mail when it is to be presented for additional processing. The commercial technologies utilized in the TMS designs to meet this requirement are automated storage and retrieval systems (ASRS), vertical carousels (towers) and linear conveyor buffers. Selection by the vendors of various combinations of this technology is solely dependant on the overall system (TMS) architecture and the response time of the system with respect to delivery of mail to subsequent operations in time to avoid operational "starvation". Starvation is defined as idle time on the automated mail processing equipment while waiting on TMS to deliver the mail required for processing. The complexities of adapting the commercial storage devices mentioned above, to this particular aspect of the TMS material handling requirement, is driven by the system response demands of TMS. TMS must deliver the exact tray of mail to the appropriate piece of mail processing automation, sufficient to meet the throughput requirements of the machine and avoid operational starvation.

Addressing the starvation issue is only part of the design problem. The issues of staging system unit cost and duty cycle had to be factored into the design requirements. To meet the system design challenges the vendors had to address the issues of balancing the performance requirements of the technology against its acquisition cost, while insuring that the technology could provide the duty cycle requirements of Postal Service operations. A unique problem presented by Postal Service operations is the 24 hours per day operations, 7 days per week. This level of operational "up-time" was greater than that of other commercial applications of the same technology. In other commercial applications the system had significant blocks of "down-time", or was only accessed periodically, such that the "idle-time" between transactions was long enough to rest the machine. The TMS application forced these machines to run continuously, and demanded high throughput to keep pace with the operation demand for mail. These operational demands were mitigated in the system designs by techniques such as load balancing between more than one machine to eliminate single points of failure, coupled with a capability to dynamically route product to avoid operational impacts. The lesson learned in the developmental phase of TMS was that there were going to be certain aspects of adapting commercial material handling technology in the operational environment of the Postal Service, particularly the discrete, random access staging, that would be a unique design challenge.

3. BATCH STAGING

In postal operations, mail will arrive at a processing and distribution center throughout the day almost continuously. The operational scenario for TMS is to induct into the system all mail that will be processed within a given 24-hour period. Much of this mail that gets advanced into the TMS does not require discrete handling, but rather needs to be batch processed when the appropriate operation that consumes this mail gets activated. Supporting this type of operation requirement was a challenge for our commercial suppliers. The equipment requirements born of supporting this type of mail processing presented a new challenge. That challenge was specifically developing a high capacity staging device, capable of batch extraction of certain types of mail, with an operational throughput four times greater than that of the discrete, random access technology. When the Postal Service initiated the TMS program, there was no commercial solution available for this specific operational need. Meeting this objective forced a different response from industry, which was to develop new technology. The solution was developed in the form of high capacity, horizontal carousels that interfaced to high throughput insert/extract devices to handle this particular mail volume.

The design objectives for this new technology were much the same as that of the discrete random access devices. The machines had to cope with the requirement for the 24-hour, 7-day per week operating scenario. One bit of relief came from the fact that the batch requirements did preclude continuous running or up time. Like with any component in the TMS designs, the batch staging devices had to perform economically as well, from the stand point of unit cost per staging location and cost to maintain the units operationally. In satisfying this design requirement for TMS, both the Postal Service and our commercial suppliers learned that it was going to take a hybrid solution of both batch and random access staging to meet the operational needs of our processing and distribution centers.

4. TRANSPORTATION

The transportation portion of the TMS designs is the technology utilized to move mail from point A to point B within the system. To meet the volume and flow demands our commercial suppliers looked into three different types of technology to address this design requirement. The three technologies employed in the TMS design to provide the transportation function are conventional tilt tray sorters, scalable tilt tray sorters and conveyor. It is safe to assume that the transportation portion of the system is perhaps the most critical technology in the TMS architecture. When there are issues with the transportation portion of the system which lead to operational downtime the entire plant operation will suffer, since no product is flowing through the system. While the staging devices mentioned above are critical to the operation, loss of one or more of these devices due to operational downtime will only degrade a certain percentage of the total TMS functionality. All of these issues made the TMS designers go after robust transportation technology, which in their own system architecture was optimized to provide the correct amount of transportation function at an optimal price to keep their system priced competitively.

The economies of scale presented to the commercial material handling community by the Postal Service's initiation of the TMS program enabled them to pursue high quality and high performance transportation technology. Unlike the systems we installed as part of earlier new facility projects, the TMS program offered the opportunity for our commercial suppliers to amortize production and developmental costs, as well as achieve economies of scale that enabled them to supply superior transportation technologies. These superior technologies provide a higher degree of uptime, and lower life cycle operational and maintenance costs that if not for the economies of scale, would not be affordable.

5. SYSTEM CONTROLS

The heart of the TMS is its control system. In all of the competing TMS designs we have designed in capabilities to both intelligently control the system and its components, while integrating the information system technology that will enable us to achieve our network and system wide objectives for better control of information on our work in process. The Postal Service is looking to the control and information system portion of TMS to be a critical enabling feature for our future strategic objectives of better control of our products and services, and more deterministic real-time information for both our internal and external customers.

Knowing that there was going to be more than one single supplier for TMS, the Postal Service made the critical decision early on in the planning of the program to standardize our database requirements on the "Oracle" database software. The particular product was proven to be robust, and widely accepted as the best commercial product of its kind. The Postal service felt that with a database standard for the information capture, we could pursue technology from one or more TMS suppliers and not risk our objective of creating an information rich network. Oracle would enable the universal sharing of this information. Having this standard simplified some aspects of the control and information system design since it did provide constraints for the designers to engineer their system solution. This standard also introduced some additional complexity to create a control system that could meet the demands of dynamic changes to the operating environment, while interfacing to a relational database. This proved to be a major design challenge to the commercial suppliers developing TMS. The operational constraints of mail processing placed demands on the material handling system to be able to support many variations of operational sort plans and schedules, that went beyond what other commercial applications had to operationally contend with. TMS had evolved into a complex control and information system, that had enough computer power to fly the space shuttle, as more than one supplier pointed out.

6. CURRENT STATUS

By May of 2000, the Postal Service expects to have deployed and fully operational the first 30 TMS's across our processing and distribution network. Preliminary financial results from the systems that are presently operating indicate that we are achieving returns on investments of 15 to 25 percent, depending on such factors as the size of the site and the cost of site preparation. These returns are based only on observed reductions in mail handling labor, without taking account of savings due to improved operating efficiency, reduced cycle times, increased accuracy, etc.

The design challenges discussed in this paper have been addressed over several years of system development as well as by gathering empirical data to assist in the development of solutions to support the Postal Service material handling needs. While TMS is one of the material handling industries most complex systems ever developed; the commercial industry has proven to be capable of taking this technology to a new level. The deployed systems are continuously monitored and remain stable, while supporting our processing and distribution operations.

The initial developmental efforts of the TMS program focused on developing a technology solution that could overlay the existing mail processing operations of the Postal Service. This is perhaps the worst case scenario for the system design, since it forces issues of system response, staging capacity and operational throughput to be critical in the system design. While future design iterations of TMS will still deal with these issues, we do expect that over time that mail processing operations will realize methods of utilizing TMS that can simplify the current distribution operations and realize a net reduction in current staging capacity as compared to today's designs. Additionally we are continuing to expand the role of material handling technologies such as TMS beyond their current scope. We are integrating dispatch technologies, such as robotics, sleevers and automated airline assignment to enhance the capabilities of TMS and provide more comprehensive coverage of our operations by material handling technology. Additionally as the Postal service moves toward a "lights out" mail processing facility, we are looking into additional opportunities to go beyond trays/tub and letter/flat mail operations. By the term "lights out", we are implying that we wish to further and more completely automate the mail processing and material handling activities within our plants. From a labor perspective, we wish to arrive at a situation where the staffing is limited to personnel whose tasks are monitoring the performance of all of the plants processing equipment and performing maintenance activities. To accomplish this we will integrate a mail cartridge system, automating the sweeping of our mail processing automation and move toward bundles, bulk, and Priority products handled on a common material handling system, derived from TMS.

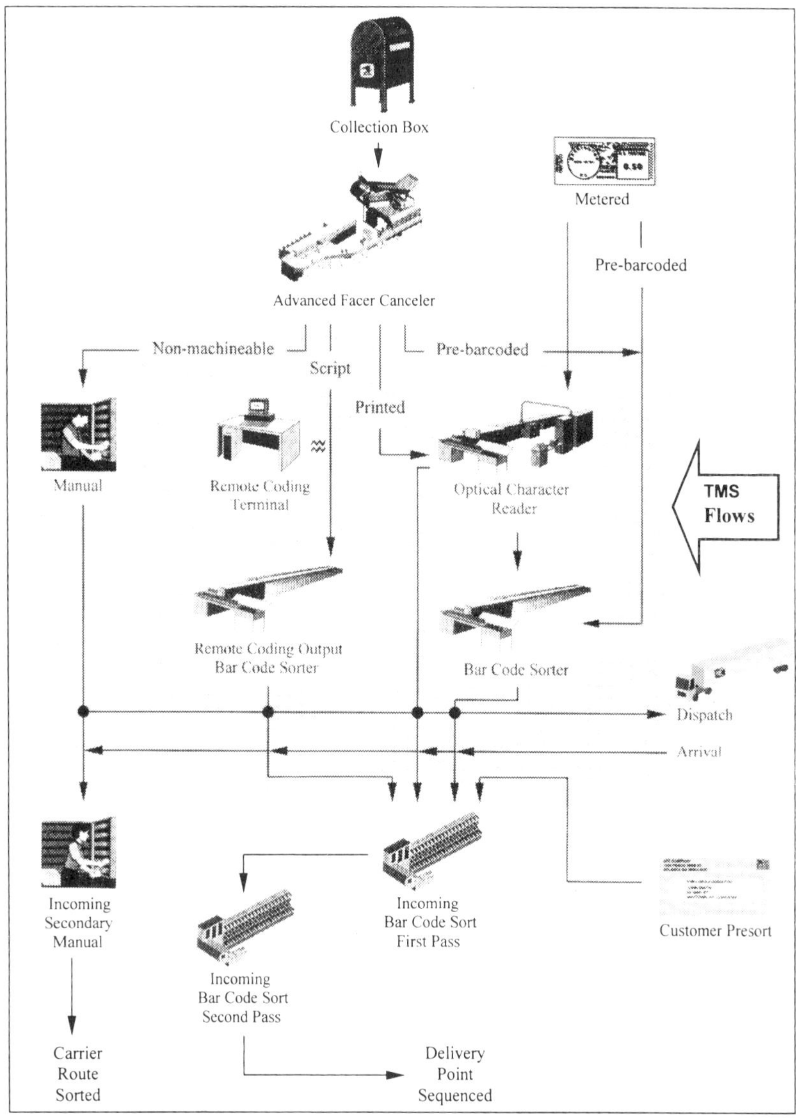

Notes: 1. TMS handles essentially all trayed flows in a processing center except dispatches of delivery point sequenced mail.
2. An analogous set of operations for flat mail is also handled by TMS.

Figure 1. Letter flows handled by TMS

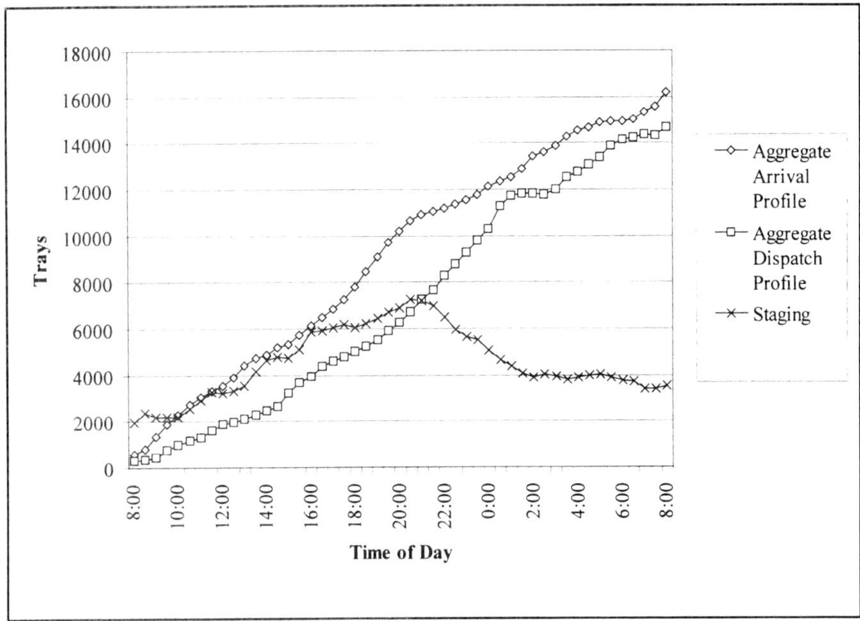

Notes: 1. This facility has a total of 24,300 square meters (262,000 square feet) of processing floor on two levels. Its TMS design has 654 meters (2,145 feet) of transport spine in two loops, 1,800 meters (5,800 feet) of powered roller conveyor, and 13 random access staging units comprising 6,977 staging locations.
2. Volumes are for an exceptionally heavy day three years in the future.

Figure 2. Overall flow for a typical facility (Westchester, NY)

Table 1. Equipment Specifications

Random Access Staging

    Suppliers:
        Daifuku
        Rapistan Demag
        Northrup Grumman
        Computer Aided Systems, Inc. (CASI)

    Throughput: 200 to 900 trays per hour

    Capacity: 130 to 2,000 tray locations

    Floor Space: 42 to 72 square meters (450 to 780 square feet)

    Height: Up to 7.3 meters (24 feet)

Batch Staging

    Throughput: 3 trays per minute

    Capacity: 43 trays (letters only) or 33 trays (letters or flats)

    Floor Space: 1.2 square meters (13 square feet)

Transportation Spine

    Suppliers:
        Crisplant
        Siemens
        Rapistan Demag

    Speed: 1.5 meters per second (295 feet per minute)

    Pitch: 0.5 to 0.8 meters (1.6 to 2.6 feet)

    Maximum Theoretical Throughput: up to 10,800 trays per hour

    Maximum Useful Throughput: Approximately 4,500 trays per hour

    Loop Length: 150 to 820 meters (500 to 2700 feet)

C568/024/99

Robotics in postal handling

A HENRIKSSON
Transman AB, Trosa, Sweden

SYNOPSIS

This paper is a short introduction to robotised materials handling in general, and robotised postal materials handling in particular. A survey of the existing robot concepts is presented, and evaluated in terms of performance.

The provisions for modern quality assurance and traceability is illuminated, and related to the present situation in postal industry.

The paper finally concludes a few future possibilities and demands in a visionary way.

1. POSTAL INDUSTRY IN THE 21:th CENTURY.

The heading above is definitely extremely pretentious. This paper will focus on the present situation and maybe the next 5-10 years. It will also delimit to the part of the postal industry that consists of materials handling, within the mail sorting centres.

To obtain some external input beside my own personal experience from materials handling in general industrial and postal industrial environments, I have made a short enquiry about the demands in postal industry. A short description of the enquiry is found in the end of this paper.

The results show that the postal administrations have found themselves in a new situation during the last years. The situation is continuously changing, but one major difference is that a lot of countries will permit competition in this area. The monopoly situation is changed or brought to its end, allowing both small local and very large global agents to take part in the postal activities. Of course this development has effected the plans of the postal administrations.

Other conditions that will direct the progress are the technical development in the information and media technologies, both in terms of the competition to the traditionally mailed letter and in terms of the possibilities these new technologies offer within the postal industry.
The general technical development in automation will also be influential, since it offers many new solutions. The automation and robot technology have previously been very concentrated to just a few industrial branches, but recently (perhaps the last ten years), the technology has matured and gained a wider popularity in most industries.

The result from the enquiry could be interpreted in many ways, but I have made a list of my own conclusions:

1. Top priority in the efforts will be to reduce the costs

2. The demands of quality and quickness are ranked higher than the preparedness of an increase in the total mail volume.

3. The number of employees will be reduced, as a result of the cost reduction process, and because of the ergonomic problems with manual work in the handling of trays or other pieces.

4. The postal administration has a tradition of long term cost analysis. This will be useful also in the future since the maintenance costs is very important component of the total costs. This is not considered with the same ambition in other branches, but it will probably be more usual in other branches, and stay important in postal industry. Life cycle cost analysis is a method growing in popularity.

5. Main contractors that take care of very extensive deliveries are getting more usual.

6. Trays, containers etc. will be further adapted to automated handling. The trays will be prepared for robot grippers, automatic labelling, and stable stacking. Generally the quality of the trays will be higher, to improve the flow and avoid problems with damaged trays.

7. The life of trays will then also be extended.

The competitors will naturally select parts from the postal administrator's total service. The selection will be made on the basis of the competitor's possibilities and on the basis of the potential to provide the selected service with a more rational approach. For instance, if a certain category of customers (senders) or a certain category of mailed items is selected, the processing could be more rational. This would probably reduce the costs and the service would be more competitive.

The postal administrators will then be forced to meet this competition and rationalise the processing, probably by automation to a high level. The provisions may be favourable, since the (national) postal administration originally has a very large market share.

The second most important criteria or aims, quality assurance (traceability) and rapid processing will also benefit from a higher degree of automation.

Another result of the radical high tech development in postal industry, is that a small number of very large companies have specialised to deliver complete, state of the art solutions, integrating all the technologies now involved in mail processing.
Altogether, this makes it sensible to expect a general development towards highly automated mail processing. This is already evident, since a massive majority of the planned new postal projects includes highly automated materials handling systems (i.e. robots). This is obvious all over the world, independently of the size of the countries, the general level of technology in the country etc.

2. INDUSTRIAL ROBOTICS

The robot as industrial and commercial machine is still rather young, and could be found in the industry from the beginning of the seventies. The first robots were typically of the multi-axis one-arm concept. Some kind of fundament on the floor had a number of structural elements linked together, operated by electrical motors and/or pneumatics.

There is probably more than one definition of a robot system, but according to ISO 8373, a manipulating industrial robot is:

> *An automatically controlled, reprogrammable, multipurpose, manipulator programmable in three or more axes, which may be either, fixed in place or mobile for use in industrial automation applications.*

The diagram below shows the most usual models of industrial robots. In this diagram the robots are classified by mechanical structure.

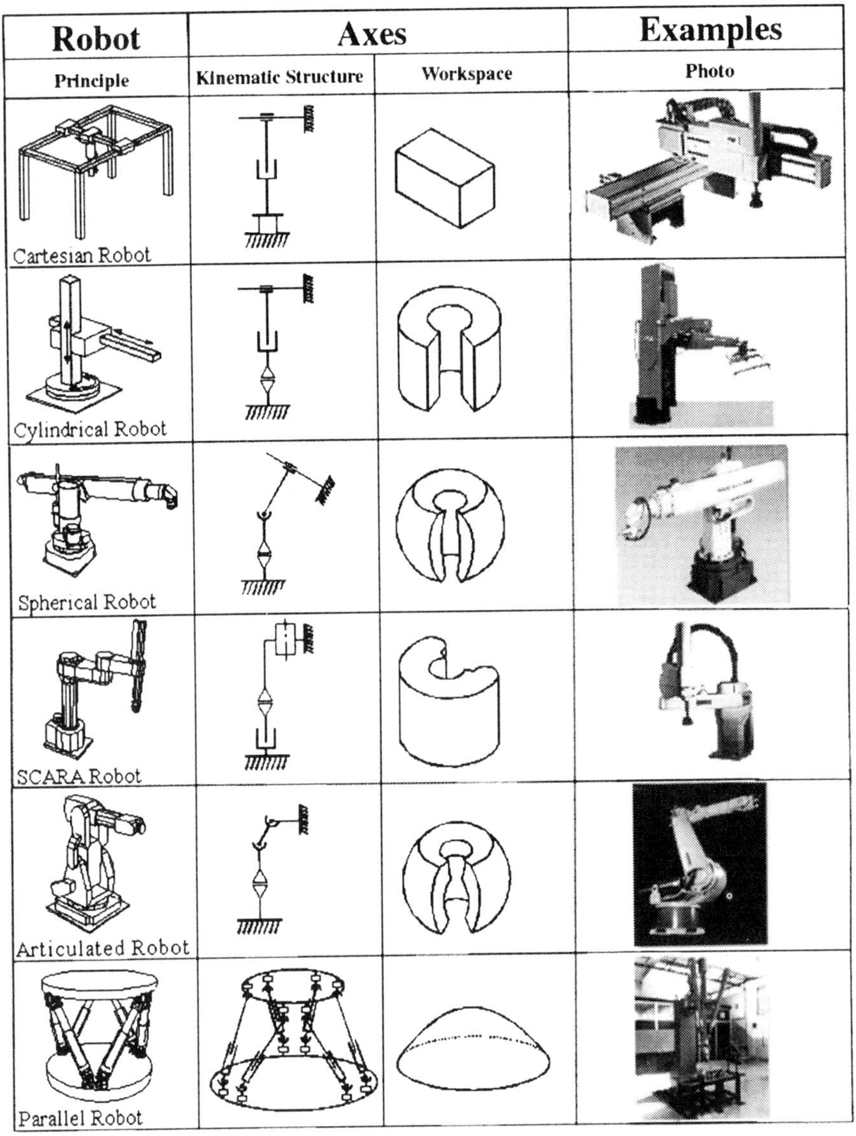

Robot	Axes		Examples
Principle	Kinematic Structure	Workspace	Photo
Cartesian Robot			
Cylindrical Robot			
Spherical Robot			
SCARA Robot			
Articulated Robot			
Parallel Robot			

Diagram No. 1 (Source IPA Fraunhofer Inst. Stuttgart)

An important consequence of the robots mechanical structure is the working area, or rather, space. In some cases it is possible and even convenient to orient the different objects within reach for the chosen robot. More often it is favourable to choose the robot type from the provisions of the application.

For the specific need, handling of postal trays, two different layouts are illustrated in diagram no 2.

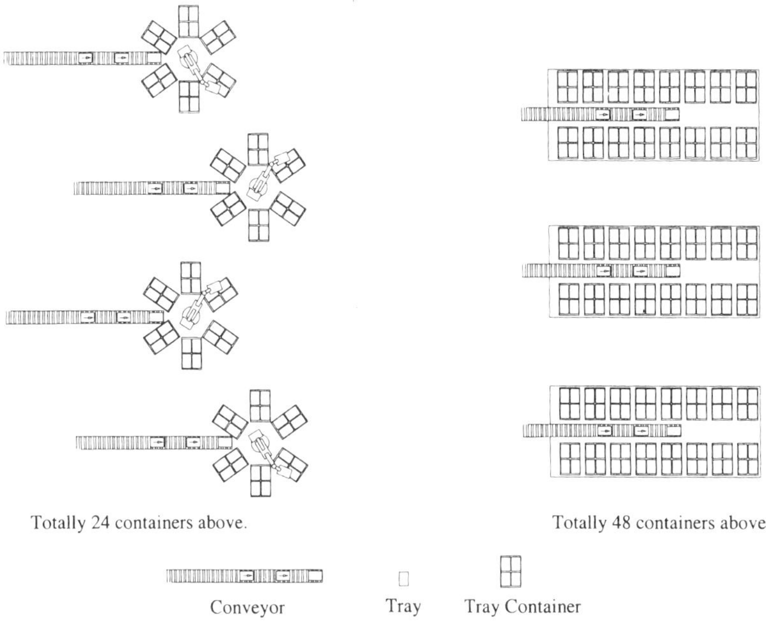

Diagram No 2: Lay-outs for articulated robots and for Cartesian robots.

1.1. Properties of different robot types.

Apart from the robots' own performance, regarding velocities and accelerations for the different axes, the layout will be very important for the capacity of the solution. Even though most modern mail centres now are situated or planned in industrial parks, less centrally than previously, the area consumed by the robot solution is a decisive factor.

Postal handling in the mail centres often means that several automatic handling systems must be integrated and adapted to each other. Even if the trays are standardised to one single, or very few sizes or models, within the systems, there are always a large number of destinations for the trays. This means that the system will occupy a large area even with a smart solution. A large and adaptable working area is certainly useful. From this point of view, the Cartesian robots comply with the demands. (The Cartesian robot is commonly called *box gantry robot* (three or more axes) or *line gantry robot* (two or three axes).) The working area is to a great extent adaptable. The area could be large and include a very large number of positions for the trays or tray containers. The right angle format often means that in combination with other machinery, the systems could be arranged very tightly together. Most containers for trays have a rectangular base, and the result will be a very dense working area.

Practically all of the other robot concepts have a more or less circular working area with the robot itself in the centre. It is difficult to arrange a larger number of tray containers within reach for the robot, and the robot must be a large model to reach through the array of trays even on one single container. The shape of the working area also makes it difficult to have a rational layout of a number of robots. It is space consuming and the distribution of the containers will be complicated. The reasons for choosing an articulated or SCARA robot is obviously not related to the working area.

But the articulated robot, and the SCARA, is favourable in other respects. The robot itself is often produced in large numbers, which results in a low price and high quality. The performance is also very good in terms of axis velocity and acceleration. The table below shows the typical performance of the robots'.

Robot	Articulated	Cartesian	SCARA
Velocity / axis	100-150°/s (4-5m/s)	3-4 m/s	200-300°/s (5-10 m/s)
General movement*	3-4 m/s	5-7 m/s	5-7 m/s
Acceleration/axis	200-600°/s^2	3-10 m/s^2	200-600°/s^2
Working area	Reach 0,5-3 m	1-8 x 1-75 x 0,5-3 m	Reach 1-3 m
Lifting capacity	0,2-150 kg	1-1500 kg	0,1-150 kg

*linear transport in a general direction, activating 3 axes

Usually the end customer is indifferent to the actual performance above, the capacity in terms of units per hour or other resulting effects are the desired assets. Some of this "applied performance" is shown in the table below. The performance will of course be dependent of the total solution; the trays, roll containers, grippers, and various important technical provisions and devices. The table is very general and the factors apart from the robot are assumed to be equal.

Robot	Articulated	Cartesian	SCARA
Trays per minute*	4-7	5-10	5-9
Tray Containers/cell	1-7	1-30	1-6
Tray Containers/sqm	0,3	1	0,3

*For the Cartesian robot the capacity is obviously effected by the working area. The higher capacity is for a 2-container solution, and the lower for a 30-container solution.

1.2. Quality assurance.

One of the most important aims in postal industry today is to raise the quality, i.e. increase the possibilities of quality assurance.

To improve the level of quality by raising the level of automation is not an obvious relation, but the errors, which usually is referred to as "the human failings" will reasonably be reduced. Of more importance, the possibilities of continuous logging and monitoring will correspond to very high demands of traceability and quality assurance.

It is subjected to the administrator to decide how much of the activities that will be registered and stored. It is already possible to register every single letter, and its physical properties, in order to trace the way and time it has been travelling from the receipt to the final delivery. Presently the first occasion for registration is at the first sorting machine. The letter could of course have its history back to the specific mailbox reconstructed and registered. If the mail-

C568/024 © IMechE 1999

piece is furnished with an identity (e.g. bar code) at this occasion it may then be possible to follow it by direct and indirect monitoring throughout the processing.

Presently most systems in mail processing offer more possibilities of this kind, than what is utilised.

If robots are used for the handling of trays, it will be certain to a very high degree that every tray is handled to the proper destination. Furthermore every operation by the robot may be registered, and the registers will be easy to use in databases. This means that even if a robot has made a mistake, this will be possible to trace.
Since the trays *cannot pass the robot cell without being actively handled by the robot*, all trays will be traceable, if registered. This is an important difference compared to most conveyor systems for sorting. In these systems it may happen that once the label is detected, the tray will go in the wrong direction because of defective sensors. Since there are small possibilities to find out when a certain sensor started to fail, it its very difficult to trace, or decide the extent of the problems, even if the defective sensor is discovered.

Of course, it has to be remembered that in all these situations, only the trays are detected and registered, and a single letter can only be traced or monitored indirectly.

3. FUTURE POSSIBILITIES AND DEMANDS.

A future perspective is exciting and complex, while including all the future possibilities and development in different areas.
Not only will the robots and the handled pieces be different, but all the machinery and the information systems within the mail centres will certainly be developed. In fact, the provided postal services themselves will definitely change. Thus my visions will be very limited and uncertain, as would any prognosis ranging beyond 10 years.

It is very hard to isolate the robots in the postal automation, and predict the future visions for them, without any consideration of the surrounding technologies. Even today every postal administration have rather specific solutions as regards the robots, as a consequence of the total solution and the layout.

The (potential in) robot technology could probably be categorised as follows.

- Mechanical components
- Servo technology
- Controller technology

Certainly each category will be developed and improved. For the mechanics and the servos, it is natural to expect a performance that will mean a higher capacity of the robots. The accuracy could also be improved. The fastest and most accurate robots of today, are however so fast and accurate that the capacity of the total solution will partly be limited of other factors.

To reach higher speeds (4 m/s today) the acceleration has to be very high. Today the acceleration is typically 1 G for a fast Cartesian robot. This means that the letters in the trays are moving around due to the inertia, the letters are sometimes damaged, and the trays are not rigid enough for the bouncing letters. Thus, the trays will have to be improved to allow higher speeds. If these problems would be solved the robot speeds could be increased.
Possibly the speed and the acceleration of these robots will be raised to 8 m/s and 15-20 m/s^2 in a ten-year perspective.
Considering the resulting forces and energies at these levels, the mechanics and servos have to be refined, very few linear systems (one-axis) can reach beyond 7 m/s today.

To improve the capacity further, and to have more options for the logistics within the robot cell, it is also possible to use configurations of the Cartesian robots that include more than one traversing unit. With two (or even three) units (Y and Z axes), the capacity could be enhanced, and in case of a situation where one unit is down for service, the other unit could take care of the whole working area.

Controller technology will probably improve more during the next ten years. Naturally the hardware is undergoing rapid development, as most computer hardware. This will make the controllers even more capable regarding data base features and sophisticated users interfaces. The users interface of the modern robots' already includes informative graphics, some times presented on large full colour touch screens. This interface can be even more sophisticated with extended on-line support from the suppliers. The user will also have extended possibilities to change the program etc. with a minimum of training. The support and maintenance functions will probably be one of the areas with high priority, when the systems get more advanced, and the demand of 99-100% accessibility is necessary.

Generally the investment costs for the systems will probably get lower. In such a situation the maintenance costs will be more important; if the investment costs are reduced and the maintenance costs stay at the same level (or even grow because of shorter upgrade intervals), the maintenance costs will make up a larger part of the life cycle cost.

This will automatically increase the demand for service and maintenance features allowing the local maintenance personnel, or even the operators, to solve most of the possible problems.
To reduce the need of special training and integration problems, standards for network communications, field-bus technology, servo communication etc., will be improved.
A general demand for further integrated systems **and** higher flexibility (compatibility) was clear in the enquiry, preceding this paper.

This development is almost impossible to quantify, but the next ten years robot development could be summarised like this:

- Capacity and performance will increase with about 50%.
- The trays and tray containers will be well adapted to the automated handling.
- The controllers will be more powerful, particularly the features for database and statistics, on-line and off-line support, diagnostics and maintenance.
- The controllers will be adapted to standards in greater extent.

To improve the materials handling at the mail centres further, the trays and tray container systems will have to be developed. A time consuming part of the handling is still the loading of the lorries or trains. This is a manual work, demanding a lot of manpower at specific moments during the 24 hours.

To reduce the time needed for the vehicle at the quay, it would be possible to automate (possibly with robots) the loading.
One method would be to forward the containers from the trays loading robots to a very large slip-sheet, or a container. When the slip-sheet or container is filled, ready for delivery, it could be loaded in less than 15 seconds. To pre-load or handle tray containers generally, larger Cartesian robots could be used, a tray container weighs typically below 300 kg. There are fast robots capable of payloads up to 1 000 kg. The advantages with robotic handling instead of monorails or conveyors are:

- non-consecutive handling is possible
- the controllers offers the same possibilities as in the tray handling

If a tray container is designed for automatic transportation within the mail centre, automatic tray loading and automatic loading on the vehicles, it is obvious that it would be quite different to a roll container optimised for manual handling throughout. For instance, the wheels, common on the tray container today all over the world would not be necessary.

To improve the logistics and reduce the handling, it would be a great advantage if the same container would be useful on lorries, trains and aeroplanes.

This illuminates another bottleneck in the handling today; the import and export of mail.
There is some co-operation today, between countries in order to use the same trays for the air transports. It will certainly *eliminate* a lot of handling if further international agreement will be settled.

Generally, the handling could be further automated with existing technology. For instance, the final sorting for each postman, including the complete address, could be made with more or less the same letter sorting machines, and the existing robots. Certainly, improvements of the capacity will make the automation more attractive and cost-beneficial.

Cost-analysis, ergonomics and quality demands will be crucial, but considering the dramatic changes some of the decisions will be of a more political character.

<u>**Postal Industry Enquiry 1999**</u>

One of the sources for the paper "Robots in Postal Technology was an enquiry delivered to a total of 20 countries all over the world in the beginning of 1999.
16 questions was formulated to illuminate the technical direction of postal automation, the prognosticated extent of automation, the reasons for this development and also the purchase methods for investments of this size and category.

50 % of the addressed postal administrations contributed to the result, forming a representative group of larger and smaller countries from all parts of the world.
Countries with obviously different geographical and infra structural constraints, as well as different populations and economies did show a surprisingly homogenous result.

Most postal administrations have extensive co-operation across the borders, and do not act as competitors. This promotes the communication between administrations all over the world, and the industrial systems and logistics develop rather similarly.

The most obvious conclusions were:

- The development of new technology and systems will be carried out of the suppliers or the main contractors, not by the technical departments of the postal administrations.

- Most postal administrators have, or will, change vital parts of the equipment (also trays and containers) to further the possibilities of advanced automation.

- The competition from private companies and other circumstances will force the postal administrations to reduce the costs dramatically.

- Postal administrations regard the ergonomics as one of the more important criteria in the development.

The result of the enquiry is distributed to the participating countries, and the enquiry is planned to be carried out yearly.

　　　　　　　　　　　　　　　　　　　　　　　C568/024 © IMechE 1999

C568/051/99

Postal business – home of the cross belt technology

M CORONA
CML Handing Technology Spa, Lonaze Pozzolo, Italy

1   EVOLUTION IN THE WAY OF CONCEIVING AN AUTOMATION CHOICE

Today, an increasing number of applications require a significant involvement in terms of automation. No doubt, the postal process is one of the most demanding in this respect. The fast growing of mail streams to be handled urges Postal Organisations toward innovation within a more and more competitive background whose aim is to meet the user's increasing request for diversified and cheap services.

In the choice of an automated system coping with organisation and architecture of the controlled process, efficiency and quality parameters shall no more be connected only with the mechanical characteristics of each component. Nowadays, competitive advantage lies in the way system components are linked and used to set up a flexible and reconfigurable system. This allows to optimise integration of the new available technologies for acquisition and processing of the data requested for a proper management of the postal process.

2   EVOLUTION OF A TECHNOLOGY BORN TO BE EFFECTIVE ON POSTAL ENVIRONMENT

Since 1982 our effort to innovate the Cross Belt technology has been supported by this point of view taking into account the above assumptions.

Unmatched sorting precision together with extremely wide handling ability and high productivity, led the SBIR® to become the 'state of the art' technology for postal bundles and packets processing: between 1986 and 1998 more than 350 systems have been deployed in this application.

Cross belt technology found its application also in the postal business of the express courier services with high throughput applications for the sorting of overnight letters and small parcels. Over the years, CML cross belt technology has been improved and its application range has been enlarged toward king-size items and has been installed in airport premises for baggage handling in a completely automatic environment, fitted with fully automatic storage and retrievable buffer line system.

For the handling of in-house mail within large companies, such as banks and insurance companies, the Cross Belt technology and, in particular, its multilevel application called PLUSSORT, has proved to be an excellent solution where a very high number of low-traffic destinations is requested.

We think the SBIR® Cross Belt sorting technology is really suitable to manage automation of the sorting process within postal centres requiring a high processing complexity.

3   MAIN CHARACTERISTICS OF THE CROSS BELT TECHNOLOGY (SBIR)

Here is a short overview of the SBIR® Cross Belt technology main characteristics:
the total development of the SBIR® sorter consists of a continuous "train" of carriers, each supporting one sorting cell. Every single cell is able to receive an item loaded from the automatic induction station, to convey it to the assigned destination and to unload it.

A belt wound on the cell allows to perform the actions connected with the sorting operations; the application of this belt entails the following advantages:

- A perfect dynamic continuity with the 45° induction line. The items are transferred on the cells at almost zero relative speed allowing a gentle handling with the most accurate loading trajectories irrespective of items size, weight and deformation liability without any damages to items due to rolling or crashes.

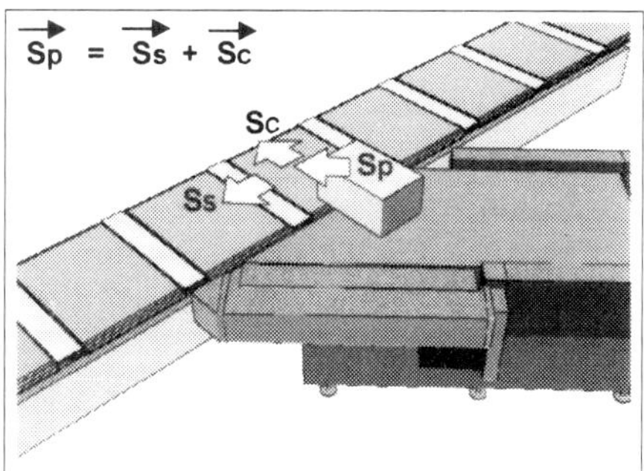

Fig. 1

- The correct items positioning on the cell after the loading action is adjusted and assured by the re-centering function.

- Very accurate off-loading trajectories are assured since items are not discharged by exploiting gravity, but "actively" by powering the cell belt under the command of the control system. In this way, the item off-loading is always constant and controlled. Different trajectories can be obtained by varying the loading speed profile or the activation point of the cell belt based on shape and weight of the items to be handled.

All these characteristics are possible thanks to the way the control system uses to manage the cell drive during the loading and unloading phases. All commands for the actions to be performed are transferred to the control units distributed on board of the sorter as software messages. Messages are transmitted at any location of the sorter loop using a bus because transmissions are not bound to any special positions where they are enabled. On-board messages are transmitted via full-duplex closed loop exchanging messages in both directions: therefore it is impossible that a wrong command is executed.

Cell activation is thus performed without using any mechanical device.

The sorting process based on the Cross belt Technology is therefore reliable, accurate and can be adapted to a large variety of item types. Thanks to the possibility of adapting the parameters controlling the item loading and unloading functions, handling accuracy and quality can be unchanged, irrespective of the items handled. If the item characteristics are known, some functions can be implemented to improve the sorting process result in terms of items distribution in the outlet devices.

Some examples:
- optimisation of the outlet filling

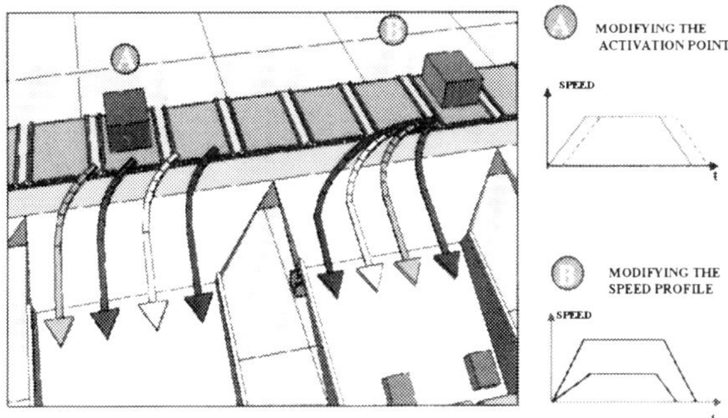

Fig. 2

- where needed, separation of heavy and light items inside the same outlet to reduce any possible risk of damage

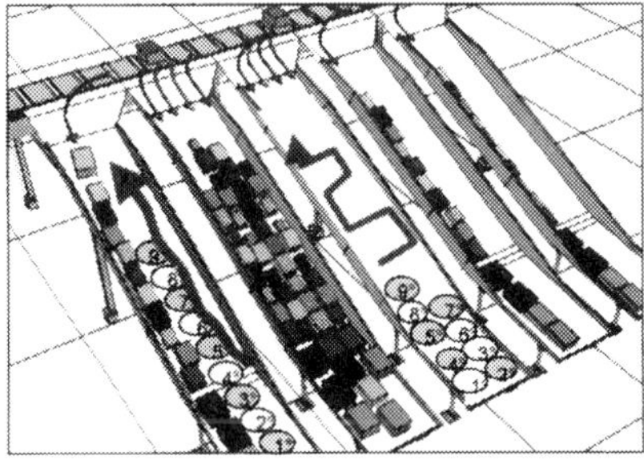

Fig. 3

4 HOW THE CROSS BELT TECHNOLOGY CAN SUPPORT THE RE-ORGANIZATION IN POSTAL PROCESS LEADING TOWARD THE CONCEPT OF "LIGHTS OUT" POSTAL CENTER

The high flexibility of the active sorting technology allows introducing some innovations in the organisation of mail streams through a sorting system inside a large postal centre. An articulated postal process requires to handle a great variety of items (bundles, packets, parcels, flats, letters, trays, bags) and therefore also different kinds of outlet devices to collect sorted items.
But the differentiation of the outlet devices is usually not enough and the use of different sorting technologies is also required.

The "vision" inside the sorting application envisages a sorter that can sort all types of items with the same care and accuracy level, but concentrating several streams in one single layout.
The proposed technological solution shall allow to organise the postal process so as to maximize automation and to minimize the personnel contribution required to manage the centre: the concept of "lights out" postal centre could even be advanced. A quick return on the investment employed in automation and further savings as a result of optimized technological choices represent the target of every postal organisation.
I will try to explain now how it is possible to reach this goal organising the sorting process based on just one Cross Belt sorter: more than one stream will be handled with one single machine but assuring sorting accuracy and efficiency in the various applications.

We will consider the implemented concepts and technical characteristics lying at the base of the system (sorter + upstream and downstream components) proposed for the application in question.

Let's start to analyse the sorter that is the heart of the system.

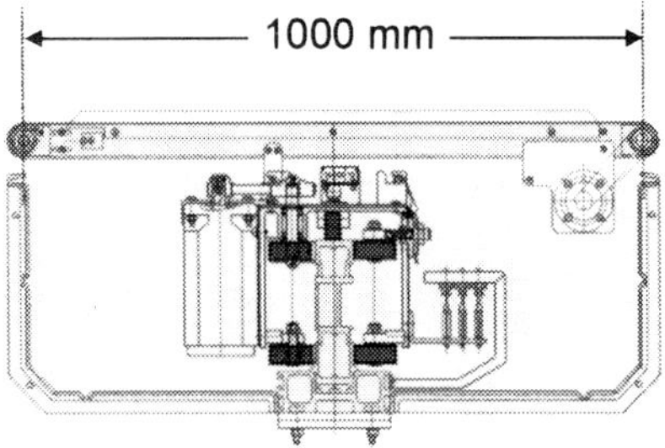

Fig. 4

The designed cell has a large size but a reasonably short pitch, thus allowing to load any item on one single cell. The resulting sorter assures 8000 cells/hour at a speed lower than 1,8 m/s: this capacity is equivalent to the hourly throughput in terms of items per hour because no cells are "missed" as a consequence of loading a large-size item on two consecutive cells. To further increase the cell loading capacity, it is also possible to load items whose length exceeds by 20% the cell dimension. Thanks to the continuous control function performed on the cells and to the availability of stations checking the shape of the items loaded, items can be repositioned in the middle of the cell at the end of the induction area. In this way the item exceeding part is distributed on the two cell ends and does not cause any problem during the conveying and sorting phase thanks to the outlets design.

In case the item shall recirculate, before the loading area, it is moved toward the side opposite to loading to avoid any interference with the induction lines.

The sorter conveying surface has been designed in such a way as to eliminate any possible discontinuity of the conveying plane. The cells belts and the connection plates represent a homogeneous conveying surface: thanks to this feature loading, sorting and re-loading operations are facilitated and more safe, especially if they are completely automatic, without any presence of personnel.

Discontinuity has been eliminated also on the sorter induction lines: side panels have been lowered so that they can neer interfre with the items in transit. Besides virtually eliminating any risky situation during operations, the induction stations dimensions have been reduced to the minimum needed with a consequent remarkable reduction of the plant footprint. In addition, the reduced cell pitch has allowed reducing also the outlet pitch imroving compactness of the outlets area. Thanks to the totally independent control of the cells, various layout combinations are possible for the outlet devices: both type and pitch can be alternated.

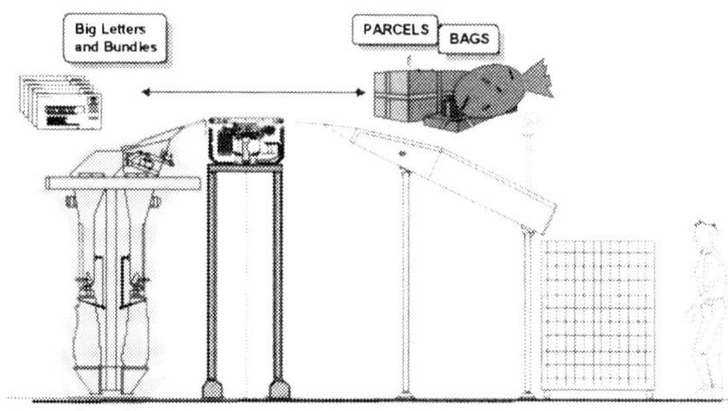

Fig. 5

A special remark shall be made for the availability of all these characteristics at different conveying speeds: in this way the real sorter throughput assured at any moment is the most suitable with respect to the real working conditions without the need of increasing the command and control devices, either in the mechanical or electronic field.

As a direct consequence of this ability to adapt throughput to the real needs the power consumption required to operate the plant are reduced.

5   THE IMPORTANCE OF THE PLANT AVAILABILITY FOR A SYSTEM PROPOSED AS BACKBONE

Thanks to the typical characteristics of "active" sorting, totally automatic lines can be arranged to carry out some special operations. For example: if the items addressed to a certain destination shall be submitted to some security checks, they can be extracted from the main stream to perform the requested screening. If the check result is negative, the items can be reloaded onto the sorter and sorted at their destination; on the contrary, in case of doubt, they can be sent to a dedicated area where a more accurate screening is performed. These operations can be carried out in a completely automatic way, which offers important advantages both in terms of logistic organisation and safety for the personnel charged with the task.

Concerning plant reliability, very high levels can be reached thanks to the redundancy of some critical components: The number of high-efficiency driving groups distributed on board of the cells can be increased in order to assure a constant sorter speed, even under extremely degraded conditions. In addition, the control system can be doubled to allow an easy switch from the main control to the back up. The possibility of having a redundancy of only some components allows obtaining almost total plant availability. This assures also to handle all streams with the same high availability level without resorting to back-ups of dedicated machines, which are not normally used, with a patent advantage from the investment point of view.

The combination of the technical characteristics peculiar to the described system and the organisation based on the handling all mail piece streams in one passage offer many direct and induced advantages.

The system layout can be considerably reduced and the interfaces for the single mail piece streams can be optimized.

Since all streams can be handled at the same time, the system is able to easily adapt to any sudden changes in the mix of each single stream. Furthermore the high flexibility of the system allows to reconfigure the quantity and arrangement of the outlet devices at a reasonable price, in order to better adapt to any possible change of the operating conditions.

Streams concentration in one single handling system allows a better utilization of the code identification equipment (OCR, laser scanners, videocoding, etc.) and of the weight and volume control devices for all kinds of handled mail pieces. The possibility of using equipments having a high unit price on a large scale makes their employment more profitable and further increases the already mentioned possibility of easily complying with composition changes in the mix of items to be handled.

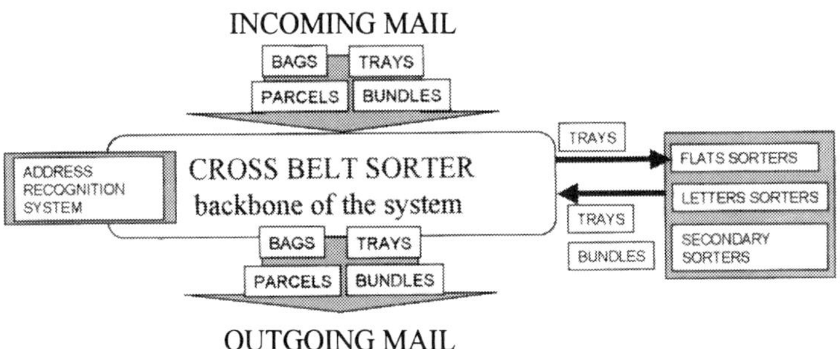

Fig. 6

Centralised control of various streams based on one single HW/SW platform allows to optimize interfaces with upper level systems, which assures a more efficacious solution to face problems connected with the management of the large data bases used for coding processes, track & trace, recording and Revenue Protection. Having few machines in a postal

centre implies as an indirect benefit the reduction of the operating costs deriving from spare
parts stocks management and allows a higher efficiency of maintenance teams.

6   CROSS BELT TECHNOLOGY APPLIED TO INTERNAL MAIL
HANDLING PROCESS

The concentration of several postal streams on one single sorter offers a lot of irrefutable
advantages. For some applications and streams an approach with dedicated technology is
necessary anyway. The handling of letters, but also of flats is a clear example of application
for which sorting technology offers dedicated solutions. Nevertheless a minimum percentage
of letters and flats shall still be handled manually as a consequence of the incompatibility of
format and contents with respect to the sorting devices which are peculiar to these
applications. In large-size centres the quantities of mail to be sorted manually could justify the
automation of the corresponding process. A similar application is represented by the
technology currently adopted to sort in-house mail inside large organisations such as banks
and insurance companies. This application is based on a multilevel Cross Belt sorter, called
PLUSSORT, which can distribute mail over a very high number of destinations. In this
particular application, destinations consist of extractable drawers having a size of
500x150mm.

Fig. 7

Drawers are arranged at several levels (up to 12) to obtain a very high destination density with
a limited footprint of the plant.
The reduced dimensions of the drawer are justified by the low traffic they are going to
receive. Thanks to the large number of available destinations (1018 at Deutsche Bank) the
receiver's assignment can be subdivided to such a point that one physical destination (drawer)
can be assigned to one single person. Greater aggregations are possible if one drawer is
assigned to an office, a department or a small branch.
The types of handled items consist of letters and "irregular flats", that means the selection
which is typical of the manual application within traditional postal centres (rest mail).
Handled dimensions range from min. 100x100x0,1mm with a weight of 20g up to max.
450x300x50mm with 5kg. For this application the sorter mechanical capacity is reduced to
4.600 cells/hour by the lightness of the items handled and by the very short pitch of the used
destinations. If the statistical increase of capacity is exploited by loading from various points
or if one single pre-sorting is performed, the installation capacity can be brought up to 9.000
or more cells/hour. Special attention must be paid to the coding procedure used for this

process and to the check of the drawers status monitoring any possible full condition or jamming situation occurring on the sliding plate connecting the sorter to the drawers.

Fig. 8

Manual coding through keyboard is supported, if necessary, by rapid access to a database which can drive the operators to find all the information required to reconstruct the mail piece address in case the data contained in the address are uncertain or unclear. Also automatic coding is possible by means of both an automatic barcode scanner installed on the line and a Vision System for OCR coding of the address.
The drawers filling degree check is based on the calculation of the accumulated volume. It is monitored and shown to the operators on displays installed on the outlets devices. The ID number of the full drawers is displayed so that the operator can see to their emptying. The same display is used to warn against any possible jamming. The jamming location is obtained combining information transmitted by some photocells installed on the outlets with the data available on the sorting real-time control system.
The multilevel application of the Cross Belt technology has been adopted also for market sectors differing from the postal one, in particular for garments distribution or mail order companies. The above-mentioned handling of in-house mail is the field where this technology has proved once again to be extremely flexible and adaptable to the different problems of a non-industrial environment, such as the postal department of a large organisation.
The restraint of the plant footprint and the reduced clearance available (height) do not allow to install any mezzanine; this kind of environment is therefore one of the most difficult to house an automation system.

The multilevel Cross Belt technology has adapted to these restrictions offering a smart solution for all issues connected with the need of having different levels for sorting/loading and available passages for operators under the sorter. All this is possible thanks to the active lifting of the sorter cell which is activated by an electrical winch completely controlled by the on-board unit. Therefore the extension/retraction of the pantograph supporting the active sorting cell can be activated at any point of the loop, which considerably reduces the layout constraints imposed by this type of application.

7   CONCLUSIONS

The two applications described above lead us to draw the conclusion that the sorting process lying at the base of any postal application requires a high flexibility degree of the selected technology. Only in this way the problems deriving from the organisation of the sorting process can be faced and solved.
However, the element which is common to all mentioned problems is the "Optimisation" of the solution based on a consolidated technology. This shall allow minimising the compromises accepted when confirming a choice of automation.
The Cross Belt active sorting technology is for sure the foundation stone of all fundamental processes within an innovative postal centre.

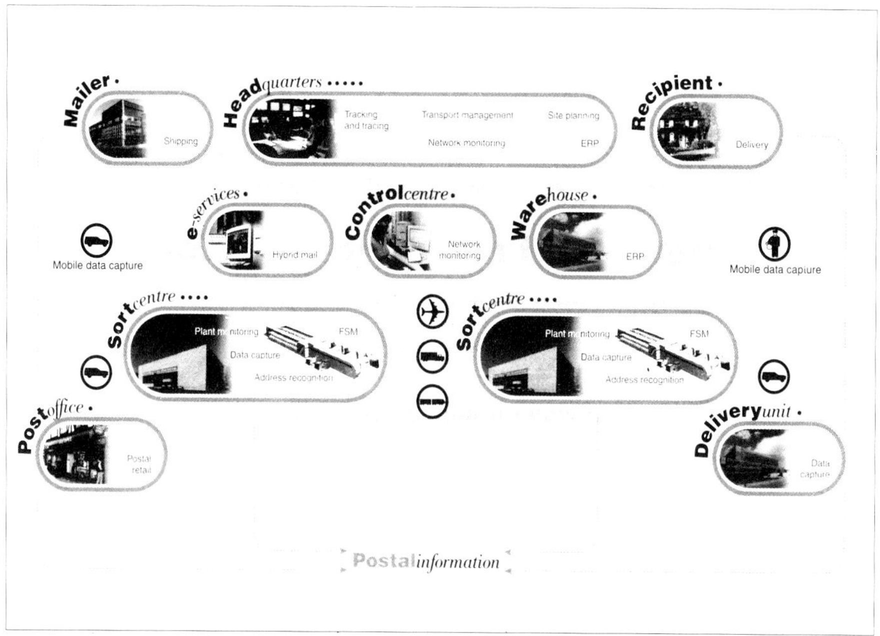

IBM's Postal Information FrameWork

More than just a pretty picture.

Need a framework for future prosperity?

Integrating your IT with the rest of your business might seem simple. But the reality is often quite different. While many postal organisations know that IT can give them competitive advantage, working out the best way to achieve this can seem a nightmare.

That's why you'll want to hear about how IBM can help.

IBM has already invested billions in being able to help you overcome the business challenges you're facing today.

And the best news is that IBM is now focusing more of its resources on developing solutions specifically for the Postal Industry. Familiar issues, such as improving mail processing, managing logistics more effectively, and enhancing retail operations can already be resolved with solutions from IBM.

Allowing you to take advantage of innovative electronic network services IBM has pioneered to help your organisation reach for greater success.

Announcing the IBM Postal Information FrameWork

The IBM Postal Information FrameWork has been designed to help postal organisations align IT with their business objectives, to plan and create future solution investments.

For more information on the IBM Postal Information FrameWork or any of our Solutions for the Postal Industry please visit our Web site at www.government.ibm.com/gov/postal or e-mail us at postal@uk.ibm.com or write to us at Global Postal Business, 31s New Square, Bedfont Lakes, Feltham, Middlesex, TW14 8HB.